Climate Variability Imp
and Livelihoods in Dr)

Mahesh K. Gaur • Victor R. Squires
Editors

Climate Variability Impacts on Land Use and Livelihoods in Drylands

Springer

Editors
Mahesh K. Gaur
ICAR-Central Arid Zone Research Institute
Jodhpur, Rajasthan, India

Victor R. Squires
Institute of Desertification Studies
Beijing, China

ISBN 978-3-319-85972-9   ISBN 978-3-319-56681-8 (eBook)
DOI 10.1007/978-3-319-56681-8

© Springer International Publishing AG 2018
Softcover reprint of the hardcover 1st edition 2017
This work is subject to copyright. All rights are reserved by the Publisher, whether the whole or part of the material is concerned, specifically the rights of translation, reprinting, reuse of illustrations, recitation, broadcasting, reproduction on microfilms or in any other physical way, and transmission or information storage and retrieval, electronic adaptation, computer software, or by similar or dissimilar methodology now known or hereafter developed.
The use of general descriptive names, registered names, trademarks, service marks, etc. in this publication does not imply, even in the absence of a specific statement, that such names are exempt from the relevant protective laws and regulations and therefore free for general use.
The publisher, the authors and the editors are safe to assume that the advice and information in this book are believed to be true and accurate at the date of publication. Neither the publisher nor the authors or the editors give a warranty, express or implied, with respect to the material contained herein or for any errors or omissions that may have been made. The publisher remains neutral with regard to jurisdictional claims in published maps and institutional affiliations.

Printed on acid-free paper

This Springer imprint is published by Springer Nature
The registered company is Springer International Publishing AG
The registered company address is: Gewerbestrasse 11, 6330 Cham, Switzerland

Foreword

Drylands are lands where soil moisture content is constrained by climatic variables, low rainfall, and high solar radiation, to the extent that annual evapotranspiration is (at least, potentially) about 1.5 higher than annual precipitation. Thus, dryland productivity is limited by water availability. Yet, these long-term average climatic metrics that qualify lands as drylands mask another, more critical attribute of drylands' climate – the extreme between-year variability in precipitation frequency, intensity, and spatial distribution that make dryland livelihoods constantly vulnerable.

It can be thus expected that the tight linkage of land use-dependent livelihoods to climate variability has made both the drylands' biodiversity and dryland people that depend on this biodiversity highly adaptable to climate variability. Since the human-induced global climate change issue is no longer a source of controversy but is of deep concern, the current state of the drylands, some 40 % of the global land and used by about one billion people, has aroused attention.

This is so because on the one hand it can be assumed that the millenia of evolving adaptations under climatic variability made the drylands' socio-ecological systems climate change resilient and thus make these systems a showcase of climate change adaptation for the rest of the world. But on the other hand, another scenario may be plausible – the resilience acquired by these socio-ecological systems through millennia of land use would succumb when grappling with the projected and, in places already detected, further and fast increase in global change-induced climatic variability. This projected outcome might generate ominous repercussions affecting the rest of the world.

These two diverging scenarios for the drylands and the significance of their impact on the post-2015 global community call for an urgent assessment and stock-taking of the current natural, human, and social capitals of the drylands and their performance under the impact of local, global, natural, and human-induced drivers of change. The products of this assessment would facilitate molding land use-relevant responses that could retain the drylands' natural productivity and at the same time also contribute to maintaining a supportive rather than destructive climate to which global biodiversity, land productivity, and land users have become adapted.

For this to happen, it is imperative to recognize that though the common denominator of all drylands is their inherent water-dependent low biological productivity, their socio-ecological systems considerably differ in the degree of aridity, climatic variations, biodiversity, land use practices, culture, and governance, depending on their geographic location and the geopolitical affiliation of their users and their countries. This book "Climate Variability Impacts on Land Use and Livelihoods in Drylands" provides updated information on socio-ecological systems embedded in drylands the world over, geographically, altitudinally, and ecologically, in the northern and southern hemispheres, and in developing and industrialized countries. Its production at the temporal crossroads in which the endorsed *UN Sustainable Development Goals* and the *Paris 2015 Climate Agreement* initiate their implementation is indeed timely. One of the thrusts of this implementation addresses neutralizing negative human impact on land and climate, both locally and globally. The information provided by each of the chapters could be used as a baseline for evaluating the progress made subregionally, regionally, and globally in sustainable dryland management for years to come.

Uriel N. Safriel, M.Sc.D.Phil (Oxon)
Professor, Hebrew University, Jerusalem
Former Director Jacob Blaustein Institute for Desert Research,
Ben Gurion University of the Negev, Israel
Chair, UNCCD Committee on Science and Technology

Biographical Notes on Editors

Dr. Mahesh K. Gaur is an Indian, who has been a Senior Scientist at the ICAR-Central Arid Zone Research Institute, Jodhpur, India, since September 2012. He specializes in *aridlands geography* and the application of satellite remote sensing, GIS and digital image processing for natural resources mapping, management and assessment. He also researches *drought, desertification, land degradation, indigenous knowledge systems, and the socio-economic milieu of the Thar Desert of India*. He has worked as an Associate Professor for more than 13 years with the Department of Higher Education, Government of Rajasthan State, India and has also worked at Sandhan, Jaipur, IIM Ahmedabad, IHS Hyderabad and IN-RIMT in various capacities. He has also worked with ICAR, ISRO, SAC, UGC, various Ministries of Government of India, and ICARDA for sponsored studies. He has published three books and is the author of a number of research papers and book chapters that have appeared in national and international research journals. He is a member of a number of national and international scientific organizations, such as the Association of American Geographers and the Society for Conservation Biology, and several editorial boards of journals including *Journal of Arid Environments* and *Journal of Ecosystem and Ecography*. He has been awarded the Indian Government Citizen Karamveer Award 2011 by iCONGO for working in the field of Higher Education, Environment and Technology Applications for

Community Upliftment, the Millennium Award for working in the field of Environment, and prestigious recognitions by the UGC of India and Scientific Assembly of the International Committee on Space Research (COSPAR).

Dr. Victor R. Squires is an Australian who as a young man studied animal husbandry and rangeland ecology. He has a Ph.D. in rangeland science from Utah State University, USA. He is former dean of the Faculty of Natural Resource Management at the University of Adelaide, Australia, where he worked for 15 years following a 22-year career in Australia's leading research organization, CSIRO. He is the author/editor of 11 books including *Livestock Management in the Arid Zone, Range and Animal Sciences and Resources Management, Rangeland Stewardship in Central Asia: Balancing Improved Livelihoods, Biodiversity Conservation and Land Protection, Rangeland Ecology, Management and Conservation Benefits*, and *Rangelands Along the Silk Road: Transformative Adaptation Under Climate and Global Change*, and has written numerous research papers on aspects of range/livestock relations. Since retirement from the University of Adelaide, Dr. Squires was a visiting fellow in the East-West Center, Hawaii, and is currently guest professor at the Institute for Desertification Studies, Beijing, China, and an adjunct professor in the University of Arizona, Tucson, USA. He has been a consultant to the World Bank, the Asian Development Bank, and various UN agencies in Africa, China, Central Asia, and the Middle East. He was awarded the 2008 International Award and Gold Medal for International Science and Technology Cooperation and in 2011 was awarded the Friendship Award by the government of China. The Gold Medal is the highest award for foreigners. In 2015, Dr. Squires was honored by the Society for Range Management (USA) with an Outstanding Achievement Award.

Editors' Preface

The early part of the twenty-first century has begun with a lot of change. Some of the change is good but regrettably there are ominous signs of things not going well. This is especially true in the world's drylands (aridlands), where human populations continue to rise and where livestock inventories are at record levels. All of this is occurring at a time when the impact of global change (including climate change) is being felt and awareness of further likely impacts have heightened. Many aridland regions are home to ethnic minorities who are marginalized not only geographically but also socially, politically, and economically. Poverty is widespread in such regions and is likely to deepen and encroach upon people who may have recently been lifted out of poverty. This reversal of fortune will come from an accelerated land and water degradation, from a restricted supply of water for basic human needs, and for irrigation to produce food crops, fodder for the burgeoning livestock populations, and cash crops that may hold some hope of allowing poverty to be averted.

In this book, we draw upon experiences from many countries and examine the implications of climate variability (including prolonged drought) and extended cold spells that wreak havoc in North Asia (in particular High Asia) (Kreutzmann 2012) and in cold arid regions in South America.

Climate change is affecting the culture, health, economies, and lifestyles of peoples in the drylands. We are already observing and experiencing the impacts of climate change. Environmental degradation exacerbated by climate change is threatening vital community infrastructure and is leading in some places to forced displacement and relocation. Reductions in precipitation and continued experiences of prolonged drought affect soil quality and herding and agricultural practices. Drought impacts have been worsened by increasing evapotranspiration rates, reduced soil moisture, and intensified stress on vegetation and local water sources. The influx of invasive species and prolonged drought are disrupting subsistence practices. These impacts threaten traditional knowledge, food security, water availability, historical homelands, and territorial existence and may undermine traditional ways of life that have persisted and adapted for thousands of years. In some areas of High Asia, permafrost melting and glacial retreat are making it more difficult for hunter/gatherers and transhumant herders to access traditional habitats and are changing the migration patterns of certain species as well as threatening the age-old seasonal migration by herds and flocks of seminomadic herders.

Insights about how these impacts are affecting herders and agropastoralists in the world's drylands can contribute to the development of policies, plans, and programs for adapting to climate change and reducing greenhouse gas (GHG) emissions. A growing body of literature examines the vulnerability, risk, resilience, and adaptation of peoples to climate change (Williams 2012; Norton-Smith et al. 2016). The knowledge and science of how climate change impacts are affecting peoples in the drylands contributes to the development of policies, plans, and programs for adapting to climate change in vulnerable locations and reducing greenhouse gas emissions.

Vulnerability is not characteristic of a community but is the product of systems of inequality. Herders and agropastoralists differ in their vulnerability to climate change based on their distinct cultural practices and economies, and the vulnerability of a particular ethnic group's sociopolitical, economic, and eco-cultural systems may differ by geography and climate regime. Although groups may face similarities in terms of how climate change may affect their socioeconomic status and dependence on natural resources, distinct cultural practices influence how climate change vulnerability is experienced. Despite this variability, similarities among dryland communities may exist in terms of the institutional barriers – including legal, and administrative, policies – that affect adaptation and resilience among land users. Government policies may have unintended consequences of limiting or removing climate adaptation options and in turn constraining, restricting, and undermining adaptation efforts within local dryland communities.

Not all drylands are unproductive, and where irrigation water is available, high levels of productivity can be realized. Concerns about glacial melt and thawing of the permafrost relate to water supplies for basic human needs and for irrigation. Already irrigators in dryland communities are adapting to the changed circumstance of reductions in the supply and quality of water.

Dryland communities are also adapting to the climatic variability by altering cropping patterns including the switch to new crop plants, changes to tillage practices that conserve moisture, water-saving agriculture, and soil conservation measures. Herders and other livestock owners are making adjustments too, e.g., a switch from raising cattle to sheep or goats (or camels) and better winter housing for livestock to conserve energy and ease the burden of supplementary feeding. These winter "barns" (often built using a greenhouse design that captures solar energy) can be used in summer/autumn for vegetable production for either subsistence (own use) or for barter or sale. Other adaptations include accepting the principle of "more from less" (Michalk et al. 2011). This involves focus on reducing livestock inventories and providing fewer animals with a high plane of nutrition and better winter housing that saves energy and leads to lower mortality, higher birth weights, and faster compensatory growth in spring and summer. The ability to turn off animals at a younger age reduces pressure on winter fodder needs and generates cash flow sooner (Squires et al. 2010).

Where land degradation is a problem (over 50 % of the world's drylands have varying degrees of land degradation), the impact of climate change, reduced precipitation (or the disruption to the seasonal distribution), rising temperatures, increase in frequency of droughts and longer duration, changes in pasture structure and botanical composition, and loss of biomass production will hamper efforts to arrest land gradation and make it even harder to reverse it.

Much can be learned from traditional knowledge held by land users in drylands, but regrettably there is a limit to how far this transfer of local ecological knowledge (LEK) can be applied to the present and future crises that land users face under the global change (including climate change) regimes. There is no doubt that LEK has a place in the planning of mitigation and adaptation strategies. Some observers point out that the sorts of challenges occurring now bear little resemblance to those dealt with in the past and that it may be asking too much of traditional herders and even long-term farming households to come up with appropriate coping and adaptation strategies to deal with challenges of global change (including climate change).

The post-2015 era has begun. The international agreement was reached on a climate accord[1] in Paris on climate change mitigation and adaptation, and the UN in New York issued its development goals that inter alia stressed arrest and reversal of land degradation.

This book is timely and should prove valuable to scholars, graduate students, UN and other aid agency personnel, NGOs, natural resource managers, and policy makers. If it can do this, we, as editors and authors, along with other contributors, will be very pleased.

Jodphur, Rajasthan, India
Beijing, China
Novermber 2016

Mahesh K. Gaur
Victor R. Squires

References

H. Kreutzmann, *Pastoral Practices in High Asia Agency of 'Development' Effected by Modernisation, Resettlement and Transformation* (Springer, Dordrecht, 2012)

D. Michalk, D. Kemp, G.D. Han, L.M. Hua, Z.B. Nan, T. Takahashi, J.P. Wu, Z. Xu, Redesigning livestock strategies to reduce stocking rates and improve incomes on western China's grasslands, in *Development of Sustainable Livestock Systems on Grasslands in North-Western China. ACIAR Proceedings No.134,* ed. by D.R. Kemp, D.L. Michalk (Eds), (2011), pp. 140–151. Available from: http://aciar.gov.au/files/node/13401/pr134_20130724_pdf_29838.pdf#page=141

---

[1] The accord was ratified and came into force in 2016. The Conference of Parties to the Climate Change Convention (COP22) met in Morocco in November 2016 to oversee development of an implementation pathway.

K. Norton-Smith, K. Lynn, K. Chief, K. Cozzetto, J. Donatuto, M. Hiza Redsteer, L.E. Kruger, J. Maldonado, C. Viles, K.P. Whyte, *Climate Change and Indigenous Peoples: A Synthesis of Current Impacts and Experiences,* Gen. Tech. Rep. PNW- GTR-944 (U.S. Department of Agriculture, Forest Service, Pacific Northwest Research Station, Portland, 2016), 136 p

V.R. Squires, L.M. Hua, D.G. Zhang, G. Li, *Towards Sustainable Use of Rangelands in NW China* (Springer, Dordrecht, 2010), 353 p

J. Williams, The impact of climate change on indigenous people –the implications for the cultural, spiritual, economic and legal rights of indigenous people. Int. J. Hum. Rights **16**, 4, 648–688 (2012)

Scope and Purpose

This book, written by 21 experts from 11 countries across the globe, seeks to provide insights into the way in which climate variability affects people, especially land users, in aridlands/drylands. We focus on the response by the myriad of herders, agropastoralists, and smallholder farmers to climate change. Consideration is given also to the measures and policies being put in place by governments at all levels. Coping strategies and adaptive transformations are discussed.

To better understand the nature and global extent of aridlands, we have commissioned chapters from selected countries from both hemispheres, countries with large populations and high densities in the arid zones (India, China) as well as countries like Australia with a large arid zone but low population densities in the arid zone. Special attention is given in this book to the less well-known problems confronting people in high-altitude, cold deserts that are widespread in Central Asia. We focus on the Qinghai-Tibet Plateau that has the highest mountains and some of the driest deserts. High-altitude regions are vulnerable to climate change, especially warming as this causes permafrost to melt and glaciers to retreat. These high-altitude places are the headwaters of major life-bringing rivers. Some of them are transboundary rivers that flow to some of the most densely populated areas in SE and SW Asia. Hundreds of millions of people in those areas (many of them living in poverty) depend on water originating in the "water tanks" of High Asia for their livelihoods.

Humans are major agents of change. The individual daily actions and decisions made on a daily basis by over one billion people who live in the world's aridlands/drylands can change the outcomes from global warming and other elements of climate change. We discuss the role of humans in what has been termed the Anthropocene (a new geological epoch in which humans bring irreversible change to the life-support systems on which we all depend). The image of a permanently changed globe that the Anthropocene represents provides compelling evidence in support of the call for the adoption of new worldviews and of the paradigm that they inform.

Apart from humans, there are countless other species whose existence is under threat as a result of anthropogenic activities that exacerbate degradation of the resource base and contribute to global warming by accelerating greenhouse gas (GHG) emissions. The burgeoning livestock inventories, especially ruminants, are contributing to GHG emissions as do the degraded aridlands that are now *net emitters* of carbon dioxide. Arrest and reversal of land degradation on the 40 % of the world's land surface that are classified as drylands could sequester at least 1 Gt[1] of carbon dioxide – a figure comparable to the total amount sequestered by the world's temperate forests. The challenge is to stabilize, and ultimately to restore, the productive capacity of the aridlands to support the human population and to preserve biodiversity of plants and animals, including invertebrates. These measures, to restore degraded lands, if taken now, will result in a "win-win" situation – better functioning aridland ecosystems and a slowing of climate change.

According to *Richard Bawden,* the world's aridlands are living examples of nature/culture dynamics in the face of global changes in both their biophysical and sociocultural environments. These water-stressed biomes are inherently vulnerable to environmental challenges on any scale and from whatever source. This vulnerability is being subjected to a veritable barrage of stresses that are truly global in their range and influences. The aridlands provide a rich focus for exploring the nature and consequences of many of these changes. The crux of the argument presented here is that the situation is now so dire that nothing less than a major change in the prevailing worldviews and thus in the dominant paradigm that they inform is essential. This shift will require a new consciousness with regard to the very concept of worldview as well as a readiness to explore alternatives. While there is some evidence of acceptance of the differences between a techno-centric and a holo-centric perspective on research, development, policy making, education, etc., it is still very much in a primitive mode. The focus of much research is changing from externally motivated strategies for increasing productivity of agricultural and pastoral enterprises in the aridland to embrace the far more systemic and inclusive notion of sustainable livelihoods and well-being achieved through what can be regarded as the adoption of a perspective of multifunctionality.

---

[1] 1 Gt is equivalent to 1^{15} tons.

Contents

Part I Background and Setting

1 **Geographic Extent and Characteristics of the World's Arid Zones and Their Peoples** 3
Mahesh K. Gaur and Victor R. Squires

2 **Recent Trends in Drylands and Future Scope for Advancement** 21
Ahmed E. Sidahmed

3 **Global Change and Its Consequences for the World's Arid Lands** 59
Richard Bawden

Part II Aridlands Under a Global Change Regime

4 **Humans as Agents of Change in Arid Lands: With Special Reference to Qinghai-Tibet Plateau, (China)** 75
Victor R. Squires and Haiying Feng

5 **Climate Variability and Impact on Livelihoods in the Cold Arid Qinghai–Tibet Plateau** 91
Haiying Feng and Victor R. Squires

6 **Integrated Watershed Management Approach in Drylands of Iran** 113
Hamid R. Solaymani and Hossein Badripour

7 **Thar Desert: Its Land Management, Livelihoods and Prospects in a Global Warming Scenario** 131
R.P. Dhir

Part III Northern Hemisphere Aridlands: Selected Examples

8 **Dryland Resources of North America with Special Reference to Regions West of the 100th Parallel** 151
Victor R. Squires

9 Desertification and Land Degradation in Indian Subcontinent: Issues, Present Status and Future Challenges 181
Ajai and P.S. Dhinwa

10 Drylands of China: Problems and Prospects 203
Shiming Ma

11 Drylands of the Mediterranean Basin: Challenges, Problems and Prospects... 223
Ahmed H. Mohamed and Victor R. Squires

Part IV Southern Hemisphere Aridlands: Selected Examples

12 Southern African Arid Lands: Current Status and Future Prospects 243
Klaus Kellner, Graham von Maltitz, Mary Seely, Julius Atlhopheng, and Lehman Lindeque

13 Arid and Semiarid Rangelands of Argentina................... 261
C.A. Busso and Osvaldo A. Fernández

14 The Impact of Climate Variability on Land Use and Livelihoods in Australia's Rangelands..................................... 293
David J. Eldridge and Genevieve Beecham

Part V Summary, Synthesis and Concluding Remarks

15 Drylands Under a Climate Change Regime: Implications for the Land and the Pastoral People They Support.............. 319
Mahesh K. Gaur and Victor R. Squires

16 Unifying Concepts, Synthesis, and Conclusions................. 335
Victor R. Squires and Mahesh K. Gaur

Contributors

Ajai E S- CSIR, Space Applications Centre, ISRO, Ahmedabad, India

Julius Atlhopheng Geomorphology and Climate Change, P/Bag UB0704, Faculty of Science, University of Botswana, Gaborone, Botswana

Hossein Badripour Forests, Range and Watershed Management Organization, Tehran, Iran

Richard Bawden Western Sydney University, Sydney, Australia

Genevieve Beecham ANU College of Law, Australian National University, Acton, ACT, Australia

C.A. Busso Departamento de Agronomía-CERZOS, (National Council for Scientific and Technological Research of Argentina: CONICET), Universidad Nacional del Sur, Bahía Blanca, Provincia de Buenos Aires, Argentina

P.S. Dhinwa TALEEM Research Foundation, Ahmedabad, India

R.P. Dhir Central Arid Zone Research Institute, Jodhpur, India

David J. Eldridge Centre for Ecosystem Science, School of Biological, Earth and Environmental Sciences, University of New South Wales, Sydney, New South Wales, Australia

Haiying Feng Institute of Administration and Management, Xining, Qinghai, China

Osvaldo A. Fernández Departamento de Agronomía-CERZOS, (National Council for Scientific and Technological Research of Argentina: CONICET), Universidad Nacional del Sur, Bahía Blanca, Provincia de Buenos Aires, Argentina

Mahesh K. Gaur ICAR-Central Arid Zone Research Institute, Jodhpur, India

Klaus Kellner Unit for Environmental Sciences and Management, North-West University, Potchefstroom, South Africa

Lehman Lindeque United Nations Development Programme (UNDP), Pretoria, South Africa

Graham von Maltitz Global Change and Ecosystems Dynamics: Natural Resources and the Environment, CSIR, Brummeria, Pretoria, South Africa

Shiming Ma Institute of Environment and Sustainable Development in Agriculture (IEDA), Chinese Academy of Agricultural Sciences (CAAS), Haidian District, Beijing, People's Republic of China

Ahmed H. Mohamed Desert Research Center, Cairo, Egypt

Hamid R. Solaymani Forests, Range and Watershed Management Organization, Tehran, Iran

Mary Seely Desert Research Foundation of Namibia (DRFN), Windhoek, Namibia

Ahmed E. Sidahmed FAO Investment Center, Africa Region TCIA, Rome, Italy

Victor R. Squires Guest Professor, Institute of Desertification Studies, Chinese Academy of Forestry, Beijing, China

List of Acronyms and Abbreviations

AfDB	African Development Bank
ASAL	Arid and semiarid lowlands
CAADP	Comprehensive Africa Agriculture Development Programme
CCAFS	Climate Change, Agriculture and Food Security
CFS	Committee on World Food Security
CGIAR	Consultative Group on International Agricultural Research
CMIP	Coupled Model Intercomparison Project
COSPAR	Committee on Space Research
CSIRO	Commonwealth Scientific and Industrial Research Organisation
DPSIR	Drivers–pressures–state–impacts–responses
EAC	East African Community
ESARO	Eastern and Southern Africa
FMNR	Farmer-managed natural regeneration
FAO	Food and Agriculture Organization
GHG	Greenhouse gases
ICAR	Indian Council of Agricultural Research
ICARDA	International Center for Agricultural Research in the Dry Areas
IDDRSI	IGAD Drought Disaster Resilience and Sustainability Initiative
IFPRI	International Food Policy Research Institute
IGAD	Intergovernmental Authority on Development
IIASA	International Institute for Applied Systems Analysis
IPCC	Intergovernmental Panel on Climate Change
ISRO	Indian Space Research Organisation
IUCN	International Union for Conservation of Nature
IWD	Irrigation water demand
LADA	Land Degradation Assessment in Dryland
LEK	Local ecological knowledge
LGP	Length of growing period
MCI	Multiple cropping intensity
NDVI	Normalized difference vegetation index
NEPAD	New Partnership for Africa's Development
NGOs	Nongovernmental organizations
ODA	Official Development Assistance
OECD	Organisation for Economic Co-operation and Development

QTP	Qinghai-Tibet Plateau
SAC	Space Applications Centre
SDGs	Sustainable Development Goals
UN	United Nations
UNCCD	United Nations Convention to Combat Desertification
UN EMG	United Nations Environment Management Group
UNEP	United Nations Environment Programme
UNFCCC	United Nations Framework Convention on Climate Change
USAID	United States Agency for International Development
WFP	World Food Programme
WMO	World Meteorological Organization

List of Figures

Fig. 1.1	Distribution of arid lands in the world	7
Fig. 1.2	Global distribution of drylands for 1961–1990 climatology derived from P/PET ratio based on observations	13
Fig. 2.1	Estimated reduction in the average annual number of drought-affected people resulting from FMNR and other technologies	44
Fig. 4.1	A simplified representation (causal diagram) of the main physical and biological mechanisms affecting ecosystem regulation in rangelands	82
Fig. 5.1	Map of Qinghai–Tibet Plateau (QTP) showing its immediate neighbors in mainland China and in adjacent countries	92
Fig. 5.2	Map showing Qinghai province and the location of the source area for the Huang He (*Yellow*) and ChangJiang (Yangtze) rivers. The Lancan Jiang (Mekong), Indus, Brahmaputra, and other important transboundary rivers also come from this area	92
Fig. 5.3	(**a**) Yaks and (**b**) Tibetan sheep and goats are the mainstay of the livestock industry on the Qinghai–Tibet Plateau. Both species can live on fibrous forage and endure severe cold	93
Fig. 5.4	The headwaters of the Three Rivers (see text) are in SW Qinghai, and the major parts of the source area are now part of the Sanjiangyuan National Nature Reserve	105
Fig. 6.1	The details of the IWM	116
Fig. 6.2	World's drylands and subtypes	118
Fig. 6.3	Iran's drylands distribution based on the climate zone	119
Fig. 6.4	Schematic diagram of a subsurface dam	125
Fig. 6.5	MENARID coordinates several levels of government and non-government groups, and local land users are represented on key committees	126

Fig. 7.1	Some artefacts of Harappan civilization: (**a**, **b**) seals; (**c**) a terracotta	132
Fig. 7.2	Growth of population over the decades in Thar, Rajasthan, and country as a whole (year 1921 = 100)	135
Fig. 7.3	District-wise net cropped area as percent of geographic area at three time periods, namely, years 1956–1957, 1981–1982 and 2010–2011. The numeral at the bottom corresponds to earliest period, i.e. 1956–1957, whereas middle and upper numeral is for successive periods, respectively	138
Fig. 7.4	Accelerated wind erosion is an outcome of current land use and management and a major manifestation of desertification problem. (**a**) Saltating sand deposits within field means a somewhat reduced fertility and a serious loss of land level. (**b**) Drift sands pile over dune flanks and crest. (**c**). These encroach also onto dwelling units	140
Fig. 8.1	Map showing the location of the "American West," including the 100th meridian	153
Fig. 8.2	Principal areas of sagebrush (*Artemisia*) shrub steppe in western USA	158
Fig. 8.3	Map showing the geographical extent of the Great Basin—a region characterized by basin and range topography and an endorheic drainage system	160
Fig. 8.4	The six major water divides in North America	162
Fig. 8.5	Federal government ownership characterizes this vast region. The public estate (shown here in percentage of each state owned by the federal government) includes national forests, national parks, and multiple-purpose public lands	163
Fig. 8.6	The western lands of USA use a lot of water for irrigation but there is a significant number of irrigated farms in the eastern states	164
Fig. 8.7	Percentage of population of each state in the contiguous western USA dependent on groundwater for domestic needs. Groundwater contributes a large proportion of total water extracted in each state in the USA west of the 100th meridian. New Mexico is especially dependent on groundwater	166
Fig. 9.1	Methodology for desertification mapping using satellite data	195
Fig. 9.2	Desertification/land degradation status map of India	196
Fig. 9.3	(**a**) Desertification Status Map for Ballia district, Uttar Pradesh, Northern India. The major process of land degradation is water logging. (**b**) The Desertification Status Map of Leh-south watershed in Jammu and Kashmir state in Northern India. Major processes of land degradation are frost shattering, frost heaving, mass wasting and water erosion	197

Fig. 9.4	Desertification change detection (Nathusari Block Sirsa district of Haryana state, India)	198
Fig. 9.5	Desertification status map for Panchkula district, Haryana state, India based on IRS images of 2001 (*left*) and 2011 (*right*)	198
Fig. 10.1	Distribution of arid and semiarid regions of China	204
Fig. 10.2	Dynamic changes of vegetation area and NDVI in the Northwest China	209
Fig. 10.3	The variations of irrigated crop-growing area in the arid areas of China	213
Fig. 10.4	Trends of temperature in the global, Northern Hemisphere, and arid region of Northwest China during 1960–2013	214
Fig. 10.5	Distributions of annual temperature change from 1971–2000 to the periods of 2016–2045 and 2046–2075 under three scenarios, respectively	215
Fig. 10.6	Distributions of annual precipitation change from 1971–2000 to the periods of 2016–2045 and 2046–2075 under three scenarios, respectively	216
Fig. 10.7	Distributions of annual irrigation water demand (IWD) change from 1971–2000 to the periods of 2016–2045 and 2046–2075 under three scenarios, respectively	218
Fig. 11.1	A typical climatograph for a site within the Mediterranean Basin	224
Fig. 11.2	There are numerous countries in the Mediterranean Basin. The drier areas are shaded in darker colours	224
Fig. 11.3	The desertification vortex: development (*left-hand boxes*) driving desertification (*solid arrows, central circle*), which generates security issues (*dotted arrows, right-hand boxes*), which drive intensified development (*broken arrow*) and further desertification	228
Fig. 12.1	Aridity map of southern Africa	244
Fig. 12.2	Broad vegetation types	245
Fig. 12.3	(a) Mean annual rainfall and (b) mean minimum temperature in July (normally the coldest month) (From World climate data)	246
Fig. 13.1	(a, b) Arid and semiarid territories	262
Fig. 13.2	Distribution of the five arid and semiarid territories in Argentina	263
Fig. 14.1	Distribution of Australia's rangelands	294
Fig. 14.2	Rainfall variability map for Australia based on data between 1900 and 1996. Variability is calculated as the difference between the 10th and 90th percentile divided by the 50th percentile	296

Fig. 16.1	World drylands..	336
Fig. 16.2	A schematic showing key elements of the DPSIR model.............	337
Fig. 16.3	Two contrasting paradigms. On the *left* is the more traditional "desertification," while on the *right* the counter paradigm where indirect drivers such as demography and sociopolitical prevail	343

List of Tables

Table 1.1	The extent of the global arid zones (expressed as a percentage of the global land area)	5
Table 1.2	Arid land climates	5
Table 1.3	Regional extent of arid zones	6
Table 1.4	Global distribution for the types of arid lands	6
Table 1.5	Population distribution in arid lands	12
Table 1.6	Human populations of the world's arid lands	15
Table 2.1	Regional extent of the drylands	22
Table 2.2	Classification of drylands by aridity index in aridity index global land area ESARO land	28
Table 2.3	Distribution of FLW along the food chain in different regions of the world	38
Table 4.1	A summary of the impacts of change agents on ecological and socioeconomic process in the Three Rivers region of Qinghai	76
Table 7.1	Land use in Thar (in %) at three points of time over more than five decades	137
Table 7.2	Vegetation cover and yield under non-degraded and degraded condition	141
Table 9.1	Indicator system for desertification monitoring and assessment	192
Table 9.2	Classification system evolved and adopted for desertification status mapping	194
Table 10.1	Annual average temperature of different decades in the different regions of the arid areas in China (in °C)	205
Table 10.2	Annual average precipitation of different decades in the different regions of the arid areas in China (in mm)	205
Table 10.3	Soil classification in the arid areas of China	210
Table 10.4	Status of land desertification in several main regions in northwest China (unit: 1,000 ha)	210

Table 11.1	Economic indicators of North African rangelands	226
Table 14.1	Summary of the short-term and long-term effects of increased climate variability on farming and grazing enterprises	305
Table 14.2	Predicted declines (%) in agricultural production in Australia for selected enterprises by 2050	308
Table 14.3	Social and economic barriers to former adaptation to climate change in rural communities	310
Table 16.1	World drylands (excluding hyper arid lands) in millions of hectares	336

Part I
Background and Setting

This Part contains three chapters. In Chap. 1, *Mahesh Gaur* and *Victor Squires* describe the global distribution of aridlands/dryland and explain something of the geography, climatology, and population patterns over every continent, except Antarctica. The areal extent of cold arid regions, hot dry regions, and a vast area of semiarid and dry subhumid together support the lives and livelihoods of about 1.3 billion people. Drylands encompass a vast mosaic of diverse and contrasting landscapes, plant and animal species, and human populations leading very different and unique lifestyles. In spite of their environmental sensitivity and perceived fragility, and despite the prevailing negative perceptions of drylands in terms of economic and livelihood potentials, these ecosystems have supported human populations for centuries. Dryland livelihoods represent a complex form of natural resource management, involving a continuous ecological balance between pastures, livestock, crops, and people. The people living in the drylands are heavily dependent upon ecosystem services directly or indirectly, for their livelihoods. But those services – from nutrient cycling, flood regulation and biodiversity to water, and food and fiber – are under threat from a variety of sources such as urban expansion, mining, and unsustainable land uses. As a result, these fragile soils are becoming increasingly degraded and unproductive.

Chapter 2 by *Ahmed Sidahmed* provides an up-to-date overview of the key issues that impinge on drylands the world over. Special attention is given to climate change and the responses by people at every level from subsistence farmers and herders through to large-scale "industrial farms" and to administrators, policy makers, politicians, and lawmakers. Drylands everywhere are under threat from a variety of sources such as urban expansion, mining, and unsustainable land uses. As a result, the resource base is becoming increasingly degraded and unproductive. Climate change is now aggravating these challenges. However, combating climate change and adapting communities to its impacts represent an opportunity for new and more sustainable investments and management choices that can also contribute to improved livelihoods and the fight against poverty among dryland communities. An ecosystem approach which includes restoration and renovation is one important direction that needs to be urgently undertaken.

Chapter 3 takes a broad view of aridlands/drylands both now and in the future as the sociology, ecology, and economics of land use and further development in the aridlands plays out. According to **Richard Bawden**, the world's aridlands are living example of nature/culture dynamics in the face of global changes in both their biophysical and sociocultural environments. These water-stressed biomes are inherently vulnerable to environmental challenges on any scale and from whatever source. This vulnerability is being subjected to a veritable barrage of stresses that are truly global in their range and influences. The aridlands provide a rich focus for exploring the nature and consequences of many of these changes. The crux of the argument presented here is that the situation is now so dire that nothing less than a major change in the prevailing worldviews and thus in the dominant paradigm that they inform is essential. This shift will require a new consciousness with regard to the very concept of worldview as well as a readiness to explore alternatives. While there is some evidence of acceptance of the differences between a techno-centric and a holo-centric perspective on research, development, policy making, education, etc., it is still very much in primitive mode. The focus of much research work is changing from externally motivated strategies for increasing productivity of agricultural and pastoral enterprises in the aridland to embrace the far more systemic and inclusive notion of sustainable livelihoods and well-being achieved through what can be regarded as the adoption of a perspective of multifunctionality.

Geographic Extent and Characteristics of the World's Arid Zones and Their Peoples

Mahesh K. Gaur and Victor R. Squires

1.1 Introduction

Arid zones are generally defined as regions where rate of evaporation is greater than precipitation. These are also characterized by persistent water scarcity, frequent drought, high climatic variability and high wind velocity and various forms of land degradation, including desertification and loss of biodiversity. As such, arid lands occupy 5.36 million km^2 or 41% area of the earth's land surface and are home to roughly more than a third of the world's population (Mortimore 2009). Large areas of arid zones are located in North and South America, North Africa, the Sahelian region, Africa South of the Equator, the Near East and the Asia and the Pacific regions between latitudes of 15 and 30° in both northern and southern hemispheres. Arid zones are home to roughly 2.5 billion people who rely directly on arid land ecosystem services for their livelihoods (UN EMG 2011) and support 50% of the world's livestock where 44% of the world's food is grown.

As per the *aridity index*, there are four categories of arid lands, hyperarid, arid, semi-arid and dry subhumid regions. So, the main distinction could be made between deserts (hyperarid and arid) and semi-deserts (semiarid). Hyperarid zones cover 6.6% of earth's land surface, arid zones cover 10.6%, while semiarid zones are more extensive, occur in all continents and cover 15.2% whereas the dry subhumid category covers 8.7% of earth's land surface (UN 2010). Lack of water, limited foodstuffs and extremes of climatic phenomenon have generally made arid zones critical places for any kind of habitation. This desperate situation is predicted to worsen even further as a result of climate change and shifting weather patterns due to anthropogenic activities.

M.K. Gaur (✉)
ICAR-Central Arid Zone Research Institute, Jodhpur-342003, India
e-mail: maheshjeegaur@yahoo.com

V.R. Squires
Institute of Desertification Studies, Beijing, China
e-mail: dryland1812@internode.on.net

From the climatic point, *aridity* is an overall moisture deficit (especially when resulting from a permanent absence of rainfall for a long period of time) under average climatic conditions. As such, the cause of aridity is significant within the context of various environmental and atmospheric factors (including global climate change), which contribute to the occurrence of aridity. In classification of aridity zones, the aridity index or the index of moisture deficit is generally taken into consideration. It is calculated by comparing the incoming moisture totals with potential moisture loss: the evapotranspiration/precipitation levels.

1.2 Definition of Arid Zones

Definitions and delimitations of arid environments and deserts abound, varying according to the purpose of the enquiry or the location of the area under consideration. Numerous literary, climatic and biological classifications have been proposed for arid lands (McGinnies 1988) and there seem to be two separate types of definition of arid zones: the one is series of *literary definitions* based on the changing meanings of '*desert*', and the other is a series of *scientific attempts* to establish a meaningful *boundary* to the arid zone. *Literary definitions*, thoroughly reviewed by Heathcote (1983), commonly employ terms such as 'inhospitable', 'barren', 'useless', 'unvegetated' or 'devegetated', and 'devoid of water'. The *scientific approaches* to the definition of the arid zones have varied according to the aims of the enquiry. Arid zones may be those areas of the world where evaporation exceeded precipitation. This boundary, or '*dry line*', is comparable in some ways to the 'snowline' in mountainous areas. *Scientific definitions* have been based on a number of criteria, including erosion processes (Penck 1894), drainage pattern (de Martonne and Aufrère 1927), climatic criteria based on plant growth (Köppen 1931) and vegetation types (Shantz 1956). A region is arid when it is characterized by a severe lack of available water, to the extent of hindering or preventing the growth and development of plant and animal life. Environments subject to arid climates tend to lack vegetation and are called *desertic*. As such, the word '*arid*' has been derived directly from Latin *aridus* meaning "dry, arid, parched," from *arere* "to be dry". Further, all schemes involve a consideration of moisture availability, at least indirectly, through the relationship between precipitation and evapotranspiration.

Meigs's (1953) classification of arid environments was produced on behalf of UNESCO. Being ultimately concerned with global food production, it is not surprising that arid areas too cold for plant growth (such as polar deserts) were excluded from the classification. Meigs based his scheme on Thornthwaite's (1948) indices of moisture availability (Im):

$$\text{Im} = \frac{(100S - 60D)}{PE}$$

where PE is potential evapotranspiration, calculated from meteorological data and S and D are, respectively, the moisture surplus and moisture deficit, aggregated on an annual basis from monthly data and taking stored soil moisture into account.

Table 1.1 The extent of the global arid zones (expressed as a percentage of the global land area)

Classification	Semiarid	Arid	Hyperarid	Total
Köppen (1931)	14.3	12.0	–	26.3
Thornthwaite's (1948)	15.3	15.3	–	30.6
Meigs (1953)	15.8	16.2	4.3	36.3
Shantz (1956)	5.2	24.8	4.7	34.7
UN (1977)	13.3	13.7	5.8	32.8

Source: Heathcote (1983)

Table 1.2 Arid land climates

		Mean temperature (°C)	
	% of arid lands	Coldest month	Warmest month
Hot	43	10–30	>30
Mild winter	18	10–20	10–30
Cool winter	15	0–10	10–30
Cold winter	24	<0	10–30

Source: Meigs (1952)

Meigs (1953) identified three types of arid environments: semiarid ($-40 \leq Im < -20$), arid ($-56 \leq Im < -40$) and extreme (or hyper–) arid ($\leq Im < -156$). Grove (1977) subsequently attached mean annual precipitation values to the first two categories (200–500 mm and 25–200 mm, respectively), though these are only approximate. Hyperarid areas have no seasonal precipitation regime and occur where 12 consecutive months without precipitation have been recorded (Meigs 1953). According to this scheme, these three environments cover about 36% of the global land area, differing only slightly from areas calculated according to this schemes (Table 1.1).

United Nations has declared 2010–2020 as *Decade for Deserts and the Fight Against Desertification*. According to it, the spread of the arid lands around the world is as follows:

Common name		Area (mill.km^2)	Percentage of total arid lands	Percentage of world land surface
Desert	Hyperarid	9.8	16.09	6.6
Semi-desert	Arid	15.7	25.78	10.6
Grassland	Semiarid	22.6	37.11	15.2
Rangelands	Dry subhumid	12.8	21.02	8.7
Total		60.9	100.00	41.3

Adapted from http://www.un.org/en/events/desertification_decade/whynow.shtml and modified

Meigs (1952) divided arid lands into those that are hot all year round and those with mild, cool and cold winters (Table 1.2). Variations in temperature affect the seasonal availability of moisture, by influencing evapotranspiration rates and affecting the form of precipitation in relatively high-latitude arid areas.

Table 1.3 Regional extent of arid zones

Region	Aridity zone							
	Arid		Semiarid		Dry subhumid		All arid lands	
	1000 km²	%	1000 km²	%	1000 km²	%	1000 km²	%
Asia (incl. Russia)	6164	13	7649	16	4588	9	18,401	39
Africa	5052	17	5073	17	2808	9	12,933	43
Oceania	3488	39	3532	39	996	11	8016	89
North America	379	2	3436	16	2081	10	5896	28
South America	401	2	2980	17	2223	13	5614	32
Central America and the Caribbean	421	18	696	30	242	10	1359	58
Europe	5	0	373	7	961	17	1359	24
World total	15,910	12	23,739	18	13,909	10	53,558	40

Source: FAO (2008)

Table 1.4 Global distribution for the types of arid lands

Arid land sub-habitat	Aridity index[a]	Share of global area (%)	Share of global population (%)	% rangeland	% cultivated	% other (incl. urban)
Hyperarid	<0.05	6.6	1.7	97	0.6	3
Arid	0.05–0.20	10.6	4.1	87	7	6
Semiarid	0.20–0.50	15.2	14.4	54	35	10
Subhumid	0.50–0.65	8.7	15.3	34	47	20
Total		41.3	35.5	65	25	10

Source: Safriel and Adeel (2005)
[a]The ratio of precipitation to potential evapotranspiration

There is no widely accepted definition of the term *arid zones*. Two of the most commonly accepted definitions are those of FAO and the United Nations Convention to Combat Desertification (UNCCD 2000). FAO has defined arid zones as those areas with a length of growing period (LGP) of 1–179 days (FAO 2000); this includes regions classified climatically as arid, semiarid and dry subhumid. The UNCCD classification employs a ratio of annual precipitation to potential evapotranspiration (P/PET). This value indicates the maximum quantity of water capable of being lost, as water vapour, in a given climate, by a continuous stretch of vegetation covering the whole ground and well supplied with water. Thus, it includes evaporation from the soil and transpiration from the vegetation from a specific region in a given time interval (WMO 1990). Under the UNCCD classification, arid zones are characterized by a P/PET of between 0.05 and 0.65.

According to both classifications, the hyperarid zones (LGP = 0 and P/PET <0.05), or true deserts, are not included in the arid zones and do not have potential for agricultural production, except where irrigation water is available. While about 47.2% of the world's total land area is considered to be arid zones (according to the UNCCD classification system), the extent of arid zones in various regions ranges from about 20% to 90% (Tables 1.3 and 1.4 and Fig. 1.1).

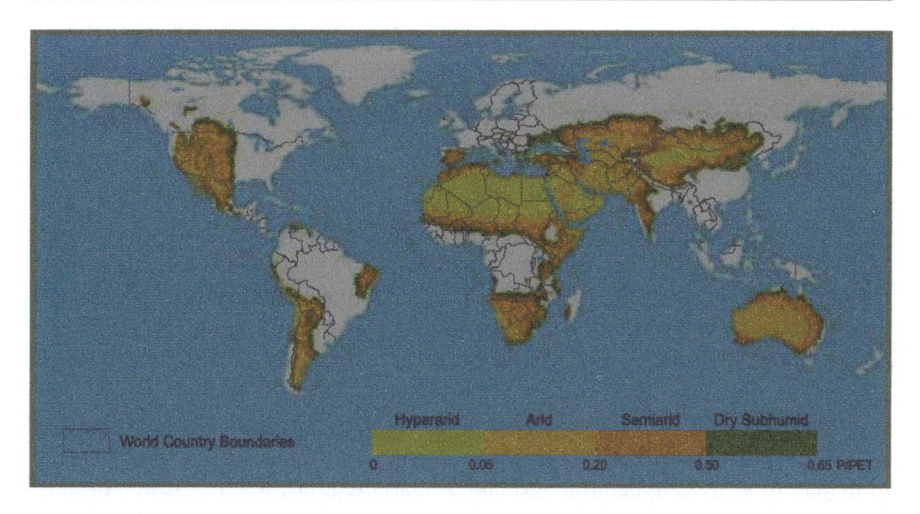

Fig. 1.1 Distribution of arid lands in the world (*Source*: IIASA/FAO 2003)

Actually, global arid zone is considered to include all elements of this threefold classification. There are several reasons for this:

1. The division between the three elements are somewhat arbitrary, often based on limited climatic data.
2. In arid lands, annual precipitation frequently varies substantially from year to year, so that in semiarid, arid and hyperarid areas, the only safe assumption is that any year could be extremely arid (Shantz 1956).
3. Distinct geomorphic thresholds in terms of processes and landforms have not been identified between the three elements of the scheme.
4. Climatic fluctuations and anthropogenic activities in the twentieth century have caused the expansion of arid physical conditions into some semiarid areas.
5. Semiarid areas are often called '*deserts*' by their inhabitants.

1.3 Characteristics of Arid Zones

Arid lands are a vital part of the earth's human and physical environments. They encompass grasslands, agricultural lands, forests and urban areas. The salient characteristics of hot arid zone are the following:

1. Aridity index is high (more than 70%).
2. High range of temperature and solar radiation, erratic and low rainfall, low atmospheric humidity and high wind velocity in summer.
3. Famine of food, fodder and water, i.e. '*Trikal*' is a permanent guest in every 3 or 5 years.
4. Soil is sandy with low water holding capacity, low organic content and deficit in nitrogen and phosphorus contents.

5. Vegetation is sparse, limited with low biomass production and xerophytic.
6. Heavy pressure of men and animals (cutting, felling, lopping, grazing, browsing, etc.) on vegetation.
7. Human population is thin except in the Indian arid zone, which is one of the most densely populated arid zones of the world.

Arid land ecosystems play a major role in global biophysical processes by reflecting and absorbing solar radiation and maintaining the balance of atmospheric constituents (Ffolliott et al. 2002). They provide much of the world's grain and livestock, forming the habitat that supports many vegetable species, fruit trees and micro-organisms. Sand covers 20% of the earth's surface. Over 50% of this area is deflated desert pavements.

Water scarcity is the predominant feature of arid lands. While heavy rains may occur, rainfall typically varies, sometimes dramatically, from season to season and from year to year. In hyperarid, arid and semiarid regions, water is scarce most of the time, and human settlements may cluster around rare sources of water such as rivers, springs, wells and oases. In such areas, traditional cultures have developed ways of finding, conserving and transporting water, including specialized land management techniques and structures to capture and retain precipitation or to encourage groundwater recharge.

Unsustainable land and water use and the impacts of climate change are driving the degradation of arid lands. Approximately 6 million km^2 (about 10%) of arid lands bear a legacy of land degradation. Such degradation – sometimes also referred to as 'desertification' – can take the form of soil erosion, nutrient depletion, water scarcity, altered salinity or the disruption of biological cycles (UNEP 2007). It has been estimated that about 1–6% of arid land human populations live in desertified areas, but a much larger percentage is under threat from further desertification (MEA 2005).

System productivity is greatly limited by inherently poor soil and/or human-induced soil degradation. On poorly managed land, the share of water that is available to plants can be as low as 40–50% of rainfall. On severely degraded land, as little as 5% of total rainfall may be used productively. '*Agricultural droughts*' can emerge even when water itself is not scarce within the landscape: when low soil fertility, poor crop and soil management and the use of poorly adapted varieties combine, the result is rainfall that is not being fully utilized for plant growth and grain filling (Humphreys et al. 2008).

In the wetter semiarid and subhumid regions, total seasonal rainfall often exceeds crop water needs. In fact, as long as appropriate levels of inputs are used, there is typically enough rainfall to double, and sometimes even quadruple, yields. In these areas, low soil fertility and a lack of inputs (particularly nitrogen) are major constraints to increasing yield and rainwater productivity – for example, most poor, smallholder farmers in sub-Saharan Africa do not apply fertilizer (Hilhorst and Muchena 2000; Morris et al. 2007; Twomlow et al. 2008). Nevertheless, there is evidence of positive trends in productivity in long-term data for certain African arid land countries.

Even in dry, semiarid temperate areas, such as Central West Asia and North Africa, seasonal rainfall of only 300–400 mm is enough to produce as much as 4 tonnes per hectare (t/ha) of wheat grain because precipitation falls during the cool winter growing season and because the growing season is longer. However, yields are typically less than half of this.

High variabilities in both rainfall amounts and intensities are characteristics of arid land regions, as are the occurrence of prolonged periods of drought. A drought is defined as a departure from the average or normal conditions, sufficiently prolonged (1–2 years – FAO 2004) as to affect the hydrological balance and adversely affect ecosystem functioning and the resident populations. There are actually four different ways that drought can be defined (National Weather Service 2004). Meteorological drought is a measure of the departure of precipitation from normal. Due to climatic differences, a drought in one location may not be a drought in another location. Agricultural drought refers to situations where the amount of soil water is no longer sufficient to meet the needs of a particular crop. Hydrological drought occurs when surface and subsurface water supplies are below normal. Socioeconomic drought describes the situation that occurs when physical water shortages begin to affect people.

The terms drought and aridity are sometimes used interchangeably, but they are different.

1.4 Causes of Aridity

Aridity is basically a comparison between water supply and water demand. Water supply in general is the amount of water received from precipitation, while water demand is measured in terms of evapotranspiration. Potential evaporation may be estimated by use of commonly observed climatological data. Aridity may be considered as an expression in a qualitative or quantitative manner of the dryness of an area. Aridity results from climatic, topographic and oceanographic factors which prevent moisture-bearing weather systems reaching an area of the land surface. Broadly, aridity arises from general causes acting individually or in combination of various other factors.

1.4.1 Continentality or Distance

Distance from the oceans prevents the penetration of rain-bearing winds into the center of large continents by topography or by distance. Precipitation and evapotranspiration are both usually lower than in arid areas owing their origins to atmospheric stability, while cold winters are common. Relatively small arid areas are surrounded by an extensive zone of semiaridity. Part of the desert area of the United States and the Monte-Patagonian Desert to the leeward of the Andes in South America is a result of the acidifying effect major mountain barriers have on air masses which move over them. One of the causes of the Takla Makan, Turkestan and Gobi deserts of Central Asia is the great distance from major moisture sources.

1.4.2 Atmospheric Stability

A general cause of aridity is the formation of dry, stable air masses that resist convective currents. The Somali-Chalbi desert probably owes its existence to a stable environment produced by large-scale atmospheric motions. Deserts dominated by the eastern portions of subtropical high-pressure cells originate in part from the stability produced by these pressure and wind systems. Tropical and subtropical deserts cover about 20% of the global land area (Glennie 1987). These are concentrated in zones of descending, sable air, the tropical high-pressure belts. In these areas, large arid zones are composed of central arid areas surrounded by relatively small, marginal, semiarid belts.

Also, aridity can also result from the lack of storm systems where precipitation is very unreliable and the mechanisms that cause convergence, create unstable environments, and provide the upward movement of air which is necessary for precipitation. The paths, frequencies and degrees of development of midlatitude cyclones or tropical cyclones are crucial factors in the production of rainfall.

The deserts of the subtropical latitudes are particularly sensitive to the climatology of cyclones. The Arabian and Australian deserts and the Sahara are examples of regions positioned between major wind belts with their associated storm systems.

1.4.3 Rain

Widespread rains are almost unknown over large parts of the hot deserts, most of the precipitation coming in violent convectional showers that do not cover extensive areas. The *wadis*, entirely without water during most of the year, may become torrents of muddy water filled with much debris after one of these flooding rains.

Because of the violence of tropical desert rains and the sparseness of the vegetation cover, temporary local runoff is excessive, and consequently less of the total fall becomes effective for vegetation or for the crops of the oasis farmer. Much of the precipitation that reaches the earth is quickly evaporated by the hot, dry desert air. Rainfall is always meagre.

In inter-ascending air, motion is associated primarily with extratropical wave cyclones, large-scale low pressure systems carried by the winds blowing from the west. This polar jet stream brings abundant precipitation to middle latitudes (40° and 60°), particularly along the west coasts of the continents.

In addition, it is extremely variable from year to year. The dependability of precipitation usually decreases with decreasing amount. No part of the earth is known for certain to be absolutely rainless, although in northern Chile, the rainfall over a period of 17 years was only 0.5 mm. During the whole 17 years, there were only three showers heavy enough to be measured.

1.4.4 Temperature

Skies are normally clear in the low latitude deserts so that sunshine is abundant. Annual ranges of temperature in the low latitude deserts are larger than in any other type of climate within the tropics. It is the excessive summer heat, rather than the winter cold, that leads to the marked differences between the seasons.

During the high-sun period, scorching, desiccating heat prevails. Midday readings of 40–45 °C are common at this season. During the period of low sun, the days still are warm, with the daily maxima usually averaging 15–20 °C and occasionally reaching 25 °C. Nights are distinctly chilly with the average minima in the neighborhood of 10 °C.

Most marked causes, however, are the large daily ranges, clear cloudless skies and relatively low humidity which permit an abundance of solar energy to reach the earth by day but also allow a rapid loss of energy at night. Large diurnal ranges in deserts are also associated with the meager vegetation cover, which permits the barren surface to become intensely heated by day.

1.4.5 Cold Ocean Currents

Cold ocean current affects the western coastal margins of South America and Southern Africa. This reinforces climatic conditions, causing low sea surface evaporation, high atmospheric humidity, low precipitation (very low rainfall, with precipitation mainly in the form of fog and dew) and a low temperature range. Aridity may result if air is cooled, and then rewarmed, prior to reaching the region, when:

(a) Cool air holds less moisture than warm air.
(b) When warm, moist air is cooled, excess water condenses and falls as precipitation. If it is subsequently rewarmed, it will be drier than it was previously.
(c) Winds that blow onshore tend to do so across cold currents produced by movement of water from high latitudes (poles) to low latitudes (equator) and associated with the upwelling of cold waters from the ocean's depth.
(d) Cold or cool winds have relatively small moisture-bearing capacity and, when warmed during their passage over the land, they become stable and, thereby, reinforce the stability produced by the global stability of these latitudes. (subtropical highs).
(e) This also occurs along coastal areas where there are cold coastal seas (*Baja California*) and in rain shadows (adiabatic heating and cooling).
(f) Sometimes, air moving across the frigid currents is cooled to a low temperature; thus, the air holds little moisture when it arrives over land, where it may provide fog or mist but rarely rain (Namib and Atacama).

1.5 Types of Aridity

Aridity refers to the average conditions of limited rainfall and water supplies not to the departures from the norm, which define a drought. All the characteristics of arid regions must be recognized in the planning and management of natural and agricultural resources (Jackson 1989). Because the soils of arid environments often cannot absorb all of the rain that falls in large storms, water is often lost as runoff (Brooks et al. 1997). At other times, water from a rainfall of low intensity can be lost through evaporation when the rain falls on a dry soil surface. Molden and Oweis (2007) state that as much as 90% of the rainfall in arid environments evaporates back into the atmosphere leaving only 10% for productive transpiration.

Ponce (1995) estimates that only 15–25% of the precipitation in semiarid regions is used for evapotranspiration and that a similar amount is lost as runoff. Evapotranspiration is the sum of transpiration and evaporation during the period a crop is grown. The remaining 50–70% is lost as evaporation during periods when beneficial crops are not growing.

During the last 20,000 years, the extension and the position of arid lands have changed to a great extent, that is, from glacial maximum 20,000 years ago, to the humid optimum 11,000–7000 years ago, and finally by drier conditions which have continued till the present day.

The problem of desertification is particularly acute in Africa, which has 37% of the world's arid zones. About 66% of its land is either desert or arid lands. The impact is also severe in Asia, which holds 33% of the world's arid zones. At present, almost 2.5 billion people inhabit in arid and semiarid lands meaning they are the home to one in three people in the world today. According to UN-Habitat, the 18.5% population growth rate in the arid lands was faster than that of any other ecological zone. Population density increases as aridity decreases. It ranges from 10 people per km^2 in the deserts to 71 people in the dry subhumid (rangelands) areas (Table 1.5). They are traditionally, hunter-gatherers, agriculturalists (rainfed or irrigated) and pastoralists. These traditional lifestyles are now rapidly fading under the influence of ever-expanding urbanization, mining, industrialization and tourism. The arid regions in general contribute significantly to the economy of the respective country in terms of employment, land holding, mineral assets, crop, forage and livestock production, including dairy products and industrial production. This is being supported despite a hostile climate and limited natural resource base.

Table 1.5 Population distribution in arid lands

Ecosystem	Total population	Share of global population
Desert	101,336	1.7
Semi-desert	242,780	4.1
Grassland	855,333	14.4
Rangelands	909,972	15.3
Total	2,109,421	35.5

Source: http://www.un.org/en/events/desertification_decade/whynow.shtml

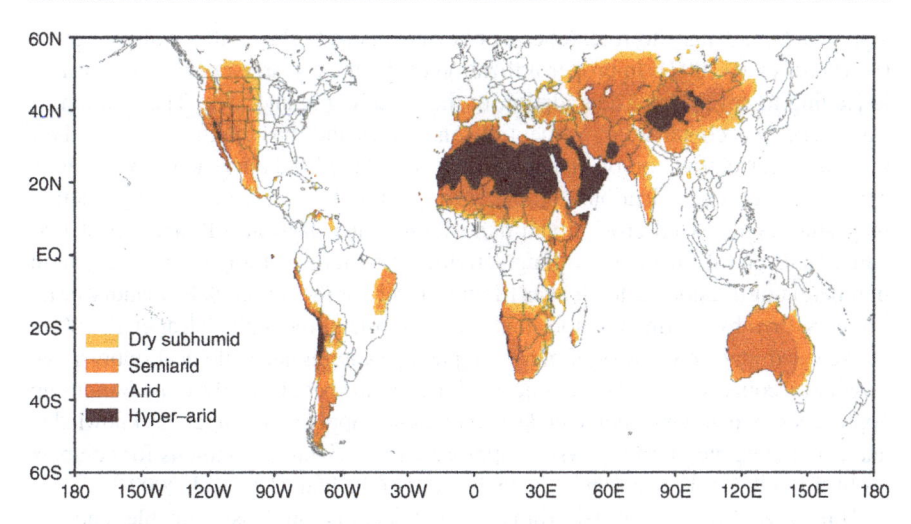

Fig. 1.2 Global distribution of drylands for 1961–1990 climatology derived from P/PET ratio based on observations (After Feng and Fu 2013)

Actually, arid and semiarid lands have been listed in Agenda 21 (UNCED 1992) among the fragile ecosystems of the world although they are described as being important ecosystems, with unique features and *resources*. These ecosystems are regional in scope, as they transcend national boundaries. Desertification is land degradation in arid, semiarid and dry subhumid areas resulting from various factors, including climatic variations and human activities. The most obvious impact of desertification, in addition to widespread poverty, is the degradation of 3.3 billion hectares of the total area of rangeland, constituting 73% of the rangeland with a low potential for human and animal carrying capacity, as well as decline in soil fertility and soil structure on about 47.2% of the arid lands (United Nations Conference on Environment & Development, Rio de Janeiro, Brazil, June 1992).

Feng and Fu (2013) have examined the changes in the areal extent of arid lands by analyzing observations for 1948–2008 and simulations for 1948–2100 from 27 global climate models participating in phase 5 of the Coupled Model Intercomparison Project (CMIP5) (Taylor et al. 2012). They have shown that global arid lands have expanded in the last 60 years and will continue to expand in the twenty-first century. By the end of this century, the arid lands under a high greenhouse gas emission scenario are projected to be 5.8×10^6 km^2 (or 10%) larger than the 1961–1990 climatology (Fig. 1.2). Expansion of arid lands associated with an increase in aridity as a result of climate change has a direct consequence on the desertification (i.e. land degradation in arid, semiarid and dry subhumid areas) and is a central issue for sustainable development, especially in the face of population growth (Reynolds et al. 2007).

Livelihood sustainability in these regions is threatened by a complex and interrelated range of social, economic, political and environmental changes that present significant challenges to researchers, policymakers and, above all, rural land users (Reynolds et al. 2007). Improper utilization of the most scarce resource, i.e. water, is leading to severe land degradation in the canal command area. The fixation of sand dunes has contained movement of sand dunes and thus making possible their utilization for cultivation or plantation. It has curtailed the apprehension of desert expansion, but desertification is still occurring through lateral routes of degradation of pasture lands, wind erosion and deposition, water erosion, salinization of irrigated lands and ill-managed mining activities. Shortage of forages for a very high population of livestock often makes life miserable for the livestock husbandry practitioners, thereby denting the traditional social fabric of the region (Kar et al. 2009).

New and innovative ways of generating and promoting technological options that take into account social and economic conditions of the people involved use bottom-up approaches by involving people in decision-making, apply vast indigenous knowledge and adopt integrated farming system approaches in arriving at solutions for complex problems of the arid ecosystem that are the need of the hour (Kar et al. 2009).

Human well-being is at risk from arid land degradation. Unsustainable land and water use and the impacts of climate change are driving the degradation of arid lands. Approximately 6 million km^2 of arid lands (about 10%) bear a legacy of land degradation. Such degradation – sometimes also referred to as 'desertification' – can take the form of soil erosion, nutrient depletion, water scarcity, altered salinity or the disruption of biological cycles. Degradation reduces biological productivity and can impact the ability of ecosystems to absorb and use rainwater. Combined with poor crop and soil management, and the use of poorly adapted varieties of crop, this can lead to 'agricultural droughts'. Climate change is already causing significant decreases in crop yields in some rainfed African agricultural systems.

This is likely to worsen by 2020 because climate change will cause grassland productivity to decline between 49% and 90% in semiarid and arid regions. It is also forecasted that high levels of desertification and soil salinization, and increasing water stress, will occur in parts of Asia, sub-Saharan Africa and Latin America. Situation is accentuated due to poor land management and irrigation practices in the drylands of the world where economy is primarily based on agriculture. Any shortage of water for agricultural use is likely to have cascading effect on various sectors of economy. Climatic fluctuations may be most pronounced in the poorest regions with high levels of chronic undernourishment and a great degree of instability. Food price fluctuations already represent a risk to vulnerable populations that is expected to increase with climate change. Arid land degradation costs developing countries an estimated 4–8% of their national gross domestic product (GDP) each year. It has been estimated that about 1–6% of arid land human populations live in desertified areas, while a much larger number is under threat from further desertification. Land degradation and poverty are mutually reinforcing, but the former has low public visibility. It is hard to deal with the problem due to cyclical swings in rainfall, land tenure which is not well adjusted to environmental conditions, and regional and global forces driving local management. Inaction would mean a cumulative addition to a long, historical legacy of degradation, from which recovery has already previously proven difficult (UN EMG 2011).

Climate change associated with changes in precipitation and PET would lead to changes in aridity (i.e. P/PET) and thus the areal extent of arid lands. The global arid lands from observations have been expanding during the past 60 years (Feng and Fu 2013). The area of global arid lands of the past 15 year (1991–2005) is ~2.4 × 10^6 km^2 (or 4%) larger than that during the 1950s. The areas of individual arid land components (except hyperarid regions) show increases. The observed maximum (minimum) area of semiarid (arid) regions during 1970s is a result of large positive precipitation anomalies in the southern hemisphere over Australia, Southern Africa and Patagonia (Hulme 1996), which led to a transition of arid to semiarid regions.

Evidence from both observations and models shows that over the past few decades, the tropics have widened (e.g. Fu et al. 2006; Hu and Fu 2007; Lu et al. 2007; Seidel et al. 2008; Johanson and Fu 2009; Zhou et al. 2011; Fu and Lin 2011), which results in shifts in precipitation patterns (Seidel et al. 2008) and especially a reduction of precipitation in the latitudes between subtropical arid zones and mid-latitude precipitation belts (Zhou et al. 2011; Scheff and Frierson 2012). Extensive land use and other human activities may further amplify these changes (Reynolds et al. 2007). Another important consequence of these changes is the desertification, i.e. the land degradation in arid, semiarid and dry subhumid regions resulting from various factors, including climatic variations and human activities (UNCCD 1994).

1.6 People and Land Use

Arid land populations are frequently some of the poorest in the world due to frequent drought, famine and internal civil wars. The population distribution patterns vary within each region and among the climate zones comprising arid lands. The population distribution patterns vary within each region and among the climate zones comprising arid lands. Regionally, Asia has the largest percentage of population living in arid lands: more than 1400 million people or 42% of the region's population. Africa has nearly the same percentage of people living in arid lands (41%) although the total number is smaller at almost 270 million. South America has 30% of its population in arid lands or about 87 million people (Table 1.6).

Table 1.6 Human populations of the world's arid lands

	Aridity zone							
	Arid		Semiarid		Dry subhumid		All arid lands	
Region	1000 km^2	%	1000 km^2	%	1000 km^2	%	1000 km^2	%
Asia (incl. Russia)	161,554	5	625,411	18	657,899	19	1,444,906	42
Africa	40,503	6	117,647	18	109,370	17	267,563	41
Oceania	275	1	1342	5	5318	19	6960	25
North America	6257	2	41,013	16	12,030	5	59,323	25
South America	6331	2	46,852	16	33,777	12	86,990	30
C. America and the Caribbean	6494	6	12,888	11	12,312	8	31,719	28
Europe	629	0	28,716	5	111,216	20	140,586	25
World total	222,043	4	873,871	4	941,922	17	2,038,047	37

Source: UNSO/UNDP (1997)

The population growth in semiarid regions has been quite rapid, and the number of people inhabiting these regions has grown tremendously during the recent decades as in the rest of the world.

Rural people living in arid lands can be grouped into nomadic, seminomadic, transhumant and sedentary smallholder agricultural populations. Nomadic people are found in pastoral groups that depend on livestock for subsistence and, whenever possible, farming as a supplement. Following the irregular distribution of rainfall, they migrate in search of pasture and water for their animals. Seminomadic people are also found in pastoral groups that depend largely on livestock and practise agricultural cultivation at a base camp, where they return for varying periods.

Transhumant populations combine farming and livestock production during favorable seasons, but seasonally they might migrate along regular routes using vegetation growth patterns of altitudinal changes when forage for grazing diminishes in the farming area. Sedentary (smallholder) farmers practise rainfed or irrigated agriculture (Ffolliott et al. 2002) often combined with livestock production. The proliferation of woody vegetation in drylands has been a consequence of heavy grazing by the cattle and small ruminants that has removed most of the herbaceous (grass) species. Simultaneously, arid and semiarid regions are being pushed across a bioclimatic threshold induced by the interaction of land management and climate variability. The drylands in Uzbekistan face a significant and growing threat of degradation, because forestry and extensive pastures have to compete for land use. It has significant direct implications for local rural populations as well as also have a significant national implications for food security and long term sustainable development because of the impact on biodiversity.

The human populations of the arid lands live in increasing insecurity due to land degradation and desertification and as the productive land per capita diminishes due to population pressure. The sustainable management of arid lands is essential to achieving food security and the conservation of biomass and biodiversity of global significance (UNEP 2000).

The rapid population growth in arid lands due to improvements in health conditions and other factors has placed tremendous pressure on the natural resource base. Often, the inevitable result of increasing population in resource-poor areas is land degradation defined as the loss of production capacity of the land (FAO 2000). The simple view of population pressure in a fragile environment causing permanent environmental degradation has been subject to re-evaluation. In the Yatenga province of Burkina Faso, farmers rescued their fields from imminent desertification by erecting low stone walls along the contours of hillsides to keep soil and water on the land. The Dogon people of eastern Mali practise some of the most intensive irrigated agriculture in Africa to feed a rapidly rising population in an era of declining rainfall – but do so without causing desertification. Elsewhere, Sahel communities have adopted rainwater harvesting methods to halt soil loss and improve the productivity of their land. Desertification makes 12 million hectares of land useless for cultivation every year. Since 1965, one sixth of the populations of Mali and Burkina Faso have lost their livelihoods and fled to cities. In Mauritania between 1965 and 1988, the proportion of the population who were nomads fell from 73% to 7%, while the proportion of the population in the capital Nouakchott rose from 9% to 41%.

But desertification is not exclusively a problem of the developing world. Commercial agriculture and livestock farming can cause as much damage to arid ecosystems as pastoralism and subsistence agriculture. Australia, one of the world's richest but least densely populated countries, has serious land degradation problems. In arid and semiarid regions, uncontrolled grazing – the single largest form of land use – takes place in areas often considered to be "commons" or "grazing lands" which are usually common property resources and over 25% of grasslands and marginal lands have been converted to cropland to meet foodgrain demand.

Important features of arid land soils for agricultural production are their water holding capacity and their ability to supply nutrients to plants. Arid zone soils which cover approximately one-third of earth's surface, are vulnerable to both wind and water erosion due to dryness. Arid and semi-arid soils are mainly found in Africa (Sahara, Namibian and Kalahari deserts), the Middle East (Arabia desert, Iran, Afganistan, Thar desert of India, etc.), North and South America (Mohave desert, Chile, etc) and Australia. As there is little deposition, accumulation or decomposition of organic material in arid environments, the organic content of the soils is low and, therefore, natural soil fertility is also low. Arid lands are inhabited by more than 2.5 billion people, nearly 30% of the world's population.

1.7 Conclusion

In spite of large-scale scientific studies, misconceptions about the arid lands still persists. *'Desert creep'*, *'rejuvenation of deserts'*, *'desert encroachment'*, etc. are various melodramatic terms keep on making headlines, without any basis or reality. Various agencies have been involved in the mapping of arid lands. Due to differences in perspectives, objectives and methodologies, differences in area are also visible. As such, climatic shifts are the main indication of any such severe problems in arid lands. Whereas, presently available scientific evidence neither substantiate it. Anthropogenic role not only in the arid lands but elsewhere is much more visible and documented. These arid lands' most important and distinguishing environmental characteristic is the lack of precipitation. Intense heat waves during summers and severe cold below freezing point in many parts of the arid world during winter are quite common and characteristic phenomenon. Further, significant achievements have been made in the technology of arid lands agriculture, water use efficiency (through highly efficient, highly sophisticated irrigation systems), livelihood and economic resources. Large-scale intensive systems for dairy, poultry and feed and fodder have been developed. Large quantities of low-cost energy, in the form of wind and solar, have been generated. Higher dependency on rainfed agriculture in the arid lands is still a cause of concern. Further, due to lack of perennial sources of water resources, dependency on groundwater resources either for irrigation or drinking purpose is increasing to cater needs of the teeming population. Existing agricultural production systems are part of a vicious cycle due to poor carrying capacities of the land resources. The social costs of continuing with these agricultural production systems are shocking and threatening. The dry lands ecosystems are so heavily stressed; the consequences of land degradation are inevitable. Human impact in arid and semiarid lands has led to soil degradation through salinization, waterlogging and wind erosion and has entailed desertification now affecting almost 70% of the region. Desertification describes a situation where a productive land becomes less and less productive due to degradation until it becomes non-productive and desertified. Desertification is a serious situation that has to be monitored, controlled and combated through local, regional, national and international efforts for it has to be viewed as a global problem (Barakat 2004).

References

K.N. Brooks, P.F. Ffolliott, H.M. Gregersen, L.F. DeBano, *Hydrology and the Management of Watersheds* (Iowa State University Press, Ames, 1997)

FAO, *Water and Cereals in Drylands* (Earthscan & FAO Publication, 2008), p. 105

FAO, Land resource potential and constraints at regional and country levels, in *World Soil Resources Report* No. 90 (Rome, 2000)

FAO, *Compendium of Agricultural-Environmental Indicators (1989–91 to 2000)* (Statistics Analysis Service, Statistics Division, Food and Agriculture Organization of United Nations, Rome, 2003)

FAO, *Carbon Sequestration in Dryland Soils. World Soil Resources Reports No. 102* (Food and Agriculture Organization for the United Nations, Rome, 2004)

S. Feng, Q. Fu, Expansion of global drylands under a warming climate. Atmos. Chem. Phys. **13**, 10081–10094 (2013)

P. Ffolliott, J.O. Dawson, J.T. Fisher, I. Moshe, T.E. Fulbright, A. Al Musa, C. Johnson, P. Verburg, Dryland environments. Ar. Land. News **52** (2002)

Q. Fu, C.M. Johason, J.M. Wallace, T. Reichler, Enhanced mid-latitude tropospheric warming in satellite measurements. Science **312**, 1179 (2006)

Q. Fu, P. Lin, Poleward shift of subtropical jets inferred from satellite-observed lower stratospheric temperatures. J. Clim. **24**, 5597–5603 (2011)

K.W. Glennie, Desert sedimentary environments, past and present – a summary. Sediment. Geol. **50**, 135–166 (1987)

A.T. Grove, The ancient ergs of Hausaland, and similar formations on the south side of the Sahara. Geogr. J. **124**, 528–533 (1958)

R.L. Heathcote, *The arid lands; Their use and abuse* (Longman, London, 1983)

T. Hilhorst, F. Muchena (eds.), *Nutrients on the Move: Soil Fertility Dynamics in African Farming Systems* (International Institute for Environment and Development, London, 2000)

Y. Hu, Q. Fu, Observed poleward expansion of the Hadley circulation since 1979. Atmos. Chem. Phys. **7**, 5229–5236 (2007). doi:10.5194/acp-7-5229-2007

M. Hulme, Recent climatic change in the world's drylands. Geophys. Res. Lett. **23**, 61–64 (1996)

E. Humphreys, D. Peden, S. Twomlow, J. Rockström, T. Oweis, A. Huber-Lee, L. Harrington, *Improving Rainwater Productivity: Topic 1 Synthesis Paper* (CGIAR Challenge Program on Water and Food, Colombo, 2008.) 19 p

IIASA/FAO, *Compendium of Agricultural-Environmental Indicators (1989-91 to 2000)* (Statistics Analysis Service, Statistics Division, Food and Agriculture Organization of United Nations, Rome, 2003)

I.J. Jackson, *Climate, Water and Agriculture in the Tropics* (Longman, Singapore, 1989)

C.M. Johanson, Q. Fu, Hadley cell expansion: model simulations versus observations, J. Climate **22**, 2713–2725 (2009)

A. Kar, B.K. Garg, M.P. Singh, S. Kathju, *Trends in Arid Zone Research in India*. (CAZRI, 2009), p. 481, CAZRI

W. Köppen, in *Die Klimate der Erde* (Berlin, 1931)

J. Lu, G.A. Vecchi, T. Reichler, Expansion of the Hadley cell under global warming. Geophys. Res. Lett. **34**, L06805 (2007). doi:10.1029/2006GL028443.

E. de Martonne, L. Aufrère, Map of interior basin drainage. Geogr. Rev. **17**, 414 (1927)

W.G. McGinnies, Climatic and biological classifications of arid lands: A comparison. pp. 61–68. *In* Arid lands today and tomorrow. E.E. Whitehead, C.F. Hutchinson, B.N. Timmermann, and R.G. Varady (eds.). Boulder, Colo: Westview Press (1988). 1435 p

P. Meigs, World distribution of arid and semi-arid homoclimates. Arid Zone Res **1**, 203–210 (1952)

P. Meigs, *World Distribution of Arid and Semi-Arid Homoclimates, Arid Zone Hydrology* (Arid Zone Research Series. UNESCO, Paris, 1953)

Millennium Ecosystem Assessment, *Ecosystems and Human Well-Being: Desertification Synthesis* (World Resources Institute, Washington, DC, 2005)

D. Molden, T.Y. Oweis, Pathways for increasing agricultural water productivity, in *Water for Food, Water for Life*, ed. By D. Molden, (Earthscan, Lsondon and International Water Management Institute, Colombo, 2007), pp. 279–310

M. Morris, V. Kelly, R. Kopicki, D. Byerlee, *Fertilizer Use in African Agriculture: Lessons Learned and Good Practice Guidelines* (Agriculture and Rural Development Division, World Bank, Washington, DC, 2007)

M. Mortimore, *Dryland Opportunities* (IUCN/HED/UNDP, Gland/London/New York, 2009)

National Weather Service, *What Is Meant by the Term Drought?* (U.S. Department of Commerce, Washington, DC, 2004.) (available at http://www.wrh.noaa.gov/fgz/science/drought.php)

A. Penck, *Morphologie der Erdoberflache* (Engelhorn, Stuttgart, 1894)

V.M. Ponce, Management of droughts and floods in the semiarid Brazilian northeast—the case for conservation. Soil Water Conserv. J. **50**, 422–431 (1995)

J.F. Reynolds, D.M. Stafford Smith, E.F. Lambin, B.L. Turner, M. Mortimore, S.P.J. Batterbury, T.E. Downing, H. Dowlatabadi, R.J. Fernandez, J.E. Herrick, E. Huber-Sannwald, H. Jiang, R. Leemans, T. Lynam, F.T. Maestre, M. Ayarza, B. Walker, Global desertification: building a science for dryland development. Science **316**, 847–851 (2007)

U. Safriel, Z. Adeel, Dryland systems, in *Ecosystems and Human Well-Being: Current State and Trends Volume 1*, eds. By R. Hassan, R. Scholes, N. Ash, (Island Press, Washington, DC, 2005), pp. 623–662

J. Scheff, D. Frierson, Twenty-first-century multimodel subtropical precipitation decline are mostly midlatitude shifts. J. Clim. **25**, 4330–4347 (2012)

D.J. Seidel, Q. Fu, W.J. Randel, T.J. Reichler, Widening of the tropical belt in a changing climate. Nat. Geosci. **1**, 21–24 (2008)

H.L. Shantz, in *The Future of Arid Lands*, ed. by G.F. White. History and Problems of Arid Lands Development, (American Association for the Advancement of Science, Washington, D.C., 1956), p. 3–25

K.E. Taylor, R.J. Stouffer, G.A. Meehl, An overview of CMIP5 and the experiment design. B. Am. Meteorol. Soc. **93**(485–498), 2012 (2012)

C.W. Thornthwaite, An approach toward a rational classification of climate. Geogr. Rev. **38**, 55–94 (1948)

S. Twomlow, D. Rohrbach, J. Dimes, J. Rusike, W. Mupangwa, B. Ncube, L. Hove, M. Moyo, N. Mashingaidze, P. Mahposa, Microdosing as a pathway to Africa's green revolution: evidence from broad-scale on-farm trials. Nutr. Cycl. Agroecosyst. **88**(1), 3–15 (2008)

UN (2010). http://www.un.org/en/events/desertification_decade/whynow.shtml

UNCCD (United Nations Convention to Combat Desertification), U. N. Doc. A/A C. 241/27, 33 I. L. M. 1328 (United Nations, 1994), UNCCD

UNCCD, An introduction to the United Nations Convention to Combat desertification (2000), available at http://www.unccd.int

UNCED. Agenda 21, Chapter 12, Paragraph 5, United Nations World Conference on Environment, Rio de Janeiro, Brazil, June 1992, (1992), p. 2

UNDP/UNSO, *Aridity zones and drylands populations* (1997). AAAS Atlas of Population & Environment, American Association for the Advancement of Science

UNEP, *UNEP-World Conservation Monitoring Centre* (Assessing forest integrity and naturalness in relation to biodiversity. Cambridge, UK, 2000), p. 75

UNEP, *The Global Environment Outlook 4. Environment for Development* (UNEP, Nairobi, 2007)

United Nations Environment Management Group, *Global Drylands: A UN System-Wide Response* (United Nations, October 2011), p. 132, UN EMG Secretariat

WMO, *Glossary of Terms Used in Agrometeorology*. CAGM No. 40. WMO/TD-No. 391 (WMO, Geneva, 1990)

Y.P. Zhou, K.M. Xu, Y.C. Sud, A.K. Betts, Recent trends of the tropical hydrological cycle inferred from global precipitation climatology project and international satellite cloud climatology project data. J. Geophys. Res. **116**, D09101 (2011). doi:10.1029/2010JD015197

Further Readings

A. Oroda, J. Agatsiva, Environmental Conditions and Farming Systems in the Southern Catchment Area of Lake Nakuru. Third Annual Report for the EU – INCO-DC (International Co-operation with Developing Countries) Project *"Sustainable Use of Natural Resources in Rural Systems of Eastern African Drylands (Ethiopia, Kenya and Tanzania): Strategies for Environmental Rehabilitation"* (2000)

M. Stuttard, G. Narciso, G. Carizi, M. Suppo, R. Catani, L. Isavwa, J. Baraza, A. Oroda, *Monitoring Lakes in Kenya: An Environmental Analysis Methodology for Developing Countries*. Final Report for the CEC DG XII STD-3 Programme, (1995), pp. 17–23 and 32–130

D.S. Thomas, Nick Middleton (eds) World Atlas of Desertification 2nd ed. A Hodder Arnold Publication (1997). p. 182

UNFPA, World population prospects: the 2006 revision and world urbanization prospects: the 2005 revision (2008), http://esa.un.org/unpp

World Atlas of Desertification, in *World Atlas of Desertification*, ed. By E. Arnold, 2nd edn., (UNEP, London, 1997) 182 p

Websites

http://www.un.org/en/events/desertification_decade/whynow.shtml. Accessed 13 Oct 2016
http://www.eolss.net/sample-chapters/c12/e1-01-06-04.pdf. Accessed 10 Oct 2016
http://www.atmos.washington.edu/~qfu/Publications/acp.ff.2013.pdf. Accessed 14 Oct 2016

Recent Trends in Drylands and Future Scope for Advancement

2

Ahmed E. Sidahmed

2.1 Introduction

2.1.1 Definition of Drylands and the Geographical Context and Distribution

There is no unique definition of the term drylands. While FAO and UNCCD define dryness based on the *length of the growing period LGP* (Box 2.1), the UN Environmental Programme (UNEP) uses the *aridity index – AI –* of 0.65[1] as the threshold below which both tropical and temperate areas are considered drylands (Millennium Ecosystem Assessment 2005). The use of aridity index has been endorsed by the 195 parties to the United Nations Convention to Combat Desertification (UNCCD) (World Bank 2015). Water scarcity in the drylands is caused by a faster rate of surface evaporation and evaporation from plants (Middleton and Thomas 1997).

In contrast the vast areas of the dryland ecosystems are underutilized, and most of the areas suitable for cropping or grazing are degraded by overuse or misuse. Rodríguez-Iturbe and Porporato (2004) described the drylands as complex, evolving structures whose characteristics and dynamic properties depend on many interrelated links between climate, soil and vegetation. The drylands are commonly classified into four subtypes: hyperarid lands, arid lands, semiarid lands and dry subhumid lands (Table 2.1). However, UNCCD considers the hyperarid lands as deserts and excludes them from dryland definition but for the purposes of this book,

---

[1] *The aridity index* measures the ratio of mean annual precipitation to mean annual potential evapotranspiration: a ratio of 0.65 means that potential evapotranspiration is 1.5 times greater than mean annual precipitation.

A.E. Sidahmed (✉)
Senior Adviser (Consultant) FAO Investment Centre, Africa Region TCIA,
Room D517 FAO, Rome, Italy
e-mail: ahmedsidahmed.contacts@gmail.com

> **Box 2.1 The Widely Accepted Definitions of the Drylands**
> Two of the most widely accepted definitions are those of FAO and the United Nations Convention to Combat Desertification (UNCCD):
>
> - FAO defined drylands in 2000 as those areas with a length of growing period (LGP) of 1–179 days. This includes regions classified climatically as arid (1–74), semiarid (75–119) and dry subhumid.
> - Under the UNCCD classification in 2000, drylands are characterized by a P/PET of between 0.05 and 0.65. The UNCCD classification employs a ratio of annual precipitation to potential evapotranspiration (P/PET). This value indicates the maximum quantity of water capable of being lost, as water vapour, in a given climate, by a continuous stretch of vegetation covering the whole ground and well supplied with water. Thus, it includes evaporation from the soil and transpiration from the vegetation from a specific region in a given time interval.
>
> Source: FAO (2008). Waters and Cereals in Drylands http://www.fao.org/docrep/012/i0372e/i0372e01.pdf

Table 2.1 Regional extent of the drylands

Region	Aridity zone							
	Arid		Semiarid		Dry-humid		All arid lands	
	1000 km²	%	km²	%	km²	%	km²	%
Asia (incl. Russia)	6164	13	7649	16	4588	9	18,401	39
Africa	5052	17	5073	17	2808	9	12,933	43
Oceania	3488	39	3532	39	996	11	8016	89
N. America	379	2	3436	16	2081	10	5896	28
S. America	401	2	2980	17	2223	13	5614	32
C. America and Caribbean	421	18	696	30	242	10	1359	58
Europe	5	0	373	7	961	17	1359	24
World total	15,910	12	23,739	18	13,909	10	53,558	40

Source: FAO (2008). Waters and Cereals in Drylands – http://www.fao.org/docrep/012/i0372e/i0372e01.pdf

we may deal with true deserts as part of the global land that is used to support people, their livestock and wild plants and animals that are part of global biodiversity.

For the purpose of management, the international research and development community resolved to categories the drylands into two: (i) marginal areas with high vulnerability in which a modest increase in productivity (10–20%) is targeted; (ii) areas with higher production potential with scope for sustainable intensification with targets to increase productivity by 20–30% (CGIAR 2012). According to the United Nations Convention to Combat Desertification (UNCCD) classification system, the drylands cover 40% of the earth's land surface. However, the areas classified as drylands vary between regions and continents. As illustrated in Table 2.1, 43% of Africa, 39% of Asia, 32% of South America, 28% of North America and 24% of Europe are dryland areas. According to Millennium Ecosystem Assessment (2005), there is a significantly greater proportion of drylands in developing countries (72%), and the proportion increases with aridity. Almost 100% of all hyperarid lands are in the developing world.

Drylands are home to some of the most unique biological diversity on the planet. Drylands are home to 17% of the global Centres of Plant Diversity, 47% of Endemic Bird Areas, 23% of Global Terrestrial Ecoregions and 26% of protected areas worldwide (White 2002). Dryland species' diversity is influenced by dryland ecology and extent of aridity and is highly resilient or tolerant to drought and salinity. For example, the dryland ecosystems of sub-Saharan Africa cover a variety of terrestrial biomes including grasslands at various altitudes and latitudes, tropical and subtropical savannahs, a variety of dry forest and woodland ecosystems and coastal areas, which are extremely heterogeneous.

The drylands are the home for about one third (about 2.5 billion people) of the global population, support 50% of the world's livestock and grow 44% of the world's food (CGIAR 2012). The contribution and future potential of the drylands to the overall global economy is significant, especially when considering nonagricultural wealth (oil, mineral, sources of renewable energy such as solar and wind). According to the Millennium Ecosystem Assessment (2005), the dryland populations on average lag far behind the rest of the world on human well-being and development indicators, and most majority of them (90%) live in the developing countries (UN 2011). Drylands have three primary economic functions: as rangelands (65% of the global drylands including deserts), as rain-fed farmland and irrigated farmland (25%), and as forest or sites for towns and cities (10%), which are growing rapidly. They include the world's driest places (hyperarid deserts such as the Atacama in Chile and the Namib in southwest Africa) as well as the polar regions.

Most of the 30% of the dryland inhabitants who depend on agriculture for their food security and livelihoods are vulnerable and marginalized small farmers and herders, and they are the most influenced by climate change. Most of the world's poor live in drylands including 400 million "poorest of poor" who survive on less than US$1 per day (CGIAR 2012). Furthermore, the drylands lose 23 ha/min to drought and desertification – a loss of 20 million tons of potential grain production every year.

The rural people living in the drylands are typically grouped into nomadic, semi-nomadic, transhumant and sedentary smallholder agricultural populations FAO (2008). The dryland farming systems are very diverse, vary according to the agroecological conditions of each region and are mostly coping with uncertainties of rain and lack of soil moisture. The major farming systems of the dryland areas vary according the agroecological conditions of each region. For example, a study conducted by the Land Degradation Assessment in the Dryland project (LADA 2008) identified the major farming systems in the drylands according to the socioeconomic information, agroecology and possibilities for irrigation.

Cereal cultivation (indigenous or improved varieties) dominates dryland farming. Livestock production is a major activity of the dryland agricultural systems and is raised by most of the rural and peri-urban households, although there are noticeable degrees of wealth among some nomadic individuals; most of the poor pastoral communities mix herding with some subsistence farming during favourable years. Also, there is considerable livelihood diversification among the dryland people (e.g. petty trade by women and charcoal collection by men) within the households (Headey and Taffese 2012). Nowadays, year long migratory pastoralism seldom exists as nomadism gave way to seasonal migration (e.g. transhumance pastoralism). Whereas part of the household and herd is migratory, the remainder (typically the elders, women and school children) tends the breeding female and the less vigorous animals. Also, there are sedentary smallholder farming households that practice rain-fed (shifting) or irrigated crop farming often combined with livestock production. This is mainly because livestock is an important user of the natural resources (land, water, nutrients and biodiversity), an important food and wealth commodity and an important convertor of large amounts of low value byproducts and waste into valuable products (The Global Agenda for Sustainable Livestock). However, the productivity is challenged by high climatic variability, various forms of land degradation, loss of biodiversity (CGIAR 2012), limited access to technology poor market linkages, weak institutions, lack of partnerships and marginalization and exclusion of rural people.

2.2 Recent Trends

2.2.1 Impact of Climate Variability on Drought Intensity

There are three major types of climates in the drylands – tropical, Mediterranean and continental. Typically the dryland seasons are hot and dry, moderate and rainy and cool and dry (FAO 2008). Frequent diurnal fluctuations restrict plant growth within the three seasons. This is further complicated by diverse structure and physiochemical properties of the soils. Large rivers originating from the highlands (e.g. *the Nile, Tigris-Euphrates, Indus, Ganges, Senegal, Niger and Colorado Rivers*) are the major sources of water in the drylands. Groundwater and rainwater are important sources. Whereas the rainwater is erratic (especially in the arid zones) and mostly lost by evapotranspiration or runoff, the recharge of the groundwater is dependent largely on the amount, intensity and duration of the rainfall and soil properties.

The past 250 years witnessed unprecedented increase in human and livestock population and massive internal and international displacements caused by industrialization, urbanization, conflicts and wars (FAO 2008). These trends were accompanied by fossil fuel combustion and land use change (including inter alia deforestation, biomass burning, uncontrolled harvesting of wildlife, draining of wetlands and concentration of very limited varieties of food crops and livestock species, ploughing and use of fertilizers). As a consequence the genetic diversity and ecological balance were severely disrupted and reduced. A major drastic consequence was the global increase in atmospheric concentrations of carbon dioxide (CO_2) and other greenhouse gases particularly methane (IPCC 2007). These changes are associated with extreme temperatures that alter the hydrological cycles and might result in extreme events such as droughts and floods (Woznicki et al. 2015). More recent studies confirmed that climate change is one of the major threats to the genetic diversity of animal resources (FAO 2015d).

Most of the international community and governments have acknowledged the substantial scientific evidence (Stern 2006, 2007) that the recent and future rapid changes in the earth's climate are human induced, caused by accumulations of CO_2 and other greenhouse gases (GHGs). Widely accepted predictions show that the ongoing pattern of climate change will not only raise temperatures across the globe but will also intensify the water cycle, reinforcing existing patterns of water scarcity and abundance, increasing the risk of droughts and floods.

Also, widely accepted are the IPCC (2007) assessment that hot extremes, heat waves and heavy precipitation events will continue to become more frequent. According to IPCC, frequent and protracted droughts in most subtropical land regions are likely caused by decreases in precipitation (see Box 2.2 for definition of Droughts). Examples of *"meteorological drought"* are reported in the results of research conducted in the semiarid agricultural basin of Khuzestan region of Iran, where drought occurrence is more frequent in the warmer and drier climates of the basin (Hosseinizadeh et al. 2015). On the other hand, examples of *"socio-economic drought"* are reported in Thessaly Region of Greece where annual drought severity

---

Box 2.2 Drought
The terms drought and aridity are sometimes used interchangeably, but they are different. Aridity refers to the average conditions of limited rainfall and water supplies, not to the departures from the norm, which define a drought. A drought is defined as a departure from the average or normal conditions, sufficiently prolonged (1–2 years) as to affect the hydrological balance and adversely affect ecosystem functioning and the resident populations. The National Weather Service, 2004, recommended four different ways for the definition of drought:

- *Meteorological drought* is a measure of the departure of precipitation from normal. Due to climatic differences, a drought in one location may not be a drought in another location.
- *Agricultural drought* refers to situations where the amount of soil water is no longer sufficient to meet the needs of particular crop.
- *Hydrological drought* occurs when surface and subsurface water supplies are below normal.
- *Socio-economic drought* describes the situation that occurs when physical water shortages begin to affect people.

Source: FAO (2008). Waters and Cereals in Drylands http://www.fao.org/docrep/012/i0372e/i0372e01.

is increased for all hydrological areas, with the socio-economic scenario being the most extreme (Loukas et al. 2008).

Agriculture is negatively affected by climate variability in the dryland areas. Inconsistent climatic patterns, erratic rain, excessive heat and overuse of the natural resources contributing to the loss of millions of hectares of the productive land each year adding to the vulnerability and poverty of the dryland farmers. A case in the point is the impact of climate change on drought intensity of the Horn of Africa and the Sahel regions. In the Horn of Africa, the livestock sector experienced five major droughts between 1998 and 2011, which killed more than 50% of the cattle in the most heavily affected areas and decimated the livelihoods of between 3 and 12 million people, depending on the year (World Bank 2015). In the Horn of Africa region, the drought of 2010/2011 was the worst in 60 years leading to a severe humanitarian and food crisis affecting over 13 million people mostly from Somalia, Kenya and Ethiopia (FSNAU 2011). The two major droughts that occurred in the 1970s and 1980s in the Sahel region led to the deaths of about one third of all cattle, sheep and goats (Lesnoff et al. 2012), where relatively mild drought between 2010 and 2012 was the cause of food insecurity of 12 million people (Oxfam 2012).

The impact of 2010 and 2011 in the Horn of Africa and the Sahel, respectively, on crop and livestock production was devastating (IGAD 2013a, b). Only 30% of the households in the Sahel and the Horn of Africa had adequate livestock assets to stay above poverty in the face of recurrent droughts. As a result of population growth, the number of households is projected to drop to 10% by 2030, while 60% of the households are likely to feel pressure to drop out of livestock-based livelihoods, with the remaining 30% of households projected to stay in the system while remaining vulnerable to drought and other shocks (World Bank 2015). Also, it was projected that by 2030, without resilience and mitigation measures, the number of farming-dependent households in the Sahel and the Horn of Africa that are poor and vulnerable to drought to increase by around 60%.

2.2.2 Impact of Recurrent Droughts on Conflict Dynamics

Some of the semiarid and dry humid zones are becoming drier and hotter (World Bank 2015). Increasing temperatures in the drylands are altering climate variability, the traditional water and land use patterns and plant growth dynamics. For example, the evapotranspiration decrease is slight likely caused by the increase in the atmospheric CO_2 concentrations (Woznicki et al. 2015). Climate change can alter the hydrological cycle, which may result in extreme events such as floods and droughts. Complications caused by the rapid increase in the human populations of the drylands inhabitants during the past century are expected to intensify by further increase in the twenty-first-century population. Moreover, higher population density in the drylands, combined with increasing interest from outside investors in large-scale commercial agriculture and/or extractive industries, will put additional pressure on a fragile natural resource base, pushing it in some cases beyond its regenerative capacity (World Bank 2015). In the absence of viable land use policies and regulations and because of the chronic land tenure problems in many developing countries, such changes are leading to intensified conflicts between the livestock herders and the cultivators (World Bank 2015).

The most recent impact of drought-triggered conflict on the agriculture sector is very well demonstrated in the Horn of Africa and the Sahel Region of West Africa. Prolonged droughts were the major cause of the ineffectiveness of every attempt made to reduce poverty and insecurity in the dryland areas of the Horn of Africa and Sahel, even in the countries that witnessed economic growth (Fosu 2009; IOA 2012). By the time of writing the IOA report, it was estimated that drought-related acute food shortage in the Sahel region of West Africa and the Horn of Africa have left about 18.7 million and 11.7 million people, respectively, in need of emergency assistance. Violent internal and cross-border conflicts and ethnic unrest involving fighting over water and grazing resources, stealing livestock and women erupted in various countries and contributed to displacement of millions of people, disruption of transportation and market transactions and subsequently hunger and starvation. According to the report, sub-Saharan Africa was responsible for 88% of the global conflict death toll between 1990 and 2007, in addition to over 9 million refugees and internally displaced people.

The geopolitical and economic implications of drought on the poor and the poor countries are exacerbated in many countries by governments who because of lack of knowledge, corruption, lack of support for the development of viable land tenure systems and the intentional disregard to implement the internationally agreed upon commitments of land use governance and equitable/responsible investment practices.

For example, the recent expansion in large-scale high-technology farming systems in some Sub-Saharan African countries – e.g. *outsourced ventures of oil wealthy but water-poor countries in the Middle East in Sudan* – is impacting negatively on communities exposed to land grabbing among other injustices. Also there is emerging evidence that conflicts over land, water and feed resources is increasing in several dryland areas triggered by intensified competition for the meagre resources.

Conflict in the dryland areas is the cause and effect of vulnerability. The diverse ecosystems of the dryland areas and the dependence of the rural population, in some dryland categories, on inconsistent climatic cycles and precarious natural resources enhance the vulnerability of the drylands dwellers. This is further exacerbated by the fierce competition on land and water to meet the needs of increasing population. For example, in spite of protracted droughts, civil strife and famine, the human population of the Horn of Africa is increasing at the rate of 2.5–3.5%, with over 60% of the people being youth (IGAD 2013a, b). The increase in population is reflected in the dramatic increase in land and water for cropping and livestock production. This is reflected in the increase in cropping intensity (both caused by expansion in arable land and multiple cropping intensity – MCI[2]) in the drylands – especially in Africa and Asia (OECD 2009). OECD recorded similar acceleration in livestock productivity caused by advances in technology, disease control, genetic improvement and stock management (OECD 2009).

---

[2]*MCI* multiple cropping intensity is the sum of area harvested for different crops during the year divided by the total harvested land.

Table 2.2 Classification of drylands by aridity index in aridity index global land area ESARO land

	Aridity index	Global land area[a] (%)	ESARO land area (%)
Arid	0.05 < AI <0.20	12.1	7
Semiarid	0.20 < AI <0.50	17.7	15
Dry subhumid	0.50 < AI <0.65	9.9	25
Total (excluding desert)		39.7	47

Source: Adapted from IUCN- ESARO (2010)
[a]Drylands without deserts account for 40% of the global land mass

Of great concern is the observation that such increases in cropping intensity and livestock production did not reduce vulnerability to droughts and poverty of the poor in the dryland areas of the developing countries. According to the World Bank (2015), the number of poor and vulnerable farming-dependent households in the Sahel and the Horn of Africa is projected to increase in the next 15 years by about 60%. This increase underscores the demographic implications that keeps the majority of farming in the drylands in the hands of the poor and subsistent small farmers who lack the resources and technology to sustain production or reduce vulnerability, and eventually leading to the eruption of conflicts and ethnic wars.

Drought-related conflicts are diverse but interrelated. There are conflicts caused by dryland communities averse to sharing traditional resources especially during drought years. There are incidents of the civil strife and quarrels that break out between communities competing on the use of livestock grazing areas or human/livestock watering points. Livestock rustling is a tradition among communities that lack cross-border policies and national security infrastructure.

2.2.2.1 Drought-Related Conflicts

Africa is the continent mostly affected by the impact of drought on food security and social stability. According to IUCN_ERASO[3] (2010), dryland climates are highly unpredictable and subject to extreme events such as droughts. The ERASO drylands occupy 47% of the land area in the 22 sub-Saharan African countries (Table 2.2) and are characterized by low economic growth, a poorly educated workforce and political weakness.

Land-Related Conflicts

Africa with its endowments of vast untapped land and water resources should be able to feed its people and to revert the food trade balance to its favour. Over 50% of the additionally available arable land is found in Africa and Latin America (OECD-FAO 2009). Only 10% of Africa's rural land is registered (AfDB 2016). This is

---

[3] ESAROEastern and Southern Africa Regional Office in this case refers to the following countries: Angola, Botswana, Comoros Islands, Djibouti, Eritrea, Ethiopia, Kenya, Lesotho, Madagascar, Malawi, Mauritius, Mozambique, Namibia, Seychelles, Somalia, South Africa, Sudan, Swaziland, Tanzania, Uganda, Zambia and Zimbabwe.

further complicated with inefficient land administration that delays transferring land title deeds and substantially increases the transaction cost (e.g. which is estimated – on average – as being twice that of the developed countries) (AfDB 2016). Furthermore, incomplete and poorly enforced land tenure laws discourage the private investors or deprive the traditional land users – especially the marginalized poor and deter any equitable options for development and poverty reduction. There are very few contested indicators, in the developing countries, that the governments pursue effective planning and management of land utilization, and wherever they do they remain voluntary and nonbinding. Weak policy and institutional frameworks are the leading cause of corruption in the agriculture sector, including land administration. Further complications result from the nature of the mostly male-dominated (in some cultures male-exclusive) titles, inheritance rights and access to land in a continent where women are the primary users of agricultural land.

The impact of failing to achieve a breakthrough in solving the chronic land tenure problems is more complicated in the dryland areas where the cyclic and unpredictable weather events create unsustainable pressure on land and water resources and increase the vulnerability of the herders and the occasional small farmers (IFPRI 2012) to drought. In the absence of such an overarching policy framework, it is not possible for governments to rationalize the conflicting land utilization needs, manage allocation of secured title and attract long-term investment in agriculture. There are African governments that treat land tenure in a top-down approach giving the president or the ruling Junta the authority to allocate land and to use it in favour of selected invasive local or foreign users. For example, the recent expansion in large-scale high-technology farming systems in some African countries – e.g. *outsourced ventures of wealthy but water-poor countries in the Middle East* – is impacting negatively on communities exposed to land grabbing among other injustices. The reversal of the Egyptian government of its support to the article in the 2014 constitution that allows the Nubians who were evicted by force as a result of building the Aswan Dam in the 1940s and the High Dam in the 1960s is just one example of poor or abusive policies that do not consider the inherent rights of the indigenous people.

Water-Related Conflicts

Water scarcity in the dryland areas is one of the major challenges to the contribution of the agriculture sector to food security (FAO 2014a, b, c, d, e, f). Pressure on water resources is aggravated by increasing and unsustainable use and heavy dependence on renewable water, which are replenished by rainfall or stored in deep aquifers. Heavy reliance on transboundary water resources is a major concern contributing to food insecurity in many dryland countries and is increasingly becoming a major trigger of regional conflicts. Moreover, the use of underground water at a speed faster than replenishment is threatening many parts of the drylands areas with desertification. Plans by some dryland countries favouring production of high water-consuming crops such as wheat and alfalfa (e.g. Saudi Arabia) are leading to water depletion (FAO 2016a).

A further pressure on water use commonly triggering conflict in several dryland area dominated countries is the shift from the traditional pastoralism to settled commercial systems (FAO 2016b). For example, In Oman and in other dryland countries, increasing human and livestock population, urbanization and infringement of roads and other infrastructure on the grazing resources have transformed livestock rearing from the community managed nomadic mobility of the pastoralists to the seminomadic-sedentary farming systems (Box 2.3). This transformation was accompanied by the increasing use of water for year-round cultivation of animal feed and for the cultivation of food and cash crops. This transformation was also accompanied by conversion of vast grazing areas into cultivated land.

Box 2.3 The Major Categories of the Livestock Producers (Meat and Dairy)
- *Nomadic (Badawi)*: The stakeholders of this very limited system are the Bedouins. The system they follow is a low-input/low-output system that prevails only in the desert and the mountainous areas. The producers rear camels, cattle, goats and some sheep.
- *Transhumance (seminomadic, traditional semimobile) – Agro-pastoral.* Very close to the Badawi in pursuit of seasonal grazing resources and supplemental feeding. Producers live in a specific area but move with livestock for feed and water and supplement seasonal grazing with sown forages, crop residues, concentrates and dried fish.
- *Sedentary mixed crop-livestock* (sheep, goats and cattle producers fattening for slaughter using mix of sown fodder, concentrates and crop byproducts). This group is the largest in the Sultanate (see Table 2.1).
- *Settled rural household* (back yard) smallholders of sheep, goats and cattle *(the most dominant production system for all livestock types)* who supplement seasonal grazing – in the neighbourhood pastures – with sown forages, crop residues, concentrates and dried fish.
- *Settled sheep, goats and cattle producers* graze commercially high-valued animals in natural rangelands during certain times of the year and provide supplemental feeding to maintain the desired requirement year-round.
- *Intensive local producers* fattening for slaughter: producers keep animals in fenced areas and manually feed animals from sown forages and purchased feed, and there is no grazing involved.
- *Intensive dairy farmers (medium and large)* with relatively fully integrated production.
- *Nomadic sheep and goats herders in the dry rangelands* supplemented depending on grazing year with imported concentrates.

Source: MAF (Oman) documents and reports (several); discussion with the Directorate General of Animal Resources senior and technical staff, limited field visit

2.2.2.2 The Climate Challenge in Agriculture and Its Impact in Food Security and Nutrition in the Dryland Areas

Climate change – typically associated with frequent weather events, heat waves, droughts and sea-level rise – is already affecting agriculture and food security globally and particularly in the dryland areas (World Bank 2015; FAO, SOFA 2016). The agriculture sector accounts for at least one fifth of the total emissions mainly from livestock and crop production as well as conversion of forests to farmlands (FAO, SOFA 2016).

The potential impact of climate change on agriculture in sub-Saharan Africa, and especially the drier climates, will be negative and would enhance poverty, malnutrition and hunger. This is because the frequency of extremely dry and wet years will increase, thus disrupting agriculture production. Also forage production will decrease remarkably, particularly in the Sahel as a result of drought and range degradation. Moreover, enteric emission of methane from ruminants in small farming systems (that make up the major livestock production) will complicate any measures to reduce global temperature. Rising temperatures will threaten wheat and maize yield in the mostly dry North Africa and near east regions.

Because agriculture is such a major contributor to global warming, the sector is expected to play a major role in any effort to reduce emission. Reduction of emission will entail unprecedented responsibilities and expenses on the governments and the farmers. The responsibility on the dryland areas will increase as adoption of application of efficient water use and cropping technologies and innovations will lead to expansion on the potentially arable lands particularly the irrigated areas.

However, reduction of emission will also contribute to sustained increase in agriculture productivity, and consequently to food security and nutrition. Adopting measures to stabilize climatic variability (e.g. Milestone Climate Change Agreements Box 2.4) would reduce the extent of food production uncertainties, avoid dramatic food shortages and – as a consequence – avert dramatic increases in food prices. According to the FAO's SOFA report, without urgent action, millions of low-income people would be directly affected in regions with already high rates of hunger and poverty. Most affected would be populations in poor areas in sub-Saharan Africa, especially those reliant on dryland agriculture (FAO, SOFA 2016).

Box 2.4 The Milestone Climate Change Agreements
The finding that the atmospheric concentration of carbon dioxide in 2005 exceeds by far the natural range (IPCC 2007) triggered a worldwide campaign to eliminate the causes of climate change and to reduce global temperature. After years of complicated negotiations, 197 nations met at the Climate Change Convention in Paris (UNFCC, COP21 2015) and declared that they have reached an agreement to: *"hold the increase in the global average temperature to well below 2°C above pre-industrial levels.... recognizing that this*

(continued)

> *would significantly reduce the risks and impacts of climate change,"* pursue efforts to limit the temperature increase even further to 1.5 °C and undertake and communicate ambitious efforts to contribute to the global response to climate change by strengthening the mobility of countries to deal with the impacts of climate change. All parties expressed their recognition of the importance of averting, minimizing and addressing loss and damage associated with the adverse effects of climate change, including extreme weather events and slow onset events, and the role of sustainable development in reducing the risk of loss and damage. Following subsequent dialogues on actions, the threshold for entry into force of the Paris Agreement was achieved when 94 of the 197 ratified the agreement in Paris in 5 October 2016. The international community affirmed its position to continue the dialogue and focus on actions to achieve the commitments of the Paris Agreement through sets of COPs starting with COP 22 in Morocco in November 2016. In Marrakech the heads of state, government and delegations called for the highest political commitment to combat climate change as an urgent priority, and the developed country partners affirmed a US$100 billion mobilization goal in support of climate projects.
>
> Sources: IPCC (2007), UNFCC – COP 21 and COP22

2.2.2.3 The Case of the Protracted Droughts in the Horn of Africa

The Horn of Africa (HOA) and the Sahel are the most environmentally vulnerable dryland areas in Africa and the world. The droughts and floods in the Horn of Africa region have been recently intensified as a result of the El Niño impact. While the El Niño triggered floods in the HoA affected more than 3.4 million people in 2006/2007, 14 million people were affected by the droughts of 2009/2010 (source: EM-DAT). Furthermore, the recent humanitarian crises caused by the 2011 drought in the Horn of Africa (IGAD 2013a, b) and the 2012 drought in Sahel attracted global attention for the magnitude of emergencies and calamities they caused (World Bank 2015). For example, over 10% (around US$4 billion) of 2011 resources allocated by the Overseas Development Assistance (ODA) to long-term development goals in sub-Saharan Africa were redirected to costly short-term responses to humanitarian crises (World Bank 2015).

The Horn of Africa is one of the most serious hotspots of structural food insecurity, malnutrition and hunger in the world (FAO 2014a, b, c, d, e, f). The impact of the 2010/2011 drought in the Horn of Africa was the worst in 60 years (IGAD 2013a, b). The drought caused a severe humanitarian and food crisis affecting over ten million people mostly from Somalia, Kenya and Ethiopia. The crisis further complicated the social, economic, political and security situation in both Intergovernmental Authority for Development (IGAD) and the East Africa Community (EAC) regions where the combined economic impact of the drought and related shocks contributed to significant instability and losses (IGAD and FAO 2016). The 2010/2011 drought highlighted the extent of vulnerability of the communities using these vast areas as livelihoods assets. In addition, pressures from increased human population, urbanization and infrastructure (roads, urban homes, factories) in the pastoral areas reduced the ability of the livestock keepers to cope and to mitigate the impacts of drought through their traditionally risk averse mobility strategies.

Understanding the vulnerable livelihoods of the drylands inhabitants is primary to finding sustainable solutions to the complicated drought-triggered conflicts in the Horn of Africa (IFPRI 2012). Although pastoralism is still the most dominant source of income and employment for the inhabitants of the arid and semiarid lowlands (ASAL) region of the Horn of Africa, a significant number of the inhabitants are crop-based farmers, and many others are engaged in non-farming activities in the rural and urban areas. For example, it was reported by (Devereux 2006) that almost 70% of the households in the Somali region of Ethiopia engage in livestock rearing, whereas 43% engage in cereal crop production, 17% in firewood production and 15% in charcoal production. Also according to the study smaller but significant numbers of households engage in other activities such as mat making, petty trading, cash crop production and salaried employment.

The *Karamoja Cluster* in the Horn of Africa is a typical illustration of the complexity of the dryland areas in the Horn of Africa and the events that lead to conflicts. The Karamoja Cluster comprises four cross-country political-administrative units: at the Ethiopian, Kenyan, South Sudanese and Ugandan sides – Box 2.5 (IGAD 2015). The cluster has been historically subject to recurrent droughts and disease epidemics that regularly decimated herds – Box 2.6 (Practical Action). The herders normally rebuild the herds through exchanging or loaning herds/flocks or by raiding during times of extreme environments stress. At present, and especially post the 2010/2011 protracted droughts, the conflicts intensified in the cluster areas specially as the traditional norms (for the use of water and grazing resources) were disrupted by the modern and the geopolitical realities (border, roads, expansion in cropping lands, etc). For example, more recently small groups of warriors clad with automatic arms perform criminal/profit-motivated raids. In Karamoja, currently cattle raid account for more than 70% of deaths among the males aged 30–39 years.

Box 2.5 Karamoja Cluster

At the Ethiopian Side: Bero, Surma and Dasenech Woredas

At the Kenyan Side: Turkana, West Pockot, Samburu and Trans Nzoia *Counties*

At the South Sudanese Side: Kapoeta, Ikwoto, Naia and Naurus in *Eastern Equatoria State*

At the Ugandan Side: Kaabong, Moroto, Amuda, Nakapiriprit, Napa, Abim and Kotido in Karamoja *Region*

Source: IGAD (2015). IDDRSI – Cluster Approach for Cross-Border Cooperation and Investments to Strengthen Drought Resilience in IGAD – Region 4th IDDRSI Platform SC meeting and GA, Addis Ababa. 25–27 March 2015

Box 2.6 The Karamoja Cluster in the Horn of Africa

The Karamoja cluster refers to an area of land that straddles the borders between southwestern Ethiopia, northwestern Kenya, southeastern Sudan and northeastern Uganda. The area is populated by 14 pastoralist tribes who share a common language, culture and way of life. The cluster is composed of semi-arid savannah grading into wooded grassland to the north and desert to the south. Rainfall is generally unpredictable and localized, making agriculture an unreliable subsistence strategy. To survive in this habitat, pastoralists have evolved management strategies that are finely tuned to the realities of their environment. Recent studies have affirmed the rationality of these strategies and have demonstrated them to be more efficient than "modern" approaches to resource utilization in these environments. Notwithstanding this, the pastoral way of life is not without risks.

Recurrent drought and disease epidemics decimate herds in the Karamoja cluster. In the past, when drought or disease decimated herds, people recouped stock and ensured their survival by exchanging or loaning stock or, in the worst cases, by raiding cattle from neighbouring tribes. Raiding was confined to times of extreme environmental stress and carried out by large groups of warriors armed with spears and arrows. Since the 1970s, however, the nature of raiding has changed. It is now a continuous activity carried out by small groups of men armed with automatic weapons and driven by criminal motivation for profit. The results are devastating. It is estimated that cattle raids currently account for more than 70% of deaths among males aged 30–39 in tribes that inhabit the region. The proliferation of modern weapons along with

(continued)

> changes in traditional rules of engagement have transformed an adaptive practice into a maladaptive and ongoing conflict that has increased poverty and famine in the area, placed vast tracts of grazing land and water sources out of reach of herders and rendered many pastoralist families destitute.
>
> While ecological disasters and livelihood dislocations from war and famine contribute significantly to endemic poverty and underdevelopment in the area, there is increasing acceptance that the root cause for the crisis lies in the political and economic marginalization of pastoralists and by the failure of governments and development agencies to devise and implement programmes aimed at sustaining pastoral production. In the past, most interventions adopted the position that pastoralism is intrinsically self-destructive and that a more progressive approach to development should steer pastoralists into other, allegedly more secure means of assuring their livelihood. After decades of failed development, planners are beginning to realize that the practices of the pastoralists make sense and that optimal use of semiarid range resources may involve continuing animal husbandry through extensive pastoralism, rather than radical shifts to new technologies of intensive commercial husbandry and dryland agriculture.
>
> Source: Practical Action; http://practicalaction.org/drought-resilience-in-karamoja-1

With further dwindling resources and increased competition by the growing population on the meagre resources, conflict intensified at unpredictable dimensions and scale. For example, insufficient, corrupt or invalid land use or tenure policies strained the situation and lead to banditry, civil strife, disruption of cross-border markets and politically motivated wars at a scale that threatened the stability of the whole region. A more recent example is the sharp decline in livestock prices in Karamoja markets between December 2013 and January 2014 when civil conflict erupted once again in South Sudan (FAO 2014a, b, c, d, e, f).

2.3 Future Scope for Advancement

2.3.1 Enhancing Coping Mechanism and Resilience Through the Adoption of Measures to Minimize Drought-Triggered Conflicts

The resilience of the marginalized small agriculturists and livestock producers have substantially eroded as they failed to cope with the competing forces of population growth, advancing urbanization, commercialization and the invasive large-scale farming. Also the expansion of foreign large-scale agriculture investment from food deficit rich countries (Malaysia, Middle East Arab countries) through outsourcing is adding a further scale to the marginalization and risks facing the smallholder agriculture in the poor dryland countries and aiding in the displacement of the indigenous Africans from their historical heritages (e.g. the progress in forced eviction by Sudan and Egyptian governments of the Nubians from south Egypt and northern Sudan).

As discussed in the previous sections of this chapter, the dryland regions are under considerable pressure caused by climate variability and climate change (see also Bawden, this volume). The negative impact on agriculture sector will increase the vulnerability and reduce the coping mechanisms of the poor agriculture and herding communities. On the bright side, the global community has understood that although most of the events leading to drought are natural and climatic, the triggers are manmade, and, therefore, mankind must invent the solutions. There are several priority actions developed at the subsectoral, sectoral, regional and global levels (CFS 2015; CGIAR 2012; IGAD 2013a, b; World Bank 2015; FAO 2016a, UNFCCC 2010). Subsequent to the Paris Agreement (UNFCCC 2015) the global leaders met in Marrakech (UNFCCC 2016) to demonstrate that agriculture is part of the solution to climate change and that *"the agriculture sector has unrivaled potential to simultaneously address poverty, hunger and climate change."*

2.3.1.1 Emerging Strategies and Polices to Enhance Resilience in the Dryland Areas

Below are brief examples of emerging measures and strategies directly affecting the agriculture sector needed to enhance the coping mechanisms resilient to minimize conflicts of the dwellers of the drylands particularly the poor farmers and livestock producers.

Global Consensus: Enhancing Resilience of Small Agriculturalists and Livestock Producers Against Risks and Variability in the Drylands

From the bright side is the fact that the Committee on World Food Security (CFS) meets occasionally at FAO HQ to discuss and endorse policy recommendations on various aspects of sustainable agriculture. For example, the High-Level Panel of Experts (HLPE) on Food Losses and Waste (HLPE 2014) (see Sect. 3.1.1.2 below), the endorsement of the CFS' Framework for Action for Food Security and Nutrition in Protracted Crises (CFS 2015) and the report on Sustainable Agricultural Development for Food Security and Nutrition: What roles for livestock (HPLE 2016; ILRI 2016)? Below is a summary of selected recommendations aiming to enhance resilience of the dryland small agriculturalists and livestock producers against risks and variability for which there is evident global consensus:

(i) Coordinating actions, including through existing mechanisms, that support national policies and actions promoting food security and nutrition
(ii) Strengthening the security of tenure rights in line with the CFS Voluntary Guidelines on the Responsible Governance of Tenure of Land, Fisheries and Forests in the Context of National Food Security, including in all cases of conflict

(iii) Developing policies and tools and improve capacity, to assess, mitigate and manage risks, and reduce excessive price volatility and their impacts on the most vulnerable
(iv) Investing in and strengthening prevention, preparedness, impact reduction and disaster risk reduction strategies for more timely and cost-effective responses
(v) Supporting responsible investments which create economic opportunities for smallholders, particularly rural women and youth, as well as for members of affected and at-risk populations, vulnerable and marginalized groups and people living in vulnerable situations
(vi) Understanding, using and supporting the existing capacities, knowledge, practice and experience of affected households and communities as entry points for policies and actions
(vii) Encouraging policies and actions aimed at strengthening sustainable local food systems and fostering access to productive resources and to markets that are remunerative and beneficial to smallholders
(viii) Facilitating the adaptation to and mitigation of climate change in agricultural systems in line with the Paris Agreement and with particular support for smallholders and pastoralists, and women's role in food systems
(ix) Supporting appropriate and sustainable social protection programmes, including through predictable, reliable, rapidly scalable safety nets, to mitigate and manage food security and nutrition risks
(x) Enhancing access to insurance for small agriculture and livestock systems, including index-based insurance
(xi) Facilitating access to markets for the vulnerable and marginalized, including through cash transfer and voucher modalities, or other solutions adapted to local contexts, and based on a thorough analysis of local risks and needs
(xii) Improving plans for animal disease prevention, control and surveillance, including through cross-border cooperation on transboundary diseases, in order to foster early warning and early action on disease control, spread and eradication, with emphasis on the Global Eradication

Global Consensus: Contribution of the Reduction of Food Losses and to the Reduction of Pressure on Natural Resources and the Emission Intensity

Sustainable agriculture sector practices could contribute to the reduction of soil degradation and water depletion in the dryland areas (FAO, SOFA 2016). In addition, smart agriculture practices could play a major role in the mitigation of the impact of climate change by reducing intensity of emission. This effort can be complemented by actions aimed at reducing food losses and waste and changing food consumption patterns.

Food loss and waste (FLW) along with both the production and supply chains is endemic worldwide inclusive of the dryland developing countries. Most of the dryland countries are food deficit and are net food importers. About one third of all food produced in the world is lost or wasted postharvest. Food crises intensify during drought years. Food loss is more of an issue in the developing countries mainly

Table 2.3 Distribution of FLW along the food chain in different regions of the world

Regions/	Distribution of FLW (9%)					
	Harvest	Postharvest	Processing and packing	Distribution	Consumption	Total
NENA and Central Asia	10.8	7.8	6.3	5.6	5.5	36
North America and Oceanica	10.5	3.5	3.4	2.4	12.6	32
Europe and Russia	11.3	3.4	3.9	2.8	10.6	31
Japan, Korea and China	9	6.6	3.1	4.4	10.3	33
Latin America	13.4	7.5	5	4.1	3.7	34
Sub-Saharan Africa	12.5	12.7	4.5	4.6	1.3	36
South and SE Asia	8.7	9.6	2.7	4.6	2.6	26

Source: HPLE (2014). Food Losses and Waste in the context of sustainable food systems, a report by The High-Level Panel

because of the weakness in the food supply chain (The Economist 2014) and inefficient use of the production resources. Food loss constrains food security by reducing the availability of nutritious foods. On the other hand, food waste, which is more common in high-income countries and urban societies, occurs during the later stages of the food supply chains (distribution and consumption stages) as a result of high level of discarded food (FAO 2016a).

According to the CFS's High-Level Panel of Experts (HLPE on Food Losses and Waste) in the context of sustainable food systems (HPLE 2014), the dryland-dominated Near East and North Africa Region (NENA region) has the highest percentage of cumulative food losses (36%) among all world regions, only matched by sub-Saharan Africa (Table 2.3). The highest level of losses is found at the early stages of the food supply chain. This high magnitude of loss contributes to reduced food availability, aggravated water scarcity, adverse environmental impacts and increased food imports, in this already highly import-dependent region.

Recommendation: Reducing food losses and waste would reduce both pressure on natural resources and the emission intensity from processing, conservation and transportation of more food quantities than needed.

Land Use Policies and Agreement Supporting Rural Communities from Encroachment of Large Farms and Land Grabbing

As discussed earlier, land use changes (in size, technology, farming systems, etc.) are leading to intensified conflicts between the settled farmers and the herders. Also government policies that favour external investors have marginalized the native and the tradition subsistence farmers and, in most cases, deprived them from their subsistence livelihoods.

On the positive side, several countries have signed to voluntary measures such as the following:

- *The Voluntary Guidelines on the Responsible Governance on Tenure of Land, Fisheries and Forests in the Context of National Food Security (VGGT) – FAO, 2012*[4]
- *The Principle for Responsible Investment in Agriculture and Food Systems (PRAI) –FAO, 2014*

Use of Renewable Energy Sources

Combustion of fossil fuels is largely the cause of atmospheric concentration of carbon dioxide, the principal greenhouse gas. Overconsumption in order to provide energy for industry, domestic use and agriculture reached a level unequaled for more than 400,000 years (Climate Action Reserve). As a result, more of the sun's heat is being trapped near the earth's surface and gradually changing the earth's climate and contributing to global warming, extreme droughts and floods. Although most of the international community is aware of this, several geopolitical realities and financial obstacles are behind the reluctance and irresponsible performance of several governments in seeking alternative and renewable sources of energy. The brief discussion below is not an exhaustive account of the alternatives to fossil fuels but are meant to present a hint of what is in store.

There are at least ten major sources of renewable and environmentally friendly sources of energy that harness natural processes: *tidal power, wave power, solar power, wind power, hydroelectricity, radiant energy, geothermal power, biomass, compressed natural gas and nuclear power* (LISTVERSE 2009). Many dryland countries import oil or invest in developing hydroelectricity, the widely used source of renewable energy. However, the hydroelectric power has always been connected with social problem (forced displacement of people) or has proved to be a liability and environmental nightmare that many countries are now planning to get rid of before they became a national hazard. For example, dams cause water logging and siltation and account for loss of biodiversity by adversely affecting fish population and other aquatic organisms. Also hydropower plants fail to produce power when water levels go down or dry up because of extended droughts. When water is not available, the hydropower plants can't produce electricity and create problems of rehabilitation and related socio-economic problems. Dams drown sites of livelihood, mobilize or disintegrate local cultures and drown major historic sites such as the case of the Aswan Dam, the High Dam, Merowe Dam and the planned Dal, Kajbar and Esheraik dams along the Nile valley. Furthermore, dams increase seismicity (earthquake frequency) due to large volume of water impounded.

---

[4] FAO (2012). The Voluntary Guidelines on the Responsible Governance of Tenure (*VGGT*). The VGGT is an internationally negotiated document by the Committee on World Food Security (CFS) under the Food and Agriculture Organization of the UN (FAO) The VGGT were endorsed by The Committee on World Food Security (CFS) at a Special Session in May 2012. www.fao.org/fileadmin/user.../reu/.../vggt/VGGT_KG_en.pdf.

Not all renewable energy has problems equal to those generated by dams. Actually the majority, and specifically *solar and wind sources* of energy, are the most dominant in the dryland areas and have long-term and sustainable benefits if appropriately constructed and maintained. While there are many large-scale renewable energy projects and production, renewable technologies are also suited to small off-grid applications, sometimes in rural and remote areas, where energy is often crucial in human development.

The sun offers an ideal energy source, unlimited in supply, does not add to the earth's total heat burden and does not produce air and water pollutants. It is powerful alternative to fossil and nuclear fuels. Solar power is one of the fastest-growing energy sources, and so many new technologies are developing with rapid speed to attain more from sun's energy. Moreover, solar cells are becoming more productive, flexible, transportable and easier to instal.

Wind power has one of the lowest environmental impacts of all energy resources. At present wind produces only 1.5% of worldwide electricity, but it is growing fast. It achieved high level of penetration in many countries. The wind power electricity production in a few countries is phenomenal such as 19% of electricity production in Denmark, 11% in Spain and Portugal and 7% in Germany and Ireland.

2.3.1.2 Emerging Practices, Strategies and Polices to Enhance Investment in the Arid Areas

Adoption of Climate-Smart Technologies and Application Measures (Water and Land Management)

The smallholder crop and livestock agriculture systems in the dryland areas have been negatively impacted and, in most situations, decimated by climate variability, droughts and climate change. Cropping practices are mostly suboptimal, leading to poor crop growth, low vegetative cover, low crop yields and serious land degradation. In most dryland areas, especially in sub-Saharan Africa, climate change and variability have made drought and water scarcity common. However, it is now evident that the smallholder agricultural systems can be resilient to recurrent droughts and can adapt to climate change by adopting climate-smart practices, diversifying on-farm crop and livestock production and diversifying into off-farm income and employment especially for the women, youths and men. Furthermore, it has been demonstrated that such objectives cannot be achieved without sustainable management of land and water resources. Also it is now being fully embraced by policymakers and development partners that such goals could not be achieved without improvements in infrastructure, extension, capacity development, climate information, market access, credit and social insurance that are needed to facilitate adaptation and diversification of smallholder livelihoods. The global community has shown commitment for achieving these objectives. There are at present instruments, though not comprehensive or sufficient, that allow funding major areas of investment in agriculture research, green agriculture and climate change adaptation. The following techniques have proven success in improving small farming in the dryland areas:

1. *Conservation agriculture (CA)* techniques are being used by farmers in the dryland areas across the world with full support from the development partners and donors such as The World Bank, FAO, IFAD, AfDB and ADB. The CA techniques are developed by the CGIAR research centers, as well as university research programmes all over the world. The advantage of conservation agriculture summarized by the Conservation Agriculture: Global Research & Resources (Cornell University 2016) is in Box 2.7.

Box 2.7 Advantages of Conservation Agriculture

Conservation agriculture is generally a "win-win" situation for both farmers and the environment. Yet many people intimately involved with worldwide food production have been slow to recognize its many advantages and few consider it to be a viable alternative to conventional agricultural practices that are having obvious negative impact on the environment. Much of this has to do with the fact that conservation agriculture requires a new way of thinking about agricultural production in order to understand how one could possibly attain higher yields with less labour, less water and fewer chemical inputs. In spite of these challenges, conservation agriculture is spreading to farmers throughout the world as its benefits become more widely recognized by farmers, researchers, scientists and extensionists alike. Specifically, conservation agriculture (CA) increases the productivity of the following:

- *Land* – Conservation agriculture improves soil structure and protects the soil against erosion and nutrient losses by maintaining a permanent soil cover and minimizing soil disturbance. Furthermore, CA practices enhance soil organic matter (SOM) levels and nutrient availability by utilizing the previous crop residues or growing green manure/cover crops (GMCC's) and keeping these residues as a surface mulch rather than burning. Thus, arable land under CA is more productive for much longer periods of time.
- *Labour* – Because land under no till is not cleared before planting and involves less weeding and pest problems following the establishment of permanent soil cover/crop rotations, farmers in Ghana reported a 22% savings in labour associated with maize production. Similar reductions in labour requirements have been reported with no-till rice-wheat systems in South Asia and various CA technologies in South America. Much of the reduced labour comes from the absence of tillage operations under CA, which use up valuable labour days during the planting season.
- *Water* – Conservation agriculture requires significantly less water use due to increased infiltration and enhanced water holding capacity from crop

(continued)

residues left on the soil surface. Mulches also protect the soil surface from extreme temperatures and greatly reduce surface evaporation, which is particularly important in tropical and subtropical climates. In sub-Saharan Africa, as with other dryland regions, the benefits of conservation agriculture are most salient during drought years, when the risk of total crop failure is significantly reduced due to enhanced water use efficiency.
- *Nutrients* – Soil nutrient supplies and cycling are enhanced by the biochemical decomposition of organic crop residues at the soil surface that are also vital for feeding the soil microbes. While much of the nitrogen needs of primary food crops can be achieved by planting nitrogen-fixing legume species, other plant essential nutrients often must be supplemented by additional chemical and/or organic fertilizer inputs. In general, soil fertility is built up over time under conservation agriculture, and fewer fertilizer amendments are required to achieve optimal yields over time.
- *Soil biota* – Insect pests and other disease causing organisms are held in check by an abundant and diverse community of beneficial soil organisms, including predatory wasps, spiders, nematodes, springtails, mites and beneficial bacteria and fungi, among other species. Furthermore, the burrowing activity of earthworms and other fauna create tiny channels or pores in the soil that facilitate the exchange of water and gases and loosen the soil for enhanced root penetration.
- *Economic benefits* – Farmers using CA technologies typically report higher yields (up to 45–48% higher) with fewer water, fertilizer and labour inputs, thereby resulting in higher overall farm profits. In Paraguay, net farm income of no-till (NT) farming on large-scale commercial farms increased from $23,467 to $32,608 more than farms using conventional tillage over a 10-year period. The economic benefits of NT and other conservation agriculture technologies, more than any other factor, have led to widespread adoption among both large- and small-scale farmers throughout the world.
- *Environmental benefits* – Conservation agriculture represents an environmentally friendly set of technologies. Because it uses resources more efficiently than conventional agriculture, these resources become available for other uses, including conserving them for future generations. The significant reduction in fossil fuel use under no-till agriculture results in fewer greenhouse gases being emitted into the atmosphere and cleaner air in general. Reduced applications of agrochemicals under CA also significantly lessens pollution levels in air, soil and water.
- *Equity considerations* – Conservation agriculture also has the benefit of being accessible to many small-scale farmers who need to obtain the highest possible yields with limited land area and inputs. Perhaps the biggest obstacle thus far for the technology spreading to more small-scale farmers

(continued)

> worldwide has been limited access in certain areas to certain specialized equipment and machinery, such as no-till planters. This problem can be remedied by available service providers renting equipment or undertaking conservation agriculture operations for farmers who would not otherwise have access to the needed equipment. Formulating policies that promote adoption of CA are also needed. As more and more small-scale farmers gain access to CA technologies, the system becomes much more "scale neutral."
>
> - *Active role for farmers* – As with any new agricultural technology, CA methods are most effective when used with skilful management and careful consideration of the many agroecological factors affecting production on any given farm or field. Rather than being a fixed technology to be adopted in blueprint-like fashion, CA should be seen as a set of sound agricultural principles and practices that can be applied either individually or together, based on resource availability and other factors. For this reason, farmers are encouraged to experiment with the methods and to evaluate the results for themselves – not just to "adopt" CA technologies. Selecting among different cover crop species, for example, needs to be determined in relation to particular agroecological conditions of the farm, including soil type, climate, topography as well as seed availability and what the primary function of the GMCC will be. Similarly, planting distances, irrigation requirements and the use of agrochemicals to control weeds and pests among other considerations must be decided based on what the farmer needs as well as the availability of these and other resources.
>
> Source: Cornell University website (2016). Conservation Agriculture: Global Research & Resources

2. *No tillage* is a major aspect of CA but warrants detailed mention. Practices tried in many countries (e.g. ICARDA domain countries) include reduced tillage practices and breeding for adaptive crop and livestock species. Farmers and herders could be encouraged through also subsidized crop insurance schemes that could respond in a timely way to climate variability. However, zero tillage is a disregarded by many producers in developing countries because of conflicting demands on crop residues for animal feed or for household fuel. Also terraces or land levelling may prevent runoff (FAO, SOFA 2016).
3. *Watershed management* It has been proven that soils and water (surface and ground) resources can be better managed and sustained using soil water conservation (SWC) techniques (summarized in FAO 2008). The watershed approach encourages the promotion of cooperation between upstream and downstream

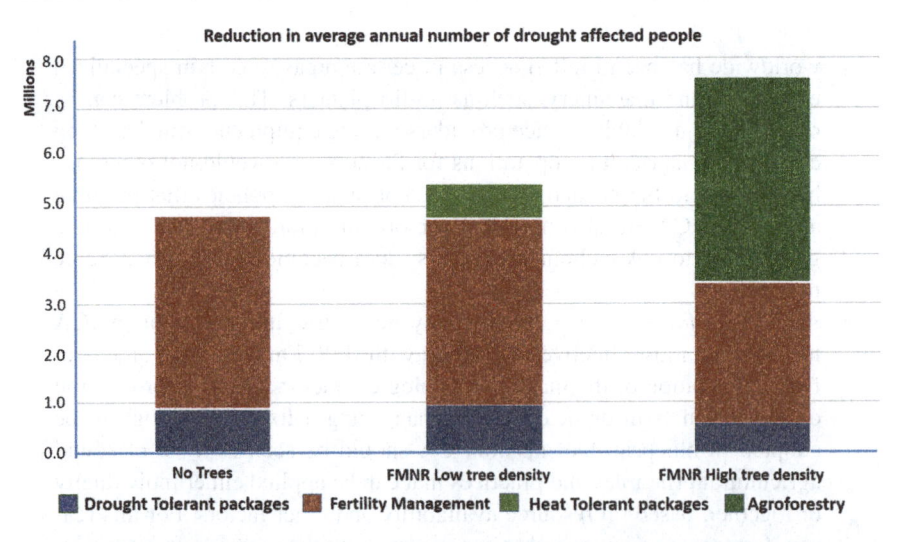

Fig. 2.1 Estimated reduction in the average annual number of drought-affected people resulting from FMNR and other technologies. *Source*: World Bank 2015 Africa Region Flagship Report: Enhancing Resilience in African Drylands. Towards a shared development agenda (Authors' estimates)

stakeholders – in an effort to minimize conflicts over land and water (FAO, SOFA 2016).

4. *Farmer-managed natural regeneration (FMNR)*: A number of modelling activities were carried out during the preparation of the Africa region flagship report "Enhancing Resilience in African Drylands: Toward a Shared Development Agenda" (World Bank 2015). The models sought either to add to the understanding of the problems or situations or to assess the impact of various approaches to mitigate drought shocks and to reduce conflicts. For example, the crop modelling helped provide orders of magnitude of the benefits of *farmer-managed natural regeneration* (FMNR) in terms of reduction of drought impacts. Impressive drought mitigation impact is possible by adding farmer-managed natural regeneration of native species to the other productivity-enhancing technologies (adoption of drought tolerant, fertility and heat-tolerant packages, agroforestry). Compared to no interventions scenario, in a group of ten countries in East and West Africa, the projected number of poor, drought-affected people living in drylands in 2030 fell by 13% with low-density tree systems and more than 50% with high-density tree systems (Fig. 2.1).

5. *CGIAR's Research Program on Climate Change, Agriculture and Food Security (CCAFS)* – CGIAR (2012).[5] Some examples are summarized in Box 2.8.

---

[5] *CGIAR, CCAFS*.https://ccafs.cgiar.org/blog/future-farming-dry-lands-smart-solutions-climate-change#.WDcL_k3ymUl.

> **Box 2.8 Examples of Climate-Smart Farming**
>
> *Niger*: Planting of 200 million nitrogen-fixing trees such as *Acacia Senegal* and *Acacia seyal* has resulted in the transformation of 5 million hectares of once infertile land. The initiative has increased crop yields and fodder availability, benefiting 2.5 million farmers. Trees increase the carbon stored in the landscape, mitigating climate change.
>
> *Tunisia*: Groups of "lead farmers" in Tunisia are being linked by mobile phone to crop and weather monitoring systems that issue alerts when irrigation is needed. The information can then be relayed to other producers.
>
> *North and sub-Saharan Africa*: The use of different planting methods, high-yielding varieties, improved water management and integrated pest management in seven countries across North and sub-Saharan Africa, new approaches tested by national research and extension systems with ICARDA has produced a 22% increase in wheat yields for Egypt and a 58% increase in Sudan – based on actual farmer experiences (not trials).
>
> Source: CGIAR. CCAFS https://ccafs.cgiar.org/blog/future-farming-drylands-smart-solutions-climate-change#.WDcL_k3ymUl

Adoption of Conflict-Sensitive Policies and Strategies

In response to the protracted droughts and the humanitarian crises they caused, the heads of state and governments of East Africa Community (EAC) and IGAD member states (*Djibouti, Eritrea, Kenya, Somalia, South Sudan, Sudan, Tanzania and Uganda*) held a summit meeting in Nairobi in 2011 that was mainly focused on the need for a strategy to address drought threats in a decisive and sustainable way. In response the Global Alliance for Action for Drought Resilience in the Horn of Africa was facilitated and funded by USAID through a technical consortium (TC) between FAO and the CGIAR with the involvement of other relevant research and academic (IGAD 2013a, b; FAO 2014a, b, c, d, e, f). Similarly 2012 drought-induced crises in the African Sahel lead to the OECD-facilitated and OECD-supported initiation of the Global Alliance for Resilience – Sahel and West Africa (AGIR) – (World Bank 2015).

To meet this challenge, the IGAD Drought Disaster Resilience and Sustainability Initiative (IDDRSI) was launched in 2013. The IDDRSI succeeded in creating a strong political momentum and commitments by the member countries and the development partners and increased investments and spurred resource mobilization. Operational frameworks and programmatic implementation mechanisms that are providing strategic direction and the operational framework for resilience-enhancing policies and investments complement the IDDRSI strategy. The common architecture of the IDDRSI is based on six strategic priority areas (natural resource management, market access and trade, livelihood and basic services support, pastoral disaster risk management, research and knowledge management and peace building and conflict resolution) that address drought resilience in a holistic manner and encompass policies and mechanisms supporting the adaptation of pastoralists and the small farmers/herders to drought shocks and uncertainties. The IDDRSI strategic framework is also aligned with the CAADP and with UN objectives to increase resilience of livelihoods to drought-triggered disasters.

Evolution of the Cross-Border Cluster Approach: A Tool for Regional and International Conflict Resolution, Early Warning and Response Mechanisms

The cross-border cluster approach is a very effective, though complicated, approach for managing conflict in the dryland areas particularly among the traditionally mobile pastoral and agro-pastoral communities (please refer to Boxes 2.5 and 2.6 above for illustrative explanations). The cluster approach can help in bridging the human development gap through specific investments projects through scaling up innovations that would enhance resilience and sustainability. These could include awareness building of the potential of these regions as well as their regional nature, early warning systems based on effective social and technological tools for measuring resilience and creating coping mechanism capable of overcoming and mitigating shocks.

Because of a multitude of geopolitical, social and climatic events, these cross-border areas are sites of conflict and competition for natural resources (specially grazing land and water). A good example presented in this chapter is the cluster approach that was developed by IGAD with the concerned member countries (Ethiopia, Kenya, Somalia, South Sudan, Sudan and Uganda) and the donors. The approach was defined (Box 2.9) during resource mobilization for the drought-resilience projects that followed the development of IDDRSI. Resources for the abovementioned countries were provided by the World Bank, African Development Bank and Germany for the design and implementation of two cross-border programmes: Drought Resilience and Sustainable Livelihoods Program DRSLP and The Regional Drought Resilience Fund RDRF. FAO provided technical support throughout the preparation of IDDRSI and the development projects (FAO 2014a, b, c, d, e, f; IGAD 2015). The following are the major requirements/features/benefits of the cross-border cluster approaches:

- An important intervention in the cluster approach is to coordinate the stakeholders in order to exchange information effectively and to synchronize activities such as vaccination campaigns, surveillance and disease control.
- There is no cross-border solution without dialogue.
- Stakeholders on multiple levels need a voice to ensure that responsibilities for long-term resilience promotion are carried out.
- Investment need to be coordinated in order to facilitate the necessary dialogue and to avoid duplication and contradiction.

> **Box 2.9 Definitions of Cross-Border Cluster**
> There are two major but related definitions for the cluster
>
> - *A geographic space cutting across a country border in which stakeholders aim to develop and implement coordinated investments to enhance resilience.*
> - *A geographic space that cuts across multiple political–administrative units within the country and international borders, where a range of resources, services, cultural values are shared by pastoral and agro-pastoral communities, and in which stakeholders aim to develop and implement coordinated investment s to enhance resilience and sustainable development.*
>
> Clusters are characterized by a set of unifying factors (e.g. social/ethnic/linguistic unity, complementary natural resources, infrastructure, trade) but also common challenges (e.g. droughts, insufficient access to water and pasture for grazing, resource conflicts). Clusters face challenges regarding natural resources (grazing land, water points), infrastructure (main road and feeder road local markets, internal and cross-border trade points), and services (veterinary and quarantine services).
>
> Sources: IGAD and FAO

- The approach helps the countries harmonize their policies related to dryland communities.
- The governments, country partners, civil society organizations and development partners should participate actively in the dialogue.

Acceleration of African Agriculture Growth and Transformation

As discussed in the above section, a large part of the global funding and programmes addressing the vulnerability of agriculture in the dryland areas is allocated to Africa. Excluding deserts, 43 % of Africa land surface is dryland areas. Furthermore, the Horn of Africa and the Sahel are the most drought-prone regions of the world where most of the global drought disaster occur. According to the report, sub-Saharan Africa was responsible for 88% of the global conflict death toll between 1990 and 2007, in addition to over 9 million refugees and internally displaced people. Therefore, the progress in the goals set for the Comprehensive Africa Agriculture Development Programme (CAADP) as continent-wide agenda for the transformation of African agriculture, economy and society will have implication on global stability and sustainability. The African heads of state endorsed the CAADP in 2003 as a New Partnership for Africa's Development (NEPAD) programme with a vision for the restoration of agricultural growth, food security and rural development. The

CAADP introduces a new approach to development focusing on both achieving economic growth and poverty reduction, entails the participation of all stakeholders and promotes regional cooperation. The CAADP covers agriculture (crops, livestock, fisheries and forestry), environment/climate change (resilience, mitigation, adaptation, adoption of innovations and new technologies), food and nutrition security, youth employment and gender mainstreaming. The CAADP's policy, planning and investment approaches are comprehensive, inclusive and multisectoral.

The CAADP's objectives are to achieve growth rates in the overall agriculture sector of minimum 6% per year in each African country. This was foreseen achievable by committing for agriculture at least 10% of government spending, developing dynamic regional markets and integrating farmers into those markets and achieving a more equitable distribution of wealth. This objective is foreseen to be achieved through measurable outcomes to be derived from four key investment priority areas: extending the area under sustainable land management and reliable water control systems; improving rural infrastructure and trade-related capacities for market access; increasing food supply, reducing hunger, and improving responses to food emergency crises; and improving agriculture research, technology dissemination and adoption.

Since 2003 most African countries developed their CAADP based National Agricultural Investment Plans (NAIPs) and a few are at present participating in the development and implementation of a second generation of NAIPs aiming to achieve further goals set in 2014 to guide African agriculture growth and transformation during the next 10 years (to 2025). The CAADP goals, and consequently the NAIPs and the Regional Agricultural Investment plans, are aligned with most of FAO's Strategic Objectives (FAO 2014a, b, c, d, e, f) and the Sustainable Development Goals (SDGs)[6] particularly those targeting reduction of poverty, hunger, food insecurity and malnutrition combating impact of climate change and desertification and enhancing resilience of livelihoods to disasters (e.g. SDG 1, 2, 3, 13 and 15).

Enhancing Resilience to Drought Shocks: Example from Somalia

The drought emergency crises in the Horn of Africa attracted the attention of the governments and the development partners to seek alternatives to the unsustainable measures of emergency humanitarian relief and food distribution program. The tripartite strategy developed by FAO, WFP and UNICEF (2012) and adopted by the development partners during the Somalia drought crises (Sidahmed 2013) is an example of how a combination of interventions to improve the current livelihoods (dryland farming, pastoral herding), to strengthen safety nets and to develop new and alternative livelihoods is being tried (Box 2.10). For example, a strategy to galvanize collective efforts aiming at laying the foundations for stability. The declaration made at the Istanbul Conference (FAO 2012a, b) recognized the urgent need to enhance resilience of the Somali households and communities, through multi-year and sectoral investments suited to each geographic location within Somalia (Box 2.10).

---

[6] UN (2015). On 25 September 2015, countries adopted a set of goals to *end poverty, protect the planet* and *ensure prosperity for all* as part of a new sustainable development agenda; http://www.un.org/sustainabledevelopment/sustainable-development-goals/.

> **Box 2.10 The Building Blocks for Enhancing Resilience in Somalia**
>
> *Strengthen productive sectors to enhance household income* for the various livelihood types (pastoralism, agro pastoralism, farming, riverine, fishing and urban) through improved access of working households to productive assets, income diversification, intensification, enhanced technologies, employment opportunities, market systems and market information.
>
> *Enhance basic services to protect human capital* access of men and women to *basic services* that enhance human capital (e.g. health, nutrition, education, sanitation, food and water safety and hygiene, adequate skill) and *support services* (extension, animal and plant health services and information and knowledge management for early warning and planning).
>
> *Promote safety nets to sustain the basic needs of the chronically destitute/at risk (for a minimum of social protection)* entails moving beyond the discontinuous cycles of short-term assistance to approaches that build resilience by providing a predictable level of assistance to those suffering from long-term destitution as well as for households that are seasonally at risk on a recurrent basis
>
> Sources: Sidahmed, A E June 2013. Somalia Country Programme Paper (CPP). Programming Framework IGAD Inititive for Drought disaster Resilience and Sustainability Initiative (IDDRSI) in the Horn of Africa (HoA). FAO Investment Centre Rome Italy

The Establishment of Family Farming Price Guarantee Programmes in Brazil

Management of drought shocks, that aim for more than humanitarian emergency rescue programmes, have proven long-lasting benefits to the vulnerable dryland farming and herding families. One successful example was achieved in Brazil under its *Bolsa Familia* (OECD-FAO 2015). The Family Farming Price Guarantee Program (PGPAF) is an insurance programme that provides discounts on credit contracts to offset drops in farm revenue owing to reductions in market prices or climate-induced crop losses. In addition, a harvest insurance fund specifically targets farmers in Brazil's semiarid region when drought causes severe crop losses for family farmers.

2.3.1.3 The Establishment of the Green Climate Fund

A major lesson learned over the past decades is the fact that climate change is a threat to the poor and the rich nations in the same manner that poverty is a threat to all. Also there is little disagreement that climate change is affecting agriculture and food security and putting millions of poor people (mainly small farmers and herders) at risk of hunger and poverty. This underscores the important role that agriculture could play as the primary investment area in the global response to climate change. For such an interrelated undertaking, sustainable agriculture production and productivity (particularly of the poor rural communities e.g. smallholder agriculturalist and herders) would require adoption of a number of appropriate technologies and innovations such as: climate-smart agriculture practices, diversification of on-farm and off-farm income and sustainable management of water and land resources However, these investments must be supported by improvements in infrastructure, extension, climate information, market access and credit and social insurance.

Therefore, the global community has increasingly understood the importance of tackling together hunger, poverty and climate change. One option, the Green Climate Fund (GCF) – a dedicated multilateral climate fund (FAO, SOFA 2016) – was established at the 16th Conference of Parties to the United Nations Framework Convention on Climate Change in 2010 (UNFCCC 2010), with a main objective of promoting low emission in agriculture (e.g. addressing emissions from land use change driven by agricultural expansion) and to help the vulnerable communities in the developing countries to adapt to the unavoidable impacts of climate change, while maintaining sustainable agricultural development.

The GCF responds to climate change threats by investing in low-emission and climate-resilient development programmes especially in sub-Saharan Africa, the most negatively impacted region of the world (deforestation, decline in water resources). For example, the common thematic areas of the GCF projects in Africa of relevance to drought management are *water and land resource management* (e.g. in Chad, Kenya, Burkina Faso, Morocco Tunisia, and Egypt), *sustainable catchment management* (e.g. in Uganda and Rwanda) and climate resilience of agrosylvo pastoral ecosystems in the DRC. As of May 2016, pledges made to the GCF reached US$10.3 billion, of which US$9.9 billion has been signed over to the fund (FAO, SOFA 2016). This sum is expected to rise to at least US$100 billion in annual climate finance to developing countries by 2020. FAO is accredited with the GCF as a grant-implementing entity for medium-sized projects (USD 50–250 million) with a medium level of environmental and social risk.

2.4 Summary and the Way Forward

2.4.1 Summary

Between 1.0 and 1.5 million of the world population live in fragile and conflict affected areas,[7] of which a major part falls within the dryland areas. Although some progress was made in reducing onset of extensive famines and in spite of two

---

[7] Kimberly Flowers (2016). The 13th Annual George McGovern Lecture "Examining linkages: The nexus between food insecurity and political instability" 22 Nov 2016; Kimberly Flowers is the Director of the Global Food Security Project, Center for Strategic and International Studies (CSIS), USA.

decades of relatively stable food market, there are some very remarkably threatening food insecurity hotspots caused by protracted droughts and floods (e.g. Horn of Africa, South Asia, respectively), and civil wars such as Syria (which have turned in 5 years from producer and exporter of food to a food deficit country), South Sudan and Somalia.

The drought-related acute food shortage in the Sahel region of West Africa and the Horn of Africa have left about 18.7 million and 11.7 million people, respectively, in need of emergency assistance.

The potential impact of climate change on agriculture in sub-Saharan Africa, and especially the drier climates, will be negative and would enhance poverty, malnutrition and hunger. This is because the frequency of extremely dry and wet years will increase, thus disrupting agriculture production. Also forage production will decrease remarkably, particularly in the Sahel as a result of drought and range degradation.

2.4.2 The Way Forward

The global community has increasingly understood the importance of tackling together hunger, poverty and climate change. Major landmark agreements include the Paris Agreement to cap the limit of global warming to below 2.0 °C or even to 1.5°. Encouraging developments include commitment of the international community to continue the dialogue and focus on actions to achieve the goals of the Paris Agreement through sets of COPs starting with COP 22 in Morocco in November 2016. In Marrakech the heads of state, government and delegations called for the highest political commitment to combat climate change as an urgent priority, and the developed country partners affirmed a US$100 billion mobilization goal in support of climate projects. In addition of commitments by the bilateral donors such as EC, DFID and USAID (e.g. the multibillion dollar Feed the Future Program), there are already committed global instruments such as the Green Climate Fund (GCF) – a dedicated multilateral climate fund – which was established at the 16th Conference of Parties to the United Nations Framework Convention on Climate Change in 2010, with a main objective of promoting low emission in agriculture (e.g. addressing emissions from land use change driven by agricultural expansion) and to help the vulnerable communities in the developing countries to adapt to the unavoidable impacts of climate change, while maintaining sustainable agricultural development.

2.4.3 Recommendations for Actions to Make Current Livelihoods More Resilient in the Drylands

Most of the summary recommendations below were derived from the analyses of the experiences in sub-Saharan African drylands. For example, the recent intensive study by a team from 16 international and regional institutions coordinated by the

World Bank and FAO (World Bank 2015). However, consideration and assessment of the validity of these recommendations to the developing country drylands areas were undertaken because of the global nature of the review as well as the global nature of the consensus in measures to address poverty, drought and climate change as extensively referenced in the text. The "research for development" recommendations were derived from global sources (CGIAR 2012).

2.4.3.1 Strategic Action Areas (SAA)

The strategic action areas listed below are adopted from the Implementation Strategy and Road Map (IS&RM) to achieve the 2025 vision on the Comprehensive Africa Agriculture Development Programme (CAADP). The action areas aim to operationalize the 2014 Declaration of the African Head of States and Governments into accelerate African agriculture growth and transformation of shared prosperity and improved livelihood.

Objective 1: Transformed agriculture and sustained inclusive growth

- Increase production and productivity
- Enhance market, trade and value chain
- Increase resilience of livelihoods and systems
- Strengthen governance of natural resources

Objective 2: Strengthened systemic capacity to implement and deliver results

- Strengthen capacities for planning
- Strengthen policies and institutions
- Strengthen leadership, coordination and partnerships
- Enhance skills, knowledge and agriculture education
- Strengthen data and statistics
- Institutionalize mutual accountability
- Increase public and private financing

2.4.3.2 Technical Action Areas (TAA)

1. *Crop production*
   - Accelerate the rate of varietal turnover and increase availability of hybrids.
   - Improve soil fertility management.
   - Improve agricultural water management.
   - Promote the development of irrigation, including both rehabilitation of existing capacity, as well as expansion, up to the viable potential (a max. of about 10 more million hectares), and focusing on small-scale systems, with good access to markets for cash crops.
2. *Natural resource management*
   - Promote farmer-managed natural regeneration (FMNR) to establish a range of beneficial trees throughout the drylands.

- Invest in tree germplasm multiplication and promote planting of location-appropriate high-value species especially in dry subhumid areas.
- Develop value-added opportunities for tree products produced in the drylands.

3. *Livestock production*
   - Increase production of meat, milk, and hides in drylands by developing sustainable delivery systems for animal health, promoting increased market integration and exploiting complementarities between drylands and higher rainfall areas.
     – In each country, public veterinary services should be strengthened for carrying out the public good services, in particular vaccination for the major contagious (transboundary) diseases, quarantine and enforcement of sanitary standards, while at the same time enabling the development of a private network of professional and para-professional animal health providers for clinical services.
     – Animal health improvements should be accompanied with increasing access to feed (water development in the arid zone and fodder crop promotion in the semiarid/subhumid zones).
     – Further increase in production can be achieved by expanding the complementarity between the drylands and nondry ecosystems by promoting stratification, with the drylands supplying younger and better quality stock for fattening in the higher rainfall areas.
   - Enhance the mobility of herds by expanding and ensuring adequate and equitable year-round access to grazing and water and by improving security in pastoral zones. The following interventions/approaches contribute to improved mobility have the potential to improve significantly the performance of livestock systems in the drylands:
     – Development of water resources to allow better access to underexploited rangelands or the organization of feed markets and feed transport. This would cover water development in underserved areas in the arid zones, clearly marked corridors between the arid and subhumid zones and reserved access to dry season grazing areas.
     – Improvement of land use planning to facilitate movement of herds and flocks (e.g. through designation of dedicated migration corridors and dry season grazing areas).
     – Engagement in comprehensive and fully inclusive dialogue with the pastoralist on priorities in development, restoring trust and convincing the pastoral population to become partners in improving security.
     – Policies that promote a more equitable access to grazing (preferential allocation of user rights to groups of smallholders, progressive taxation or grazing fees).
     – Mobility can further be increased with regional integration through harmonization of sanitary standards and the removal of informal and unnecessary administrative border procedures.

- Develop Livestock Early Warning Systems (LEWSs) and early response systems to reduce the adverse impacts of shocks. There are a number of early warning systems using modern technology that can be introduced or strengthened, the key focus being in their sustainability. There is also a number of early response actions such as subsided destocking at the onset of drought, provision of additional feed supplies, etc. The focus should be on cost-efficiency.
- Public-private partnerships (PPPs) are needed to develop stratification, with the public sector (Ministries of Infrastructure, Commerce) having the prime responsibility for the infrastructure (roads, markets) and the private sector for market development
- Identify additional and alternative livelihood strategies, including through systems of payment for environmental services, through value adding in the value chain (fattening, processing, local leather industries) and through credit and education, in particular vocational skill training, employment opportunities need to be created outside the livestock sector.

4. *Social protection*
   - Establish and gradually expand the coverage of national adaptive safety net programmes that promote resilience of the poorest.
   - Use social protection programmes to build capacity of vulnerable households to climb out of poverty but maintain the ability to provide humanitarian assistance in the short run.
   - Respond to emergencies by scaling up existing programmes, rather than relying on appeals for humanitarian assistance.
   - Tailor social protection programmes to address the unique circumstances of dryland populations.

Research for Development
1. Crop yield-increasing technology packages, which include the use of tied ridging for water conservation, improved crop varieties, use of manure and micro-dosing.
2. Training farmers in soil conservation terracing techniques.
3. Employing farming methods that adapt to changing climatic conditions and boost productivity while maintaining the sustainability of natural resources.
4. Combining the principles of Integrated Soil Fertility Management (ISFM), conservation agriculture and water management.
5. Large-scale testing of research developed disease-resistant, high temperature-tolerant varieties that are also fast growing.
6. Science and technology parks fostering innovative entrepreneurship in the Asia Pacific region
7. Community-based organizations trained and facilitated to promote climate-smart agriculture.

References

AfDB, Feed Africa: strategy for agricultural transformation in Africa 2016–2026. (2016). http://www.afdb.org/fileadmin/uploads/afdb/Documents/Generic-Documents/Feed_Africa-_Strategy_for_Agricultural_Transformation_in_Africa_2016-2025.pdf

Agriculture and Rural Development Strategy towards 2040 Investment Plan, FAO Strategy report TCIC. (2016–2020)

CFS, Framework for action for food security and nutrition in protracted crises (FFA). (2015). www.fao.org/cfs

CGIAR, *CRP Programme- Integrated Agricultural Production Systems for Improved Food Security and Livelihoods in Dry Areas* ("Dryland Systems"). Inception Phase Report. CRP Programme. (2012). www.cgiar.org

CGIAR, CCAFS, (n.d.). https://ccafs.cgiar.org/blog/future-farming-dry-lands-smart-solutions-climate-change#.WDcL_k3ymUl

Climate Action Reserve, Climate action facts. (n.d.). http://www.climateactionreserve.org/resources/climate-change-facts/. Last visit 21 Nov 2016

Cornell University website, Conservation agriculture: global research & resources. (2016). http://conservationagriculture.mannlib.cornell.edu/pages/aboutca/advantages.html

S. Devereux, *Vulnerable livelihoods in Somali Region, Ethiopia*, ODI Research Report No. 57 (Overseas Development Institute, Sussex, 2006)

EM-DAT, The International Disaster Database, maintained by the Center for Research on the Epidemiology of Disasters (CRED), Université Catholique de Louvain. (n.d.). (www.emdat.be)

FAO, Chapter 1: Drylands, people and land use, in *Waters and Cereals in Drylands* (2008). http://www.fao.org/docrep/012/i0372e/i0372e01.pdf

FAO, The Voluntary Guidelines on the Responsible Governance of Tenure (VGGT). The VGGT is an internationally negotiated document by the Committee on World Food Security (CFS) under the Food and Agriculture Organization of the UN (FAO) The VGGT were endorsed by The Committee on World Food Security (CFS) at a Special Session in May 2012. (2012a). http://www.fao.org/fileadmin/user_upload/reu/europe/documents/Events2015/vggt/VGGT_KG_en.pdf

FAO, Proceedings of Istanbul conference-partnership for resilience. (2012b). May 2012

FAO, FAO new strategic framework and the role of statistics. – AFCAS 23, 2013. (2014a) http://www.fao.org/fileadmin/templates/ess/documents/afcas23/Presentations/AFCAS_2_StrategicFramework_JS.pdf

FAO, The Technical Consortium (TC) for building resilience to drought in the Horn of Africa – completion report FAO, CGIAR and ILLRI. (2014b)

FAO, The principles for responsible investment in agriculture and food systems (PRAI) was endorsed by CFS endorsed on October 15th, 2014. (2014c)

FAO, The technical consortium for building resilience to drought in the Horn of Africa. USAID, FAO and CGIAR Completion Report May 2014. (2014d)

FAO, Special report: FAO/GIEWS livestock and market assessment mission to Karamoja Region, Uganda April 2014 1367 4E/1/03.14. (2014e). http://resilience.igad.int/attachments/article/283/160804_Uganda

FAO, Pasquale Steduto. The regional initiative on water scarcity. National Drought management Policy workshop. (2014f)

FAO, Commission on genetic resources for Food and Agriculture FAO. (2015d). Website: http://www.fao.org/nr/cgrfa/en/

FAO, RNE. Ahmed E. Sidahmed, Brian Perry and Nancy Morgan. The contribution of livestock to food security in the Neareast and North Africa Region. Primary background paper for the Discussion Paper NERC/16/4/ Rev.2 33rd Session of FAO Regional Conference Rome, Italy May 2016. (2016a)

FAO, Ahmed E Sidahmed working paper: NATIONAL PROGRAMME: sustainable livestock development for food security in Oman, in *Guidelines for Land Use System Mapping*, Technical Report # 8, ed. by ed. by Sustainable LADA, (FAO, Rome, 2016b)

FAO, SOFA, The state of food and agriculture; climate change, Agriculture and Food Security. (2016)

FAO, UNICEF and WFP Somalia, A strategy for enhancing resilience in Somalia. (2012)

Kimberly Flowers, The 13th annual George McGovern lecture "Examining linkages: The nexus between food insecurity and political instability". 22 Nov 21016 Kimberly Flowers is the Director of the Global Food Security Project, Center for Strategic and International Studies (CSIS); USA. (2016)

A.K. Fosu, Inequality and the impact of growth on poverty: Comparative evidence for Sub-Saharan Africa. J. Dev. Stud. **45**(5), 726–745 (2009)

FSNAU, Technical series report no VI 36 March 4, 2011. (2011)

D. Headey, A.S. Taffese, *Enhancing Resilience in the Horn of Africa: An Exploration into Alternative Investment Options*, IFPRI Discussion paper 01176 (IFPRI, Washington, DC, 2012)

HLPE, Report on sustainable agricultural development for food security and nutrition: what roles for livestock? Extract from the Report: Summary and Recommendations (23 June 2016). (2016). http://www.fao.org/fileadmin/user_upload/hlpe/hlpe_documents/HLPE_S_and_R/HLPE_2016_Sust-Agr-Dev-FSN-Livestock_S-R_EN.pdf

A. Hosseinizadeh et al., Impact of climate change on the severity, duration, and frequency of drought in a semi-arid agricultural basin. Geoenviron. Disaster. 2015 **2**, 23 (2015)

HPLE, Food losses and waste in the context of sustainable food systems, a report by the – The High Level Panel on Food Security and Nutrition (HPLE). (2014)

IFPRI, Derek Headey, A.D Taffesse and L You. Enhancing resilience in the Horn of Africa: an exploration into alternative investment options. IFPRI Discussion paper 01176. (2012)

IGAD, IGAD Drought Disaster Resilience and Sustainability Imitative (IDDRSI). (2013a)

IGAD, The Intergovernmental Authority for Development (IGAD) Drought Disaster Resilience and Sustainability Imitative (IDDRSI). (2013b)

IGAD, IDDRSI. Cluster approach for cross border cooperation and investments to strengthen drought resilience in IGAD Region. (2015)

IGAD & FAO, IGAD Regional Agriculture Investment Plan (IGAD-RAIP). (2016)

ILRI, UN endorses recommendations on sustainable agricultural development for food security and nutrition, including the role of livestock; Round Table of the Committee on World Food Security (CFS), meeting in Rome on 17 Oct 2016. (2016). https://news.ilri.org/2016/10/20/un-endorses-recommendations-on-sustainable-agricultural-development-for-food-security-and--nutrition-including-the-role-of-livestock/

IOA, Food insecurity and malnutrition in Africa: current trends, causes and consequences CONSULTANCY AFRICA INTELLIGENCE (CAI) Published by In *One Africa*, on 19 Sep 2012. (2012). https://polity.org.za

IPCC, *Climate Change 2007: Synthesis Report. International Panel on Climate Change Report* (Cambridge University Press, Cambridge, UK, 2007)

IUCN, IUCN_ESARO dryland situation analysis IUCN Eastern and Southern Africa Regional Office (ESARO). (2010). Jonathan.davies@iucn.org

LADA, *Guidelines for Land Use Systems Mapping. Technical Repoer #8* (FAO, Rome, 2008)

M. Lesnoff, C. Corniaux, P. Hiernaux, Sensitivity analysis of the recovery dynamics of a cattle population following drought in Sahel. Ecol. Model. **232**(2012), 28–39 (2012)

LISTVERSE, Top 10 renewable energy sources. Listserve Staff May 1, 2009. (2009). http://listverse.com/2009/05/01/top-10-renewable-energy-sources/

A. Loukas, L. Vasiliades, J. Tzabiras, Climate change effects on drought severity. Adv. Geosci. **17**, 23–29 (2008.) 2008 www.adv-geosci.net/17/23/2008/

MCI, Multiple cropping Intensity is the sum of area harvested for different crops during the year divided by the total harvested land). (n.d.)

Middleton, Thomas, The world atlas of desertification millennium ecosystem assessment (2005a). Climate change. Chapter 13, in *Ecosystems and Human Wellbeing: Current State and Trends, Volume 1*.(Island Press, Arnold, London, 1997)

Millennium Ecosystem Assessment, *Chapter 22 Drylands Systems: Ecosystems and Human Wellbeing: Volume 1. Current State and Trends* (Island Press, Washington, DC, 2005.) http://www.millenniumassessment.org/en/Global.html

OECD, Agricultural outlook 2009–2018. (2009)

OECD-FAO, OECD-FAO agricultural outlook 2015–2024. (2015)

OXFAM, (2012)

Practical Action.Org, (n.d.). http://practicalaction.org/drought-resilience-in-karamoja-1--practicalaction.org

I. Rodríguez-Iturbe, A. Porporato, *Ecohydrology of Water-Controlled Ecosystems: Soil Moisture and Plant Dynamics*. (Cambridge University Press, Cambridge UK, 2004)

N. Stern, *The Stern Review on the Economics of Climate Change* (Cambridge University Press, Cambridge, 2006)

N. Stern, *The Economics of Climate Change* (Cambridge University Press, Cambridge, UK, 2007)

Summit, The Nairobi strategy: enhanced partnership to EDE. Adopted at the Summit on HoA Crisis. 9 Sept 2011. (2011)

The Economist, Food Loss and its interaction with food security. A Special Report by the Economist Intelligence Unit. (2014).

The Global Agenda for Sustainable Livestock, (n.d.). http://www.livestockdialogue.org/

UN, Global drylands: A UN system-wide response first published in October 2011 by the United Nations Environment Management Group. (2011). http://www.unccd.int/Lists/SiteDocumentLibrary/Publications/Global_Drylands_Full_Report.pdf

UN, On September 25th 2015, countries adopted a set of goals to end poverty, protect the planet, and ensure prosperity for all as part of a new sustainable development agenda. (2015). http://www.un.org/sustainabledevelopment/sustainable-development-goals/

UNCCD, (n.d.). http://www.unccd.int/en/Pages/default.aspx

UNFCCC, *United Nations framework convention on climate change. (COP 16)*. (2010)

UNFCCC, Conference of parties 21 (COP 21). (2015). https://unfccc.int/files/meetings/paris_nov_2015/application/pdf/paris_agreement_english_.pdf

UNFCCC, Marrakech partnership for global CC and Marrakech action proclamation. (2016). http://www.cop22-morocco.com/COP22_FINAL

R.P. White, An ecosystems approach to drylands: building support for new development polices WRI.2002. (2002). http://summit.wri.org/publication/ecosystem-approach-drylands-building-support-new-development-policies

World Bank, Africa region flagship report: enhancing resilience in African drylands: toward a shared development agenda. (2015)

S.A. Woznicki, A.P. Nejadhashemi, M. Parsinejad, Climate change and irrigation demand: uncertainty and adaptation. J. Hydrol. Reg. Stud. **3**, 247–264 (2015)

Global Change and Its Consequences for the World's Arid Lands

Richard Bawden

3.1 Introduction

It is difficult to imagine a more relevant or useful focus for exploring the consequences of global change in all of their socioecological complexity, than that provided by the world's arid lands. As a living example of nature/culture dynamics in the face of global changes in both their biophysical and sociocultural environments, they present far more appropriate images than the exotic hungry polar bears or bleached tropical coral reefs or remote retreating glaciers that dominate the media. By virtue of the very conditions of the criteria of aridity that define them, these water-stressed biomes are inherently vulnerable to environmental challenges on any scale and from whatever source. This vulnerability is being subjected to a veritable barrage of stresses that are truly global in their range and influences. It is now clearly recognised that the impacts of climate change will be channelled primarily through the global water cycle (World Bank 2016) with consequences that will include increasingly erratic and uncertain water supplies.

Distributed typically in great swathes across the different continents, the arid lands, their biomic diversity notwithstanding, present both a poignant as well as a pragmatic focus for exploring the local consequences of changing global relationships between human beings and their water-scarce environment. At least five forces of profound global influence can be usefully recognised in this context, with each having an influence on nature/culture relationships, both alone and in combination with some or all of the others. While the consequences of all of these forces of influence can have some positive features, their potential for negative impacts presents challenges that at times appear literally overwhelming. Those who live in the arid lands are facing conditions in the environments about them that are as unpredictable and uncertain as they are complex and systemic. They face

R. Bawden (✉)
Western Sydney University, Sydney, Australia
e-mail: R.Bawden@westernsydney.edu.au

situations where natural, social, political, economic, cultural and technological factors are all interacting together.

3.2 Five Forces of Global Change

Global warming, as the main driver of climate change, provides the first illustration in support of this contention. The consequences of this phenomenon, in terms of its impact on local weather patterns in particular, are already impacting on those who live in the arid lands, the vast majority of whom rely on agriculture, pastoralism or forests. All too familiar with weather events that are erratic, often extreme, and invariably unpredictable, the marginalised people in these dry areas are already having to face significant shifts in the moisture and temperature regime of their environment (El-Beltagy and Madkour 2012). They are equally aware of the limitations and the consequences of their past attempts to adapt to such circumstances especially those associated with the adoption of technologies and management practices on advice from (well intentioned, if not fully informed) 'outside technical experts'.

Yet even while they are attempting to rectify the consequences of these past efforts to increase the productivity of their lands, or at least to adapt to them in the face of a changing climate, they are coming under pressure to respond to further global challenges. Commensurate with the continuing growth of the world's population, as well as their own changing demographics, plus the changing aspirations of an increasingly large segment of world's population, is an ever-increasing demand for critical resources that include water, food and energy. There are growing pressures, essentially by those who still believe that the arid lands represent underexploited resources for agriculture, for attention to be redirected at increasing the productivity of the lands, as one of the crucial strategies for responding to the challenges of the growing global food crisis – the 'coming famine' as it has been labelled (Cribb 2010).

A third global change that has been impacting upon all of the world's population and that has particularly significant consequences for the arid lands is the globalisation of trade in goods, including both inputs and commodities for agriculture, for instance, and for associated services particularly of finance, banking and the free movement of investment capital (Soros 2002). There is ample and long-standing evidence to implicate distortions to these flows that can be identified as consequences of particular behaviours of certain multinational including agribusiness corporations (Sargent 1985).

The fourth source of global change which has had consequences for those in the arid zones might best be expressed as the ever-increasing reach and intensity of what might be referred to, borrowing from Kuhn (1962) as the 'normal' or 'dominant paradigm' which is globally pervasive. In the present context, paradigms represent the practical expression of sets of particular beliefs and assumptions that constitute particular approaches to problematic situations in the arid lands by research scientists, policy makers, regulators, educators, NGOs and so-called development specialists. In essence they reflect particular views of the world (worldviews) on such matters as the nature of nature, the nature of knowledge (and also of knowing) and the nature of human nature specifically as it is related to human values such as aesthetics and ethics. Adopting philosophical epistemic labels, while straying somewhat for the rigours of that discipline, they can be identified as ontological, epistemological and axiological beliefs, respectively (Lincoln and Guba 1985). As is increasingly apparent to all save the most insensitive observers, a specific set of beliefs and assumptions have come to dominate approaches to development, not just specifically to agriculture and rural environments but to the whole matter of 'human progress'. These worldviews are so global in their adoption and so influential with respect to the paradigm that they have effectively nullified any attempts to replace them even under circumstances, as now, where their relevance is proving to be singularly inadequate and paradigmatically severely limiting (Norgaard 1994).

The central assertion of what follows is that the consequences of the current global changes are dictating the need for fundamental shifts in the globally prevailing worldviews and thus also of the paradigm that they inform. Of necessity, such epistemic transformations will need to involve all relevant stakeholders, 'outsiders' and 'insiders' alike, and of course any potential transformers will meet extraordinary levels of resistance from those with a variety of vested interests who will strongly resist such 'interventions'.

It is important at this point to mention communication and the world wide flow of information, as the fifth aspect of sources of global change with profound consequences. It is impossible to overestimate the growth of the significance of electronic communication over the five decades or so since the invention of the Internet and the launch of the first communication satellites. Suffice it to say that this has revolutionised communication on a genuinely global scale, and the influence of the compound media with which it is now associated is implicit in everything else that follows: with capabilities that range from amplification to attenuation of negative impacts on the one hand and a capacity to facilitate positive effects on the other.

It might be said that it is through the ubiquity of communication technologies that the complexities that are involved with all of the global challenges and changes from the warming of the planet through to geopolitical instabilities are coming to be known and appreciated for their complexity.

And again the arid lands provide a condensed example of this situation. The limiting conditions of the arid land environments have long fostered endemic poverty which has been, in turn, associated with persistent and pervasive malnourishment and disease (Morton and Anderson 2008). These conditions are also characterised by chronic underemployment and psychosocial depression that not infrequently lead to outbursts of violence. These situations have inevitably led to significantly amplified conflictual social unrest as characterised by a host of different pathologies that include violent demonstrations of defiance, armed insurgencies, acts of terrorism, and civil war. They are almost always also associated with institutional dysfunctions that are expressed in everything from political instabilities, lawlessness, crime, human rights abuses and illicit and corrupted trade practices, through to gross inequalities and inequities with respect to the distribution of both wealth and power. While certainly not confined by any means to the arid zones of the world, such fractured social relationships as these are significantly and disproportionately represented in nations that have significant areas of drylands within them particularly those that lie within less developed economies.

The exodus of refugees, both displaced persons and those seeking 'better' livelihoods, along with the export of terrorism and an associated climate of 'fear', are further examples of how locally sourced forces for change can contribute to global change of such a scale that they trigger gross geopolitical instabilities.

3.3   Complexity

So in the complexity of both their biophysical and sociocultural heterogeneity as well as the challenges that must now be addressed, the arid lands represent a veritable microcosm of what might be termed the contemporary global problematique (Blanchard 2010). These are conditions that are characterised by greatly 'mixed up' 'messy' clusters of problems that are often occurring on a truly global scale. Those who live in the arid lands are facing conditions in the environments about them that are as unpredictable and uncertain as they are complex and systemic. They face situations where natural, social, political, economic, cultural and technological factors are all interacting together: where, to paraphrase Latour (1993), 'culture and nature are churned up together every day'.

What happens in the arid lands also reflects a veritable kaleidoscope of differing human intentions and dispositions, beliefs and assumptions, moods and emotions and, most importantly, the distribution of power and hence the nature of governance. The diversity that characterises land types and patterns of land use across such a vast area of the globe is matched by an extraordinary variety of cultures, economies, and institutions of governance that are found within the boundaries of those countries that have significant arid areas within them. In this manner, a contiguous arid zone might contain a number of different nations with political systems that range across a spectrum that includes democracy, autocracy and theocracy and in some instances associated with the complete failure of a nation state, anarchy. This poses considerable challenges with respect to the development of policies and, even more relevant, forms of governance that are appropriate to sustainable improvements in the livelihoods and well-being of arid land populations across the world.

This increasing appreciation of the complexity and dynamics that the arid lands present is leading to what can be regarded as the major positive consequence of global change. With a fresh perspective, lessons can be learned from revisiting past errors, current circumstances and potential scenarios of the future.

As good fortune would have it in this regard, a very comprehensive and substantial commentary exists from a wide variety of different perspectives and sources that have extensively explored and reported on the challenges and changes to the arid lands over a period that extends back to the 1950s. This richly diverse yet focused chronicle is particularly notable as a record of international endeavours in this regard and especially the involvement of a number of different United Nations agencies. The edited report of a number of scientific conference papers, entitled *The Future of Arid Lands* (White 1957) which highlighted the involvement of UNESCO, can conveniently be regarded as one of the seminal texts of this historical commentary. Of equal fortune is the fact that half a century after its publication, *The Future of Arid Lands*, was revisited by two scholars whose objective was to use that seminal text as an opportunity to 'look back in time and see how we have progressed' (Hutchinson and Hermann 2008). As these authors submit, this retrospective allowed them to note which issues were considered critical at that time as well as to identify the context in which those issues were framed. It also provided opportunities, as they explained, to explore how science and technology had changed over the intervening years as well as to identify the origin of many of the problematic issues which remained patent.

The same might now be claimed from an exploration of the many research publications, books, conference proceedings, policy notes, etc. that constitute what can be usefully regarded as an 'arid land commentary'. Such a meta-study reveals significant changes in the character of the critical and dynamic flux between actions on the ground and the nature of the ideas, concepts, and theories that have informed them over a period of extending more than six decades.

The Great Acceleration and the Holocene

As it happens, this particular period in human history has been coincident with a number of quite extraordinary changes in (as well as to) the world – again both biophysical and sociocultural – that have been on such an acute trajectory that has been referred to as the 'great acceleration' (Steffen et al. 2004). As has recently been reported (Steffen et al. 2015) the world population over this period has increased threefold to reach in excess of seven billion from a 1950s baseline of just over two billion (which, staggeringly, is currently the approximate population of the arid lands across the globe!). Over this time the atmospheric concentration of carbon dioxide has increased from around 310 ppm to now just in excess of 400 ppm, while the related phenomenon of global warming has seen a temperature rise of 0.6 °C or so. Meanwhile the real global gross domestic product has increased from $US3 trillion to more than $US60 trillion, the global consumption of water has increased from just over 1,000 to 4,000 km^3, and as an indication of the industrialisation of agriculture, global fertiliser consumption has grown from less than 1 million tonnes to over 160 million!

The impacts of all of this have been so severe that it is now being suggested that they have actually triggered the emergence of a new geological epoch – the Anthropocene (Crutzen and Stoermer 2000). While there is still considerable debate about whether or not the changes that can be associated with anthropogenic activities on the globe have actually resulted in stratigraphic change (Williams et al. 2015), the image of a permanently changed globe that the Anthropocene represents provides compelling evidence in support of the call for the adoption of new worldviews and of the paradigm that they inform.

3.4 Shifting Paradigms

Echoes of this call can be found in the arid land commentary. For instance, in regard to the impact of climate change on agriculture specifically in the arid lands, El-Beltagy and Madkour (2012) argue that there is a need for a new paradigm for agricultural research and technology transfer. A key aspect of their submission is that such a new paradigms will require much more investment by international agencies and national governments alike, for supporting research and sustainable development efforts which include the full participation of who they refer to as the 'target communities'.

Responses to calls for shifts in paradigms with specific reference to the future of the arid lands have also begun to appear in the relevant literature. Mortimer et al (2009), for instance, in exploring new opportunities for the drylands, provides considerable detail of 'a new paradigms for people, ecosystems and development'. This focus on ecosystems is also the basis for the details provided by White, Tunstall and Henninger (2002) of what they refer to as an 'ecosystem approach' for building support for new development policies for the drylands.

Two papers published in 2008 by the same two scholars provide graphic evidence of the significance of paradigmatic differences to the way drylands can be perceived with respect to different development scenarios. They differentiate between an approach to development that continues to reinforce the vicious cycle of interdependencies between overuse of land resources and poverty on the one hand and a paradigmatic approach, based on encouragement of sociotechnical ingenuity and informed adaptation, that leads to sustainability and security. They refer to these, respectively, as the 'desertification paradigm' and the 'counter paradigm' on the other where this latter focuses particularly on a virtuous cycle that embraces the need for a multifunctional perspective on arid land development (Safriel and Adeel 2008). Citing Reynolds et al. (2007), they indicated that the first paradigmatic approach should be amended to 'dryland livelihood paradigm' to highlight the cycle in terms of the connections between poverty, conflict and national security. In a companion paper, these two workers emphasise the advantages of shifting from a focus on the lands themselves and of the potentiality (or otherwise) of them as productive resources, onto the sustainability of alternative livelihoods and well-being of the ever-growing populations in the arid zones (Adeel and Safriel 2008). They also highlight the need for inclusion of all key stakeholders in the process of development as well as the integration of the biophysical with the sociocultural.

Most significantly, the reports of these recent workers do illustrate a fundamental change in focus from the long persistent objective of improving the agricultural productivity of the arid lands through the adoption of 'technical expert advice' generated exclusively by 'externals'. In strong contrast, the emergent paradigm involves a much more participatory process that is inclusive of the inhabitants along with a range of other stakeholders (EMG 2011; Mortimer et al. 2009). The key focus now is on the cogeneration of strategies to achieve sustainable improvements in the inclusive well-being of people and 'the rest of nature' alike. Anderson, Morton and Toulmin (2009) provide further support for such a shift arguing that 'a people-centred approach' that embraces the need for improved livelihoods and highlights the need for empowerment should be taken. This is in contrast to an approach that continues to focus on the situation solely from the perspective of environmental crises. As they saw it the much more positive approach that they were recommending needed to be long term and 'holistic' with a particular emphasis on linkages between dryland livelihoods and the wider economy.

3.5 Worldview Foundations

While the concept of paradigm has been embraced by the arid land researching and development community, it is of considerable significance to note that the worldviews that provide their foundations have failed to attract rigorous critical attention.

Perhaps this lack of familiarity with the worldview concept is, as Williams (2016) has recently submitted, in large part due to the fact that many different words and phrases are used to convey the same idea. These include 'world outlook', 'world hypothesis', 'personal construct', 'episteme', 'mental model', 'meaning perspective', 'mental framework', 'root metaphor', 'philosophy of life', and even 'culture' itself. And as a further element of confusion, the words paradigm and worldview are frequently used interchangeably. But in addition to this plethora of synonyms, there is the issue of an all too frequent unfamiliarity with the basic philosophical notions that are central to the worldview concept that refer to those sets of beliefs and assumptions that 'shape' the way that we understand the world about us, how we experience it and how we interpret those experiences in ways that we can then used to inform our actions. In his comprehensive review of the psychology of worldviews, Koltko-Rivera (2004) defines them as sets of beliefs and assumptions that describe reality. They represent ways of describing the universe and life within it. As he sees it, our worldviews dictate what it is that we hold as truth and beauty, how we differentiate between what is right what is wrong and what it is that we ought to do as well as what it is that we can do. Worldviews define what can be known and how it can be known. They include affective, cognitive and cosmic dimensions that together contribute to the presuppositions that individuals and social collectives alike make about the nature of those things that they use to both makes sense of and generate order for the ways they live their everyday lives (Hiebert 2008). As Williams (2016) concludes, while our worldviews are constructions of our individual psyches, they are profoundly influenced by the cultures of which we are part with each indeed contributing to the development of the other.

The signal importance of worldviews is all too easily underestimated. Yet as Klein (2014) argues, the task in the face of the ecological crisis in which the world now finds itself is not to just develop different policy proposals but to articulate an alternative worldview which replaces 'hyper individualism' with interdependence, dominance with reciprocity and hierarchy with cooperation.

The first item in each of these three characteristics reflect the foundations of a superordinate worldview, referred to as progressivism (Foster 2015), that is globally pervasive and represents precisely the discussion earlier of both dominance and yet profound inadequacy. It refers to the obsessive pursuit of material 'progress' achieved essentially through the very same processes of industrialisation that have quite literally fuelled changes to the entire global socioecological environment. It is the commitment to material progress that has allowed multinational corporations to develop such a pervasive global presence and powerfully distorting influence (Klein 2014). It has also been responsible for the assumption of financial systems of so much globalised significance that is capable of triggering the collapse of entire economies across the globe while also facilitating grossly inequitable distribution of wealth (Steiglitz 2012). It is progressivism that has led to the degradation of so many public interest institutions (Hutton and Giddens 2000) and to the diminution of so many national, and even international, mechanisms of governance. Perhaps most significantly of all, the dominant development paradigm is so powerful that its advocates aggressively prevent any attempts aimed even at a critical review of itself let alone a rigorous intellectual and moral challenge (Hamilton 2010).

There are two epistemic aspects of this superordinate worldview that are of particular relevance to this focus on the arid lands. The first of these, 'scientism' (Foster 2015) is the unshakeable belief in the power of science and technology to identify and solve all of the problems of the world as they emerge. The second, 'hyper capitalism' (Rifkin 2001) is the unassailable assumption that everything has a price and can therefore be traded. The all too pervasive global ideology of neoliberalism with its focus on unfettered free trade, on privatisation and on globalisation (Monibot 2016) can be regarded as the explicit face of such hyper capitalism.

The inadequacies of scientism and neoliberal capitalism as foundations for paradigms for the development of arid lands does not, by any means, imply that science and economics are irrelevant. Nothing could be further from the truth and yet from the perspectives of the conventions of each that have prevailed almost throughout the entire six decade period under review here, both deserve critical examination with respect to their ontological, epistemological and axiological foundations.

3.6 Ecosystems

The almost universal embrace of the ecosystem concept by researchers, development professionals, policymakers and educators – and even these days by bureaucrats, industrialists, bankers and (especially) popular commentators like journalists – provides a useful focus for exploring this contention. This is in spite of the fact that, as Golley (1993) argued, it remains a significantly contestable issue particularly within the discipline of ecology. In the present context, the concept does provide an opportunity to reinforce the arguments being presented through comparing and contrasting two distinctly different sets of worldview beliefs that are pertinent to the present context. The narrative here relating to differences between what might be called a technocentric worldview and a holo-centric one (Bawden 1998) is presented in a form that deliberately exaggerates distinctions in order to emphasise differences.

Some 20 years after the appearance of *The Future of Arid Lands* (White 1957), an edited monograph was published (Goodall and Perry 1979) which explicitly focused on the structure, functioning and management of what were explicitly referred to as arid land ecosystems. As a publication set within a general series on international biological programmes, the focus, unsurprisingly, was on 'biological systems'. The fundamental ontological belief here – the assumed nature of nature as it were – is that the biosphere is actually composed of myriads of bounded biotic communities of different species of living organisms that in some manner or another are interrelated with each other in some form of relatively integrated whole entity. Each system includes the abiotic components with which the biotic component also interacts. Quantitative studies of the components of the system along with sets of different relationships between them allow mathematical computations to be generated that can then be used as the basis for simulation models. By varying conditions and relationships within these models allow them to be used as the basis for 'virtual' experiments. Even though ecosystems are regarded somehow or another as whole entities, their analysis is conducted through a reductionist approach that focuses on the study of the component parts. The whole system is actually assumed to have no properties that any thorough study of the component parts would not reveal. The whole is the same as a sum of its parts, and all cause/effect relationships are considered to be linear!

This realist ontology can be contrasted with the position that such whole entities don't 'actually' exist in reality but are very useful conceptual frameworks that, through their constructions, provide useful ways of thinking about biotic/abiotic relationships in a manner akin to the economy of economists or the societies of sociologists. It represents a difference between an interpreted world and a perceived world (Milton 1996). The assumption that whole entities (even construed ones) can be understood through a study of their component parts is rejected through the assumption that all systems, by their very nature, have so-called emergent properties which are unique to their wholeness. As such properties cannot be revealed or even predicted through a reductionist study irrespective of its rigour all systems, in the first instance at least must be approached in a way that respects their systemic wholeness, that is, holistically.

Examination of the Goodall and Perry work also reveals the conventional objectivist epistemology of conventional biological science. On this view, only the knowledge that can be generated through empirical study and validated through experimentation can be regarded as 'reliable truth'. All other forms of knowledge generated through any other method of enquiry or way of knowing cannot be regarded as verifiable truth and are therefore regarded as less valuable (if not entirely without value). Contextualism, or contextual relativism, lies in direct contrast to such objectivism. Whether or not any particular knowledge is valuable is a function of the context in which it is set and the particular appropriateness of its methods of generation.

And finally there is the matter of the nature of human nature with particular regard to ethics and aesthetics as expressions of human values. A fundamental assumption of objectivist reductionist scientific research is that it is essentially a 'value-free' process. That is to say, the particular values that are held by any individual scientist have no bearing whatsoever on the outcome of any biological research. The extent of any moral considerations is limited solely to the ethical defensibility of any research procedures that are in contravention of accepted norms of research behaviour. The contrary position of course is provided by the situation where ethical considerations and other assumptions related to human behaviours that include empathy, care, generosity and so on are absolutely central to considerations of matters at hand.

In the context of the arid landss, it is very interesting to note that Kassas (1977) in a paper discussing problems and prospects of the arid lands had earlier introduced the significance of sociopolitical systems, sociocultural systems, systems of education, ecoregions and so on. He made no mention however of how a study of these could be conceptually integrated together or operationalised. Much more recently White, Tunstall and Henninger (2002) have attempted to address these issues with respect to their explicitly nominated ecosystem approach to drylands where their particular objective is to build support for new development policies. This approach 'which includes people' would, they submitted, provide decision-makers 'with a policy tool for creating and implementing more effective drylands policy'.

3.7 Policy

This issue of policy within the context of the future of arid lands, and in particular with respect to its relationships with science, is beginning to attract increasing attention. While writing specifically about the situation as they relate to rangeland management, the observations of Holechek (2013) related to the need for major changes in policy in response to emerging global challenges are entirely relevant to those in the arid lands. Under many situations these are the same zones. As he concludes the need for policy responses that range from taxation, investment and debt relief through to land tenure, water management and renewable energy in the face of population pressures, growing food demand and climate change will all be needed 'to avoid declining living conditions at the global level' (Holechek 2013). A new or renewed focus on alternative livelihood strategies such as ecosystem services and tourism provides further impetus for the need of the development of relevant policies for rangeland and arid land ecosystems alike.

As is increasingly recognised, it is more relevant from a holo-centric perspective to explore the nature of governance within particular systems as the basis for developing mutually acceptable strategies for rules and regulations, rather than having to deal exclusively with policies imposed 'from above'. Examples are emerging in the commentary (e.g. Barrow 2014).

It is important at this point to emphasise that the centrality of the ecosystem concept at large has provided one of the key sources of impetus for the development of a host of other related concepts that themselves have become central to development. These include sustainability, sustainable development, resilience, self-organisation and adaptation. It is also central to theories related to ecology including panarchy, hierarchy, chaos and complexity; second- and third-order systems theories have all gained considerable currency. Most of these are regarded as applicable to both the biophysical and sociocultural systems and indeed to the fusion of both. The past six decades have seen a growth in interest by researchers from a range of different social sciences in matters associated with the arid lands and climate change and increasing evidence of collaborative work that is multi and interdisciplinary. Just a decade ago saw the birth of attempts to integrate the social sciences with the natural sciences within the context of sustainability with the establishment of what are being referred to as the sustainability sciences (Clark and Dickson 2003). The emergence of environmental ethics (Belshaw 2001) and, more recently sustainability ethics (Selinger et al. 2010), has greatly enriched understanding of the ethical dimensions of global challenges, as well as indicating the need for research and development approaches that can be classified as truly transdisciplinary.

All of this scholarship notwithstanding, the arid land commentary continues to be characterised by a wide range of conceptual inadequacies particularly with respect to the culture/nature duality along with the 'actual' structure, functions and 'behaviour' of ecosystems – both biophysical and sociocultural. The increasingly multi- and transdisciplinary nature of investigations into the arid lands is also resulting in language which is conceptually imprecise, like 'the balance of nature', 'self- organising systems', 'environmental damage', 'ecological degradation', 'human interventions', 'landscape deterioration' and so on. Even matters like sustainability and resilience, in spite of their near universal adoption, remain highly contestable with respect to their precise meanings and indeed epistemic foundations (Davison 2001).

And this is even further evidence in support of the submission that far too little critical attention is paid to the intellectual and moral nature of worldviews and of their significance as the foundations for paradigmatic approaches to research, policy making and so on that are relevant to investigating the consequences for the arid lands of global changes.

3.8 Conclusion

The 'great acceleration' that has led to the emergence of the Anthropocene has resulted in extraordinary changes in, across and to the globe – and the forces responsible are, of course, still continuing. The arid lands provide a rich focus for exploring the nature and consequences of many of these changes. Global warming, increased demand for quantity and quality of food as well as for other resources and global patterns of trade that are frequently distorted are all having profound consequences that challenge the development of sustainable livelihoods for those who live in them. They also challenge the well-being of the arid land inhabitants through their association with poor nutrition, disease and countless social pathologies.

The crux of the argument presented here is that the situation is now so dire that nothing less than a major change in the prevailing worldviews and thus in the dominant paradigm that they inform is essential. This shift will require a new consciousness with regard to the very concept of worldview as well as a readiness to explore alternatives. While there is some evidence of acceptance of the differences between a technocentric and a holo-centric perspective on research, development, policy making, education, etc., it is still very much in primitive mode. Certainly research and development initiatives are adopting much more participatory processes. And certainly the focus of much of this work is changing from externally motivated strategies for increasing productivity of agricultural and pastoral enterprises in the arid land to embrace the far more systemic and inclusive notion of sustainable livelihoods and well-being achieved through what can be regarded as the adoption of a perspective of multifunctionality.

As the great acceleration continues, it is worth reflecting on potential role for UNESCO to now concentrate considerable effort on what might be termed worldview or epistemic education at every single opportunity.

The arid lands represent the canary in the biosphere of the Anthropocene. Their song is getting fainter by the day.

References

Z. Adeel, U. Safriel, Achieving sustainability by introducing. Altern Livelihoods Sustain. Sci. **3**(1), 5–133 (2008)

S. Anderson, J. Morton, C. Toulmin, in *Climate Change for Agrarian Societies in Drylands: Implications and Future Pathways*, ed. by R. Mearns, A. Norton. Social Dimensions of Climate Change: Equity and Vulnerability in a Warming World, (The World Bank, Washington D.C., 2009)

E. Barrow, *Governance: Linchpin of dryland natural resource management. Policy Briefing* **90**, 1–4 (2014)

R.J. Bawden, The community challenge the learning response. New Horiz. **99**, 40–59 (1998.) Reprinted in 2010 as Chapter 3 in C Blackmore (ed.). Social learning systems and communities of practice the Open University: published in association with Springer Verlag London

C. Belshaw, *Environmental Philosophy: Reason, Nature and Human Concern* (McGill-Queens University, Montreal, 2001)

E.V. Blanchard, Modeling the future; an overview of the limits to growth debate. Centaurus **52**, 91–116 (2010)

W. Clark, L. Dickson, Sustainability science: the emerging research program. PNAS **100**, 8059–8061 (2003)

J. Cribb, *The Coming Famine: The Global Food Crisis and What We Can do to Avoid It* (University of California Press, Berkeley, 2010)

P.J. Crutzen, E.F. Stoermer, The anthropocene. Glob. Chang. News **41**, 17–18 (2000)

A. Davison, *Technology and the Contested Meaning of Sustainability* (State University of New York Press, New York, 2001)

A. El-Beltagy, M. Madkour, Impact of climate change on arid lands agriculture. Agric. Food Secur. **1**, 3–12 (2012)

EMG *Global Drylands: A UN System-Wide Response* (United Nations Environmental Management Group, Geneva, Switzerland, 2011).

J. Foster, *After Sustainability; Denial Hope Retrieval* (Earthscan, Routledge, 2015)

F.B. Golley, *A History of the Ecosystem Concept in Ecology: More than the Sum of Its Parts* (Yale University Press, New Haven, 1993)

D. W. Goodall, R. A. Parry (eds.), *Arid Land Ecosystems: Structure Function and Management* (Cambridge University press, Cambridge, 1979)

Hamilton, *Requiem for a Species: Why We Resist the Truth About Climate Change* (Earthscan, London, 2010)

P.G. Hiebert, *Transforming Worldviews: An Anthropological Understanding of How People Change* (Baker Academic, Grand Rapids, 2008)

J.L. Holechek, Global trends in population, energy use and climate: implications for policy development, rangeland management and rangeland users. Rangel. J. **30**(5117), 1–9 (2013)

C.F. Hutchinson, S.M. Hermann, *The Future of Arid Lands Revisited* (Springer, London, 2008)

W. Hutton, A. Giddens (eds.), *Global Capitalism* (The New Press, New York, 2000)

M. Kassas, Arid and semi-arid lands: problems and prospects. Agro-Ecosystems **3**, 185–204 (1977)

N. Klein, *This Changes Everything: Capitalism vs the Climate* (Simon and Schuster, New York, 2014)

M.E. Koltko-Rivera, The psychology of worldviews. Rev. Gen. Psychol. **8**, 3–58 (2004.) New York University, NY

T. Kuhn, *Nature of Scientific Revolutions* (University of Chicago Press, Chicago, 1962)

B. Latour, *We Have Never Been Modern* (Harvard University Press, Cambridge, MA, 1993)

Y. Lincoln, E.G. Guba, *Naturalistic Inquiry* (Sage Publications, Thousand Oaks, 1985)

K. Milton, *Environmentalism and Cultural Theory; Exploring the Role of Anthropology in Environmental Discourse* (Routledge, London, 1996)

G. Monibot, *How Did We Get Into This Mess?* (Verso, London, 2016)

M. Mortimer et al., *Dryland Opportunities: A New Paradigms for People Ecosystems and Development* (IUCN Gland, Switzerland, 2009)

J. Morton, S. Anderson, *Climate Change in Agrarian Societies in Drylands Workshop on Social Dimensions of Climate Change* (World Bank, Washington, DC, 2008)

R.B. Norgaard, *Development Betrayed: The End of Progress and a Co-evolutionary Revisioning of the Future* (Routledge, London, 1994)

Reynolds et al., Global desertification: building a science for dryland development. Science **316**, 847–851 (2007)

J. Rifkin, *The Age of Excess: The New Age of Capitalism where All of Life is a Paid for Experience* (Jeremy Tarcher, Putnam, 2001)

U. Safriel, Z. Adeel, Development paths of drylands: thresholds and sustainability. Sustain. Sci. **3**, 117–123 (2008)

S. Sargent, *The Food Makers* (Penguin books, Ringwood, 1985)

E. Selinger, R. Raffelle, W. Robinson (eds.), *Sustainability Ethics* (Automatic/VIP Press, New York, 2010)

G. Soros, *George Soros on Globalisation* (Public Affairs, New York, 2002)

W. Steffen, P.J. Crutzen, J.R. McNeill, The Anthropocene: are humans now overwhelming the great forces of nature? Ambio **36**, 614–621 (2004)

W. Steffen, W. Broadgate, Deutsch, L. Gaffney, C. Ludwig, The trajectory of the Anthropocene: the great acceleration. Anthropol. Rev. **1**, 18–36 (2015)

J.E. Steiglitz, *The Price of Inequality: How Today's Divided Society Endangers Our Future* (W.W. Norton and Company, London, 2012)

G. F. White (ed.), *The Future of Arid Lands* (American Association for the Advancement of Science, Washington, DC, 1957)

R.P. White, D. Tunstall, N. Henniger, *An Ecosystem Approach to Drylands: Building support for new development policies* (World Resources Institute. Information Policy Brief No. 1, Washington D.C., 2002)

M. Williams *Learning Scenarios Transforming Worldviews*. Unpublished PhD Thesis, University of Melbourne 2016.

M. Williams, J. A. Zalasiewicz, P. K. Haff, C. Schwägerl, A. D. Barnosky, and E. C. Ellis et al., The Anthropocene biosphere. Anthropol. Rev. **2**, 196–219 (2015)

World Bank, *High and Dry: Climate Change, Water and the Economy* (World Bank Group, Washington, DC, 2016)

Part II
Aridlands Under a Global Change Regime

The three chapters here deal with the important nexus between people and the often harsh environment that aridlands/drylands present.

Victor Squires and *Haiying Feng* analyze the impact of humans on the environment and their role as "agents of change," "for better or worse." They begin with a discussion about the nature of change and provide an introduction to some of the theoretical underpinnings to the current thinking about how change occurs. Adaptive practices used by local land users and by officials on the cold and arid Qinghai-Tibet Plateau to cope with environmental change are assessed. They use historical records to trace collective responses to environmental extremes. They conclude that traditional ecological knowledge and shared systems of beliefs can facilitate collective responses to crises and contribute to the maintenance of long-term resilience of social-ecological systems.

Haiying Feng and *Victor Squires* team up again to explore the attitudes and responses of herders, agropastoralists, and government officials on the cold and arid Qinghai-Tibet Plateau to global change, including climate change. Ecological resettlement and relocation of herders and agropastoralists is the government's preferred option in a bid to bring about ecological restoration to degraded lands in the headwaters of three major rivers. Livelihoods of all land users are under threat as climates change, glaciers melt, and permafrost thaws.

Iran's aridland watersheds are the focus of the chapter by **Hossein Badripour** and **Hamid Reza Soleymani.** Iran has a large arid zone and scarce water resources. A National Action plan is being implemented to improve the health of rangeland ecosystems in the catchment/watersheds of the major rivers. Special attention is given to mitigation and adaptation, in the face of the projected further worsening of the climate under the influence of global warming.

Humans as Agents of Change in Arid Lands: With Special Reference to Qinghai-Tibet Plateau, (China)

4

Victor R. Squires and Haiying Feng

4.1 Introduction

There are natural and human influences on landscape evolution in drylands. These influences sometimes accelerate or slow down the process of landscape evolution. Human activities such as deforestation, destruction of natural vegetation to make irrigated oases or to permit infrastructure developments to be built (road, rail etc.), and mining have impacted the Earth's changing surface. These activities are not new processes but often extend or accentuate the natural processes. When natural processes are accelerated, humans are often affected by the changes in their environment. Melting of glaciers as a result of global warming is one such example that has relevance to the Qinghai–Tibet Plateau.

Table 4.1 is a summary of some ways in which natural and human-induced changes occur.

Since the first prehistoric people started to dig for stone to make implements, rather than pick up loose material, humans have modified the landscape through excavation of rock and soil, generation of waste, and creation of artificial ground. Change is a natural, essential feature of arid lands. Being able to identify different change processes, understand the differences between natural and human influences on change and recognize a new level of concern are all critical for effective arid land management. Change, however, can be very complicated to understand and manage when many human and natural causes of change interact as they often do. The rate of change is escalating. The pressure is intense on anybody connected with the arid lands to focus time and attention on understanding the forces driving the changing environment and develop or implement the information systems needed to support the altered environment.

V.R. Squires (✉)
Institute of Desertification Studies, Beijing, China
e-mail: dryland1812@internode.on.net

H. Feng
Institute of Administration and Management, Xining, Qinghai, China

© Springer International Publishing AG 2018
M.K. Gaur, V.R. Squires (eds.), *Climate Variability Impacts on Land Use and Livelihoods in Drylands*, DOI 10.1007/978-3-319-56681-8_4

Table 4.1 A summary of the impacts of change agents on ecological and socioeconomic process in the Three Rivers region of Qinghai

Type of change agent	Ecological effects	Socioeconomic process most vulnerable to change	Remarks	References
Natural agents				
1. Earthquake	Items 1, 2, 3 Soil deterioration	Items 1, 2, 3, 4 Asset (potential) loss, Herder's income decline, poverty, indebtedness, depression, despair		Cao Yanyan (2013)
2. Landslide	Land-use value reduced	Changes to governance	Vicious cycle: more eco-disasters as a follow on	Dong (2005)
3. Flooding	Reduced land management capability		More ecological refugees	Shang et al. (2016)
4. Drought	Items 4, 5 No forage, animals die; lack of yak manure for fuel to cook food. Lack of milk for making cheese, yoghurt, etc, poor nutrition of householders		More outward migration of younger people creating problems of inheritance and intergenerational transfer	Xu et al. (2008)
5. Dzud or kengshi	5. High level of mortality in livestock (yak, sheep)	Items 4, 5 Poverty, livelihood hardship, indebtedness, mental stress, outward migration		Zhang (2015)
6. Melting glaciers	6. Floods, snow line changes, biodiversity reduction, quality reduction of grassland resource, animal habitats lost			Zhao (2015)
				Zheng (2011)
Human-made agents				
1. Overgrazing	1. Grassland degradation 2. Loss of woody plants and lower level of soil protection	1. Impacts on cash flow ultimately leads to lower income, hardship in households. Soil loss follows, C-sequestration negated	Item 1. Poverty alleviation measures taken, aid provided (cash, grain, fodder)	Cao et al. (2013)

4 Humans as Agents of Change in Arid Lands...

		2. Saving cash expense	Relocation, resettlement may follow	Tsering (2014)
2. Fuel collection				Wang (2006a, b)
3. Conversion of grassland	Item 3. Conversion of better rangeland to cropland, increases grazing pressure on remaining grazing areas, if cropland is later abandoned soil loss increases and productivity declines	3. Helpful with livelihood and sustainable grassland utilization	Govt. response may be training of herders on how to conserve/restore and improve ecological conditions	
4. Medicinal plant harvest			5. The current way of harvesting *Cordyceps sinensis* is destructive. More control on digging, especially by nonlocal people. Advanced technology can improve the search and avoid unnecessary large-scale digging	
5. *Cordiceps sinensis* collection	Items 4, 5. Uncontrolled harvest leads to soil disturbance, loss of soil C, destruction of forage	4.5. Income and cash flow (Sales can augment subsistence)		
6. Infrastructure	Item 6. Rangeland area reduction to make way for roads, rail, pipeline. Structures may impede migration of wild species and interference with transhumance and restrict access to water or to grazing	6. Transformative adaptation to new threats and opportunities		
7. Global warming (CO_2 emission)	Item 7 Degraded rangelands are net emitters of CO_2	7. Developing new emerging industry and eco-friendly animal husbandry and more sustainable ecological rangeland and forest industries		

Dzud is severe cold spell that is accompanied by extreme low temperatures and often blizzard conditions. Many livestock die

4.2 The Nature of Change

Social change is an elusive concept. It is inevitable and yet, paradoxically, it depends on the will and the actions of ordinary individuals. We embrace change, yet something in our nature fiercely resists it. People structure social movements, political campaigns and investment strategies around the need for change, yet most of them hardly understand how it works. There is increased attention to social change. There are several key questions. How do cultures evolve? Is it possible for organizations/societies to maintain stability under conditions of constant change? How do groups and individuals consciously bring about adaptive change? These questions all revolve around the crucial function of change in the maintenance of dynamic ecological systems. There seems to be little consensus about what constitutes adaptive change (Squires and Feng 2017). The causes, the processes, and the effects of change are often spoken of as if they are one and the same thing. Change occurs on many levels—cultural, social, institutional, and individual—and it is often hard to draw clear distinctions between them. Sometimes change occurs spontaneously on several levels at once. But how to recognize when a change has taken place?

The change process begins with a shift away from cultural harmony, a change that shows up first in the form of increased individual stress. A growing number of individuals find that they are unable to meet certain cultural expectations. At first this is perceived by both the individual and the society at large as an individual problem. But as the number of these individual deviations grows, it begins to weaken the social fabric, eventually to the point where the society must acknowledge that the problem is more than personal. At this stage, it is difficult for the society to return to a state of equilibrium without undergoing a process of revitalization. This process depends on a number of variables:

- *The formulation of a code.* An individual or small group—typically people who have been directly affected by the stresses in question—builds a new idealized image of a "goal culture" that stands in attractive contrast to the existing situation.
- *Communication.* The formulators then communicate their vision to others, beginning with those most affected by the stresses at the root of the problem.
- *Organization.* Once the vision begins to attract converts, some form of organization is required to manage the group and implement a plan.
- *Adaptation.* As the new vision gets broader exposure, it generally grows and changes. This happens for various reasons: the initial vision is usually incomplete, especially in its practical details; certain accommodations may be made to broaden the appeal of the vision; and the original vision may include such things as predictions that fail to materialize, thus requiring re-explanation.
- *Cultural transformation.* If the movement is able to gain enough support within the society, the thrust shifts from communications to implementation. If the "goal culture" cannot be immediately established, then a "transfer culture" is adopted as the route to get to the full vision. If this cultural shift is successful, the stress experienced by individuals declines dramatically.

- *Routinization.* Once the initial shift of cultural transformation has taken place, the next stage is to establish the new vision as the new steady state, which generally means institutionalizing it in various ways. Those in the vanguard of the transformation process may find this last stage difficult and disappointing, but the majority of the population is glad for a return to normalcy so that they can get on with their lives.

This kind of revitalization can be either reactionary or innovative in its basic thrust. The reactionary mode is characterized by a belief that present problems can be resolved by "doing the old way harder" and generally tries to undo or suppress recent changes that are seen as the cause of the problem. The innovative mode, on the other hand, attempts to get "lagging" parts of the culture to catch up to recent changes that are seen by the innovators as either positive or unchangeable.

In complex societies under stress such as among land users in the "Three Rivers" region, there are usually many pressures causing the culture to begin to disintegrate. It is common for a culture to attempt a "let's do the old way harder" revitalization as the first response to realizing that something must be done to get society back in track. It is only after the failure of a reactionary revitalization attempt that a culture is willing to risk fundamental change. This shift from innovation to reaction and back again is often described in common parlance in terms of a swinging pendulum. The metaphor suggests that social change follows predictable laws of motion and that movement is static in one direction or another. Society tends to lurch, often mindlessly, in a new direction, and it is only after a period of reaction that integration takes place. We suggest that things are at this stage in the study area on the Qinghai–Tibet Plateau.

4.2.1 Microchanges and Megachanges

We can divide changes into microchanges and megachanges. We make no great attempt at elaborate definitions. As a first approximation, the following scheme can be used to differentiate between the two:

- *Microchanges*—differences in degree
- *Megachanges*—differences in kind

As an elaboration we can say that modifications, enhancements, improvements, and upgrades in the existing system would typically be microchanges, while a new system or a very major revision of an existing one (new paradigm) would be a megachange. We must remember that one person's microchange is often another person's megachange. So when designing the total "people" strategy for any system, it is important to involve a variety of people from the very beginning, to clearly understand how groups function in the organization and how the work is really done.

4.2.2 The Key Stakeholders

For any given change (policy change, climate change, economic change) people can occupy a wide range of roles that will strongly influence their perceptions of the change and their reactions to it. These are roles such as champion, end user, watchful observer, obstructionist, and such. As on the stage, some people may occasionally play more than one role. In other cases, the roles are unique. Unless we clearly identify both the players and their roles in any change situation, we risk making decisions and taking action based on generalizations that are not true for some of the key players.

An overview term often applied to the various roles is stakeholders. The stakeholders have some interest or stake in the quality of both the change and the change implementation process. The roles of the stakeholders are subject to change, especially during a change process (like the onset of global warming) that extends over some time.

For those implementing change, the following steps are critical:

- To identify what roles they themselves are occupying in the process
- To identify what roles the others involved in the process are playing, being careful to recognize multiple roles
- To identify carefully which role is speaking when one is communicating with those playing multiple roles
- To monitor throughout the process whether any roles are changing

4.3 Resistance to Change

> It is easy to change the things that nobody cares about. It becomes difficult when you start to change the things that people do care about—or when they start to care about the things that you are changing. Lorenzi and Riley (2000)

Resistance to change is an ongoing problem. At both the individual and the organizational levels, resistance to change impairs concerted efforts to improve the status quo. Many change efforts have been initiated in a top-down process at tremendous cost only to be halted by resistance among the local administrative bureaus and the land users. Local government organizations as a whole also manifest behavior similar to that of individuals when faced with the need to change, i.e., there is reluctance to accept the directive or considerable inertia in the attempts to make the policy measures operational.

The relationship between individual and organizational resistance to change is important. An organization is a complex system of relationships between people, leaders, technologies, and work processes. From this interaction emerges organizational behavior, culture, and performance. Socialization, as a lifelong interactive process of cultural learning, involves different types of social actors. Agents of socialization are the individuals, groups, and institutions that create the social context in which socialization takes place. It is through agents of socialization that individuals learn and incorporate the values and norms of their culture as well as their various positions in the social structure in such terms as class, ethnicity, and gender.

These emergent properties and behaviors are tightly linked in two directions to the lower-level interactions. Organizational resistance to change is an emergent property, and individual resistance to change can give rise to organizational resistance. A self-reinforcing loop of increasing resistance can develop as individuals create an environment in which resistance to change is the norm. That environment in turn encourages increased resistance to change among individuals. The self-reinforcing nature of this loop can be tremendously powerful, defeating repeated attempts to break out of it. Studies of system dynamics frequently reveal that major problems that everyone thought were external are actually the unintended consequences of internal policies. The basic dynamic behind this phenomenon is that the organization is made up of a network of circular causal processes: A influences B, which then influences C, which in turn influences A, i.e., the snake bites its own tail. Figure 4.1 is a causal diagram showing these relationships in a semiarid rangeland.

Understanding these internal organizational dynamics is a prerequisite for leading effective change processes.

4.3.1 Rituals of Transition

All change involves loss. In some cases, the loss is substantial. The strategies for overcoming the barriers to change are quite diverse. No organization or societal group can begin using all the strategies at the same time or even in a short period of time. A better approach is to focus on one or two until they become part of the normal way of operating, i.e., until they become engrained in people's habits. Only then is it time to introduce another strategy. In this way, over time, the stakeholders gradually improve their ability to learn rapidly, to adapt to new conditions, and to embrace change. The immediacy of the threat is also important. The importance of the pace of change is critical. The difficulty many face with global climate change is that it is a "slow-onset" phenomenon. Contrast this with a landslide that buries a whole village or cuts off a major access road.

4.3.1.1 Technology Is Not Enough

Many projects designed to assist in adaptation to climate change focus exclusively on technology and fail to address the human and organizational aspects of work. Too many technically good applications have failed because of sabotage by those (bureaucrats, administrators, and land users) who like the old ways in which things were done. The challenge is to manage the natural resistance to change and help convert that resistance into commitment and enthusiasm as part of a planned process. New systems should enhance the quality of life and increase responsibility, empowerment, and motivation. But nothing works without people. Thus, human issues become magnified in the process of redesigning adaptations to change.

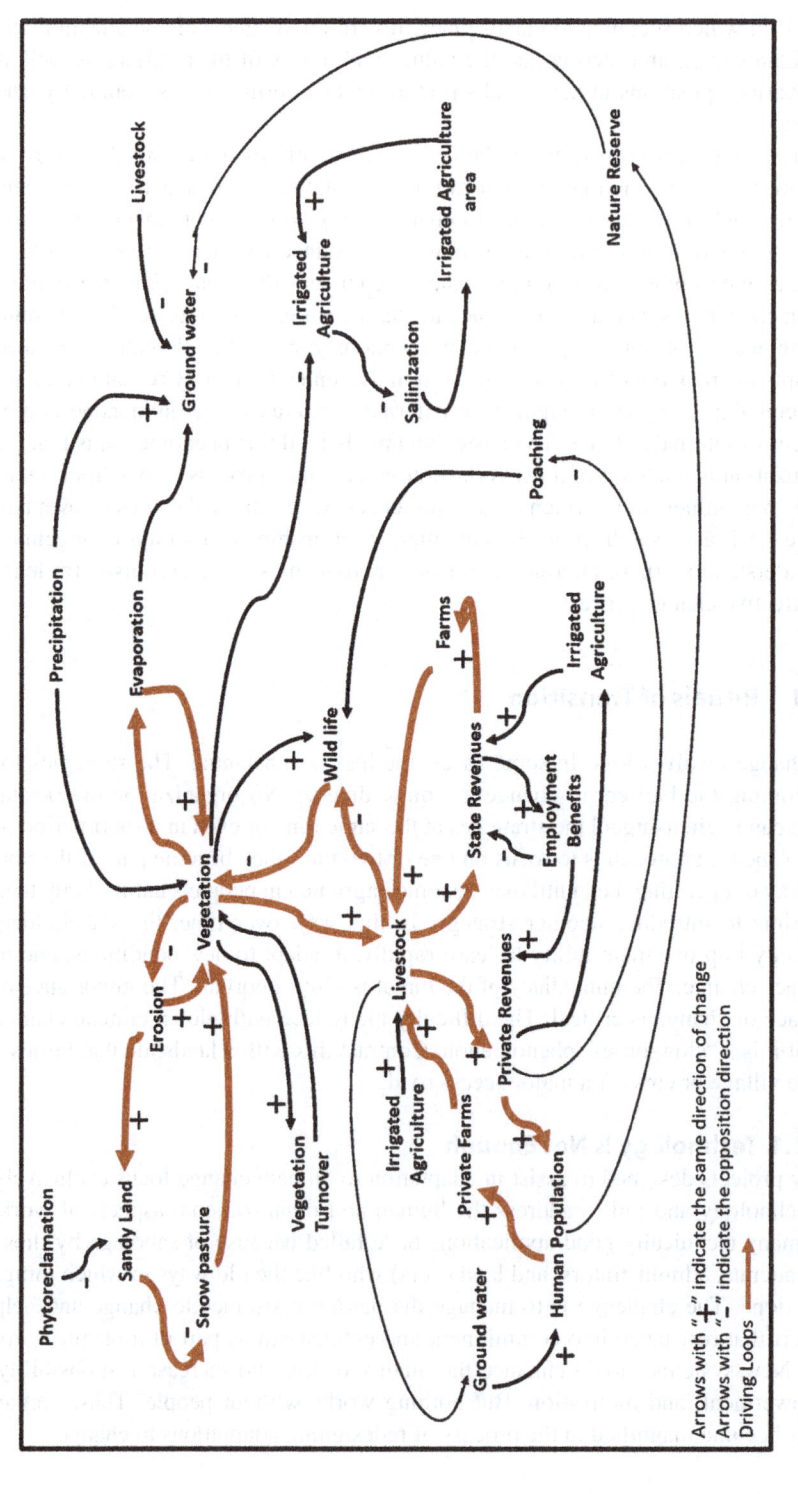

Fig. 4.1 A simplified representation (causal diagram) of the main physical and biological mechanisms affecting ecosystem regulation in rangelands

4.4 The Unique Role of Humans as Agents of Change

Humans have evolved in a changing world and show remarkable adaptability. Humans sculpt and transform the landscape through the physical modification of the shape and properties of the ground. As such, humans are geological and geomorphological agents and the dominant factor in landscape evolution through settlement and widespread industrialization and urbanization. The most significant impact of this has been since the onset of the Industrial Revolution in the eighteenth century, coincident with increased release of greenhouse gases to the atmosphere.

For further discussion of the impact of climate change on peoples of the Qinghai–Tibet Plateau and their perceptions of what it means to them, see Feng and Squires (2017).

4.4.1 Change Management

The phrase *change management* has come into common use in the literature. This leads us to ask—what is meant by change management, how did it evolve, and why has this concept become so important? Change management is the process by which an organization/institutions and human societies get to its future state, its vision. While traditional planning processes delineate the steps on the journey, change management attempts to facilitate that journey. Therefore, creating change starts with creating a vision for change and then empowering individuals to act as change agents to attain that vision. The empowered change management agents need (i) plans that provide a total system approach, (ii) are realistic, and (iii) are future oriented. Change management encompasses the effective strategies and programs to enable those change agents to achieve the new vision. Today's change management strategies and techniques derive from the theoretical work of a number of early researchers.

4.4.2 Types of Change

Changes in an organization can often be identified as one of four types, with the definite possibility of overlap among them:

- *Operational changes* affect the way the ongoing operations are conducted.
- *Strategic changes* occur in the strategic direction,
- *Cultural changes* affect the basic organizational philosophies by which the adaptation occurs.
- *Political changes* occur primarily for political and policy reasons of various types, such as those that occur at top patronage job levels in government agencies.

These four different types of change typically have their greatest impacts at different levels of society. For example, operational changes tend to have their greatest impacts at the lower levels, those land users who are key stakeholders. People working at the upper levels may never notice changes that cause significant stress and turmoil to those attempting to implement the changes. On the other hand, the impact of political changes is typically felt most at the higher organizational levels. As the name implies, these changes are typically made not for result-oriented reasons but for reasons such as partisan politics or internal power struggles. When these changes occur in a relatively bureaucratic organization, as they often do, those working at the bottom often hardly notice the changes at the top. The key point is that performance was not the basis of the change; therefore, the performers are not much affected.

4.4.3 The Key Stakeholders

For any given change (policy change, climate change, economic change) people can occupy a wide range of roles that will strongly influence their perceptions of the change and their reactions to it. These are roles such as champion, end user, watchful observer, obstructionist, and such. As on the stage, some people may occasionally play more than one role. In other cases, the roles are unique. Unless we clearly identify both the players and their roles in any change situation, we risk making decisions and taking action based on generalizations that are not true for some of the key players.

An overview term often applied to the various roles is stakeholders. The stakeholders have some interest or stake in the quality of both the change and the change implementation process. The roles of the stakeholders are subject to change, especially during a change process (like the onset of global warming) that extends over some time.

For those implementing change, the following steps are critical:

- To identify what roles they themselves are occupying in the process
- To identify what roles the others involved in the process are playing, being careful to recognize multiple roles
- To identify carefully which role is speaking when one is communicating with those playing multiple roles
- To monitor throughout the process whether any roles are changing

4.4.4 Traditional Knowledge and Shared Belief Systems in Qinghai

It has been argued by Ellis (1998) and others for Africa and Goldstein et al. (1990, 1992) for Tibet and Mongolia that traditional nomadic pastoral societies operated under a system of non-homeostatic balance—non-equilibrium systems as they are more typically referred to in ecology (Benhke et al. 1993). Rather than homeostasis being reached between populations and their resources and then maintained over time, these pastoral systems experienced great fluctuations in the number of livestock over time as a result of random, unpredictable, and uncontrollable natural calamities that periodically decimated herds. Among Tibetan nomads, these included disease and climate (heavy snows in winter that covered vegetation and prevented grazing and low rainfall summers which produced insufficient overall vegetation). Typically, herds increased rapidly for some period of time and then declined precipitously as a result of an external disaster or, more usually, a series of bad years. However, although the number of livestock in an area varied widely at different times, over long periods increases in stocking rates in Tibet appear to have been limited or at most moderate. The recurrent episodes of livestock decimation appear to have been frequent enough to create a stable, nonequilibrium system in which the grasslands were not systematically destroyed despite continuous utilization for at least one and perhaps two or more millennia.

The unpredictableness of natural disasters over large expanses of territory was replicated at the microlevel in the sense that the areas utilized by individual herders were also subject to random calamities even though the larger territorial unit of which they were a part was not severely affected. Consequently, the rational choice for the individual herder was to maximize the size of his herd at all times since there was no way for him to predict when a natural disaster would decimate his herd and certainly no way for him to prevent it. If one faces the possible loss of a large proportion of one's herd due to a random event one can neither predict or prevent, it is clearly more advantageous to possess 500 sheep rather than only 50 or 100, since the ultimate danger for any pastoralists is to fall below the minimum number needed to survive a disaster (or a series of bad years) and recoup herd size during the more advantageous years. To fall below this level in traditional Tibetan society inevitably meant losing one's autonomy and status as an independent herder and being forced to subsist by becoming a laborer for a wealthy herder. Thus, the traditional Tibetan pastoralist's emphasis on maximizing livestock numbers was actually a strategy for minimizing risk in an uncertain, nonequilibrium ecological system, although wealth and prestige were also measured by livestock numbers.

The role of traditional ecological knowledge and shared systems of beliefs in building long-term social–ecological resilience to environmental extremes has been the focus of investigations in the Qinghai–Tibet Plateau. Data were collected from seven villages of the "Three Rivers" region (see map in Chap. 5), through interviews, focus groups, and systematic reviews of historical archives. First, we assess adaptive practices to cope with environmental change. Then, we use historical records to trace collective responses to environmental extremes. Our results (1) show how environmental extremes could induce social and economic crises through declines in ecosystem services and (2) identify practices to cope with recurrent disturbance and institutional devices developed in response to environmental extremes. We conclude that traditional ecological knowledge and shared systems of beliefs can facilitate collective responses to crises and contribute to the maintenance of long-term resilience of social–ecological systems.

Traditional knowledge and related institutions increase capacity to cope with change. Shared belief systems can enhance resilience to hazards by promoting cohesion. Collective memory to cope with extremes is often coded in rituals and institutions. Traditional knowledge and beliefs tend to erode with adoption of modern technology. The emergence of new global economic centers is inducing a major expansion in the global social metabolism—the flows of energy and materials into the world economy— a transformation in the systems for the extraction and provision of natural resources, as well as setting the conditions for socioenvironmental conflicts at the commodity frontiers, particularly in areas with a dense human occupation of the territory. We point out that we are currently experiencing global transformations that constitute the beginning of a new historical phase of modern capitalism, even in the hitherto "insulated" Qinghai–Tibet Plateau region. Given the inevitability of the transition, the land-use change is likely to be persistent. Massive changes in the use of and access to land can be expected for the near future.

Change strategies fall into three categories: (1) rational–empirical, (2) normative–reeducative, and (3) power coercive.

The *rational–empirical* approach assumes that men and women are rational and practical and will change on their own given the appropriate conditions. These strategies include:

- Provide the right information, education, or training to allow individuals to change of their own volition.
- Ensure that the "right" people are in the right "place" to bring about needed changes.
- Invite the perspectives or expertise of outsiders.
- Engage in research and development.
- Promote utopian thinking to stimulate creativity and "best-case" scenarios.
- Clarify the issues and/or reconceptualize the situation in order to bring about greater overall understanding among members of the group.

The second category of strategies—the *normative–reeducative*—is predicated on the view that change begins from the bottom up, not the top down. That is to say, it focuses on changing the individuals that make up a social system.

The *power-coercive* approach to effecting change (including adaptive strategies) is the one most commonly associated with political movements and social activism. Strategies in this category include:

- Using political institutions to achieve change
- Shifting the balance of power between social groups, especially ruling elites
- Weakening or dividing the opposition through moral coercion or strategies of nonviolence

A list of community-based change strategies include consensus planning, bargaining, protest movements, research demonstrations, social action, nonviolence, organizations of client populations, community development, conflict, elite planning, organization of indigenous groups, and civil disobedience. we classify these under four headings: (1) collaborative strategies, (2) campaign strategies, (3) contest strategies, and (4) a combination of strategies.

There are two modes of viewing change: the reactive and the proactive. From one perspective, individuals and groups are the objects of change. They are at the receiving end, in the sense that change happens to them. From the other perspective, individuals and groups are the initiators of change and change follows from human volition. Both perspectives have their validity, of course, and they are closely interrelated. For instance, when one social group actively tries to bring about change, there are invariably other groups who feel put upon and try to resist the change.

All Cultures are inherently predisposed to change and, at the same time, to resist change. There are dynamic processes operating that encourage the acceptance of new ideas and things, while there are others that encourage changeless stability. It is likely that social and psychological chaos would result if there were not the conservative forces resisting change.

There are three general sources of influence or pressure that are responsible for both change and resistance to it: (i) forces at work within a society (ii) contact between societies and (iii) changes in the natural environment. All three sources of influence apply in the study area.

Within a society, processes leading to change include invention and culture loss. Inventions may be either technological or ideological. The latter includes such things as the invention of algebra and calculus or the creation of a representative parliament as a replacement for rule by royal decree. Technological inventions include new tools, energy sources, and transportation methods as well as more frivolous and ephemeral things such as style of dress and bodily adornment.

Within a society, processes that result in the resistance to change include habit and the integration of culture traits. Older people, in particular, are often reluctant to replace their comfortable, long familiar cultural patterns. Habitual behavior provides emotional security in a threatening world of change. Religion also often provides strong moral justification and support for maintaining traditional ways. Culture loss is an inevitable result of old cultural patterns being replaced by new ones. For instance, not many people today know how to care for a horse. A century ago, this was common knowledge, except in a few large urban centers. Since then, vehicles with internal combustion engines have replaced horses as our primary means of transportation and horse care knowledge lost its importance. As a result, children are rarely taught these skills. Instead, they are trained in the use of the new technologies of automobiles, televisions, stereos, cellular phones, computers, and iPods.

Community-based change strategies include consensus planning, bargaining, protest movements, research demonstrations, social action, nonviolence, organizations of client populations, community development, conflict, elite planning, organization of indigenous groups, and civil disobedience. We can classify these under four headings: (1) collaborative strategies, (2) campaign strategies, (3) contest strategies, and (4) a combination of strategies.

The processes leading to change that occur as a result of contact between societies or different ethnic groups are:

Diffusion is the movement of things and ideas from one culture to another. When diffusion occurs, the form of a trait may move from one society to another but not its original cultural meaning.

Acculturation is what happens to an entire culture when alien traits diffuse in on a large scale and substantially replace traditional cultural patterns. After several decades of relentless pressure from Han Chinese to adopt their ways, many ethnic minority cultures have been largely acculturated. As a result, the vast majority of them now speak Chinese as well as their ancestral language, wear western style clothes, go to school to learn about the world from a Chinese perspective, and see themselves as being a part of the broader Chinese society. As ethnic minority societies continue to acculturate, most are experiencing a corresponding loss of their traditional cultures despite efforts of preservationists in their communities (Feng 2012).

While acculturation is what happens to an entire culture when alien traits overwhelm it, *transculturation* is what happens to an individual when he or she moves to another society and adopts its culture. This process is exacerbated in China by the phenomenon of migrant workers who travel to cities for work in construction jobs or to become seasonal workers for picking fruit or cotton. Migrant workers who successfully learn the language and social norms of their "new homes" and accept as their own the cultural patterns of their adopted society have transculturated. In contrast, people who live as socially isolated groups in their traditional homeland without desiring or expecting to become assimilated participants in the dominant culture are not transculturating.

There is one last process leading to change that occurs within a society as a result of an idea that diffuses from another. This is *stimulus diffusion*—a genuine innovation that is sparked by an idea from another culture.

4.5 Summary and Conclusions

Our research approach combined literature review with expert judgment by researchers working in each landscape. For each landscape we described land use, rural livelihoods, and attitudes of social actors toward global change (including climate change and globalization of markets and the monetary system). Principal component analysis of indicators of natural, human, social, financial, and physical capital for the landscapes showed a loss of household income and increased poverty associated with global warming, glacial melt, and so on. High levels of natural capital (e.g., indicators of attitudes toward wildland biodiversity conservation and preservation of ethnic cultures) were positively associated with indicators of human capital, including local ecological knowledge, identification of the flora and fauna, and knowledge-sharing among land users. This exploratory study suggests that indicators of knowledge systems should receive greater emphasis in the monitoring of biodiversity and ecosystem services and that inventories of assets at the landscape level can inform adaptive management of interventions to cope with global change.

References

R. Benhke, I. Scoones, C. Kerven (eds.), *Range Ecology at Disequilibrium: New Models of Natural Variability and Pastoral Adaptation in African Savannas* (Overseas Development Institute, London, 1993)

J.J. Cao, E.H. Yeh, N.M. Holden, Y.Y. Qinn, Z.W. Ren, The role of overgrazing, climate change and policy as drivers of degradation of natural grassland in China. Nomadic Peoples **2**, 82–101 (2013)

S. Dong, Characteristics and problems of western China's eco-economic regions. Resour. Sci. **27–6**, 104 (2005.) [In Chinese with English Abstract]

F. Ellis, Household strategies and rural livelihood diversification. J. Dev. Stud. **35**(1), 1–3 (1998)

H.Y. Feng. *In the process of developing folk lore culture in tourism: Tu people's cultural reconstruction*. PhD dissertation, (China Agricultural University, Beijing, 2012) [In Chinese with English abstract]

H.Y. Feng, V.R. Squires, Climate variability and impact on livelihoods in the cold arid Tibet Plateau aspirations and prospects for herders on the Qinghai-Tibet Plateau, China , in *Climate Variability,Land-Use Change and Impact on Livelihoods in the World's Dry Lands*, ed. By M. Gaur, V. R. Squires, (Springer, New York, 2017)

M.C. Goldstein, C.M. Beall, R.P. Cincotta, Traditional nomadic pastoralism and ecological conservation on Tibet's Northern Plateau. Natl. Geogr. Res. Rep. **6**, 139–156 (1990)

N.M. Lorenzi, R.T. Riley, Managing change: an overview. J. Am. Med. Inform. Assoc. **7**(2), 116–124 (2000)

H. Ma, Explanation about the causes of ecological degradation in the "Sanjiangyuan" district under the perspectives of new institutional economics. Tibet. Stud. **3**, 88 (2007)

V.R. Squires, H.Y. Feng, Rangeland & Grassland in the region of the former Soviet Union: future implications for silk road countries , in *Rangelands of the Silk Road*, ed. By V. R. Squires, Z. Shang, A. Ariapour, (Nova Publishers, New York, 2017)

Z.H. Shang, S. Dong, A. Degen, R.J. Long, Ecological Restoration on the Qinghai-Tibet Plateau: Problems, Strategies and Prospects in *Ecological Restoration: Global Challanges, Social Aspects and Environmentl Benefits*, ed. by V.R. Squires. (Nova Science Publishers Inc, New York, 2016), p.151–176

D. Tsering, The theoretical relations between the poverty oriented resettlement and environmental migration. Tibetan Stud. **5**, 45 (2014.) [In Chinese with English Abstract]

X. Wang, On the formation reasons of eco-refugee from a geography perspective in the source of Yellow river, Lancang river and Yangtze river, ecological area. Environment 9, 131–142 [In Chinese with English Abstract] (2006a)

X. Wang, Studies on ecological refugee in the source of Yellow river, Lancang river and Yangtze river, J. Qinghai National. Univ. 17, 23–31 [In Chinese] (2006b)

X.D. Xu, G.G. Lu, X.H. Shi, WorldWater tower: an atmospheric perspective. Geophys. Res. Lett. **35**(20), L20815 (2008)

A.R. Zhang, Ecological fragility assessment of the IEFA in Qinghai Tibetan area,-a case study of the ecological function area in Sanjiangyuan. J. Tibet Univ. **30–1**, 1 (2015.) [In Chinese]

X. Zhao, Climatic characteristics of heavy precipitation events during summer half year over the Eastern Tibetan Plateau in recent 50 years. Arid Land Geogr. **38–4**, 675 (2015.) [In Chinese with English Abstract]

R. Zheng, Analysis on resettlement in ecologically vulnerable areas of western China: resettlement policy and practices aiming at ecological environment protection and poverty alleviation. Yangtze River **42–5**, 93 (2011)

Further Readings

F. Du, Ecological resettlement of Tibetan herders in the Sanjiangyuan: a case study in Madoi county of Qinghai. Nomadic Peoples **16**(12), 116–133 (2012)

M.C. Goldstein, C.M. Beall, Changing patterns of Tibetan nomadic pastoralism, in *Human Biology of Pastoral Populations*, ed. By W.R. Leonard, M.H. Crawford, (Cambridge University Press, Cambridge, 2002), pp. 131–150

P. Wang, J.P. Lassoie, S.L. Morreale, S.K. Dong, A critical review of socioeconomic and natural factors in ecological degradation in the Qinghai-Tibetan Plateau. China. Rangel. J. **37**, 1–9 (2015)

Climate Variability and Impact on Livelihoods in the Cold Arid Qinghai–Tibet Plateau

Haiying Feng and Victor R. Squires

5.1 Introduction to the Qinghai–Tibet Plateau

The area of the Qinghai–Tibetan Plateau (QTP) accounts for 23.4% of China's land area. All of it is dry land where the aridity index ranges from 0.03 to 0.35 with a growing period of less than 90 days and environmental risk coefficient of variation greater than 25%. The QTP is known as "the third pole of the world." It has impressive landscape, heritage, and biodiversity. The QTP has long been a melting pot for a number of ethnic groups including the Han, Tibetans, Hui, Tu, Mongols, and Salars The people of Asian–Hindu–Kush–Himalayan region live in the mountains and basin of the Qinghai–Tibetan Plateau. However, China's population in the Qinghai–Tibetan Plateau region is only ten million people, accounting for 0.8% of China's total population, while this region accounts for only 0.5% of China's total GDP. Therefore, the key role of the QTP lies in its ecologically strategic position to China, Asia, and perhaps globally (Sun et al. 2012) (Figs. 5.1 and 5.2).

The average elevation of QTP is over 3000 m above sea level. Mountain ranges include the Tanggula Mountains and Kunlun Mountains. The Qaidam Basin lies in northwestern Qinghai. About a third of this resource-rich basin is desert. The basin has an altitude between 3000 and 3500 m. Its average temperature is approximately −5 to 8 °C (23–46 °F), with January temperatures ranging from −18 to −7 °C (−0 to 19 °F) and July temperatures ranging from 15 to 21 °C (59–70 °F). It is also prone to heavy winds as well as sandstorms from February to April.

The Qinghai–Tibetan Plateau is one of China's five major pastoral areas with the yak being the main livestock species. The current number of yaks in the Qinghai–Tibetan Plateau is about 14 million or 90% of the total in the world. The number of

H. Feng
Institute of Administration and Management, Xining, Qinghai, China

V.R. Squires (✉)
Institute of Desertification Studies, Beijing, China
e-mail: dryland1812@internode.on.net

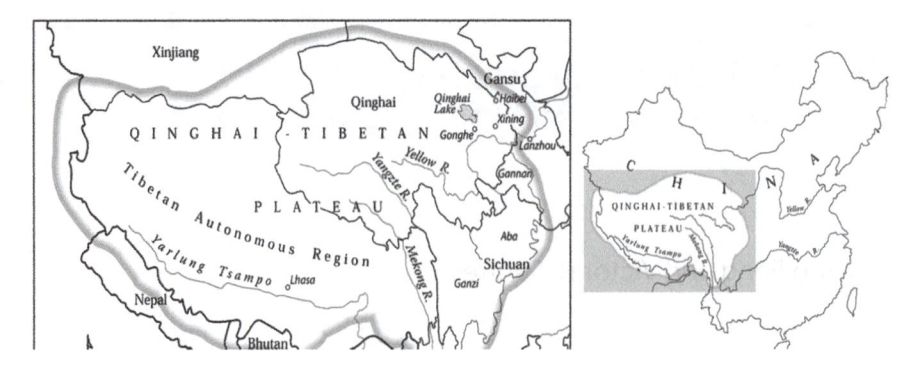

Fig. 5.1 Map of Qinghai–Tibet Plateau (QTP) showing its immediate neighbors in mainland China and in adjacent countries

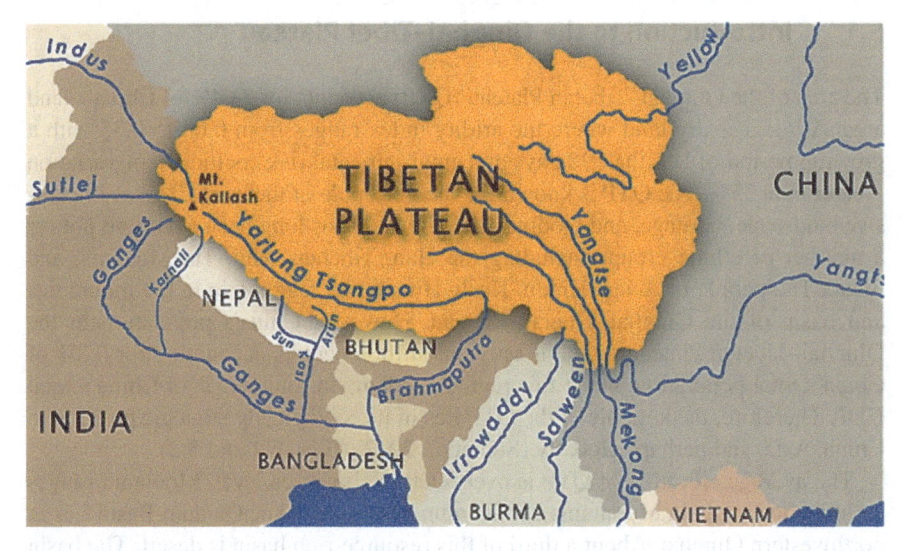

Fig. 5.2 Map showing Qinghai province and the location of the source area for the Huang He (*Yellow*) and Changjiang (Yangtze) rivers. The Lancan Jiang (Mekong), Indus, Brahmaputra, and other important transboundary rivers also come from this area

Tibetan sheep, goats, and other livestock totals slightly more than 50 million (Fig. 5.3). These livestock provide products not only for people living in Qinghai–Tibetan Plateau but also for people all over the world (Long et al. 2008). The threats to biodiversity of the special (some unique) animals and plants is very serious because of habitat fragmentation and serious reductions in populations of key species. Because of rangeland degradation, warming, and drying, some households have given up traditional livestock farming.

Livestock, namely, yak and Tibetan sheep, are the core of livelihood resources for the vast majority of land users who are engaged in pastoralism. These two livestock species are the foundation of survival of dwellers on the QTP for thousands

Fig. 5.3 (**a**) Yaks and (**b**) Tibetan sheep and goats are the mainstay of the livestock industry on the Qinghai–Tibet Plateau. Both species can live on fibrous forage and endure severe cold

of years. Yaks and Tibetan sheep provide many resources and create incomes (Yu et al. 2013). In terms of livelihood, rangelands provide less potential than farmland in other regions. So it is very important to develop diversified livelihoods in the rangeland ecosystem for sustaining a household. Although many households are in subsistence mode and technically are classified as "below the poverty line," resources of the rangelands in the QTP are not poor and sources of livelihood are considerable, so the livelihood system is not as fragile as many might imagine. Furthermore, under current policy, great importance is attached by the government to ensuring proper compensation for livestock producers (Yan and Qian 2004; Wang et al. 2015).

The Qinghai–Tibetan Plateau nurtures nine major rivers in Asia, providing 1.3 billion people with freshwater. The rangelands are the core of the water cycle, and their health directly affects the flux of water and water conservation capacity. The runoff that ends up in rivers, lakes, and other water bodies is largely through the rangeland vegetation. In terms of glacier water, the glacier area is about 49,873 km^2 and ice reserves about 4561 km^3 (Yao et al. 2017). The frozen soil area is 126 km^2, accounting for 56% of the total Qinghai–Tibetan Plateau area. The groundwater in the Qinghai–Tibetan Plateau is the source of runoff water of many surrounding low-altitude areas. The lake census showed that there are 32,843 lakes with a total area of 43,151 km^2, which makes up 1.4% of the total area of the Plateau. Around 96% of all lakes are small, with an area <1 km^2, while the 1204 large lakes (>1 km^2) account for 96% of the total lake area (Zhang et al. 2013).

5.2 Climate Change and Land Use Systems on the QTP

5.2.1 Historical Trends

Similar to global trends, QTP warmed by about 0.5 °C in the twentieth century with the most rapid warming occurring after 1970 (Du et al. 2004). Precipitation trends and patterns are more difficult to determine, and significant differences are evident because of altitude, aspect, and proximity to urban areas. Anecdotal evidence suggests marked inter-decadal variability, which adds to the difficulty of managing complex risks. A closer alignment of historical climate data with comprehensive land-use histories is urgently needed to understand the full extent of changes that have occurred in QTP rangelands over the course of the past 70 years. Many people live in drought-prone areas. Changes in the frequency and magnitude of drought could add to the complex risk management portfolios that many people use to sustain their livelihoods, and such changes will make recovery more difficult because periods between significant events will likely be shortened.

5.2.2 Coping and Adaptation to Climate Change

The concept of coping has its origins in development studies literature, particularly the sustainable livelihoods framework. The sustainable livelihood framework recognizes that livelihoods are carved out by households who choose strategies to earn a living depending on the balance of various capitals (human, social, natural, physical, and financial) made available to them by institutions and entitlements, which are themselves embedded in the wider political environment.

The sustainable livelihood approach opens up the possibility of exploring how people in developing countries modify their livelihood strategies within the context of various covariate and idiosyncratic risks. Covariate risks are the ones to which a number of people may be exposed at any one time and include many climate-related risks such as droughts and floods. Idiosyncratic risks tend to be more spatially restricted and affect one household at a time, for example, in the case of death or critical illness. Diversification is accepted as a common strategy within this framework (e.g., Ellis 1998). Coping behavior is a component of diversification.

The effects of climate change on agricultural systems *sens. lat.* in developing regions such QTP will depend on location and people's adaptive capacity. But adapting to and coping with a changing climate are not infinitely plastic, and it may be envisaged that in some places climate change may push agroecological conditions beyond the "coping range," such that current adaptation measures may not be longer be viable.

In such places, livelihood options may have to change. The mixed crop-livestock rainfed arid and semiarid systems of cropping will become increasingly risky, and this could lead to increased dependence on livestock keeping or increasing diversification. Such livelihood changes could be seen as antithetical to an evolutionary process of agricultural intensification, in which increasing human population pressure on relatively fixed land resources is seen as the driving force of agricultural intensification. Nevertheless, this reversal of an evolutionary process is entirely plausible; the ability of householders in regions of high climatic risk to adapt, using blends of old and new techniques as well as a host of methods to extensify and/or diversify the production system, has long been the subject of study (Tache and Oba 2010). If climate change in the coming decades in QTP does induce an extensive reversal to agriculture dominated by mobility of the means of production and of residence, the social implications would be profound. As for many other types of widespread livelihood transition, there would be social, environmental, economic, and political effects at local, national, and even regional levels, and these effects would need to be appropriately managed and facilitated. A wide range of strategies have been adopted in order to respond to climate variability and change. These strategies include crisis responses, modifying farming practices, modifying crop types and varieties, resource management, and diversification. Coping typically refers to short-term strategies designed to maintain survival, but the long-term nature of many of the responses suggests that they do, in fact, constitute adaptations to current variability and change. However, determining whether or not the observed strategies are examples of coping or adaptation is dependent on the particular context in which they were observed and also requires a consideration of the scale of interest. This has implications for how policies and programs are designed to support adaptation in the future.

Livelihood transitions mediated via changes in climate vis-à-vis changes caused by other drivers (e.g., immigration, conflicts for natural resources, and changing economies) need to be elucidated to disentangle the impacts of climate change on QTP rural households. In this study we tested the hypothesis that sedentary farmers who currently keep livestock in transition zones that are becoming warmer and possibly drier in the future may ultimately be forced to increase their reliance on livestock vis-à-vis cropping in the future, despite other potential driving forces shaping their livelihoods. We analyze past and current responses of farming households to climate variability and regional change in marginal lowland cropping areas of Qinghai and assess impacts on household income, food security, and food self-sufficiency while at the same time providing evidence on future coping and adaptation mechanisms. Householders were all concerned about water shortages and drought, but not because of crop failure. Water for livestock could be found by moving livestock. Livestock keepers were not willing to reduce livestock numbers to manage risk of drought. This strategy appears to be most sensible when livestock keepers live with climate uncertainty and lack support to restock after massive animal losses. The purpose (for them) of increasing livestock numbers was to increase livestock sales and not consumption of animal products. Increasing livestock productivity was also on the list of adaptation options. In the wetter areas where livestock are fewer and of more value, householders think of improved breeding to increase productivity. In the drier areas, householders plan to increase productivity through better animal health and feeding.

5.2.3 Change in Rangeland Productivity, Forage Production, and Species Composition

Rainfall and temperature are key determinants of rangeland productivity. The effect of future climate change projections on the length of the growing period (LGP), which integrates the influence of temperature and rainfall on productivity, results in a number of potential impacts, including changes in the length of the growing season for certain agricultural crops. The combined impact of changes in temperature and rainfall will result in a decrease of LGP in much of QTP, and in some cases this decrease will be severe. Temperature, rainfall, and atmospheric CO_2 concentration interact with grazing and land cover change to influence rangeland quality and composition. Increased temperature, for example, not only increases drought stress in plants but also increases lignification of their tissues, which affects both its digestibility as well as its rate of decomposition.

The amount and timing of precipitation on its own also has an important influence on rangeland species composition in both the short and long term, primarily through its differential effect on the growth and reproduction of key forage species. An extended drought can result in the death of perennial plants and the switch to an annual-dominated flora or to an increase in woody or toxic plants (including invasive species (Limbu 2017).

5.2.4 Change in Land Use Systems and Rangeland-Based Livelihoods

The general reduction in productivity projected for QTP's rangelands will have important negative consequences for development potential and likely will result in a shift in agricultural activities. Some projections suggest that in marginal crop production areas, the decrease in the length of the growth period (LGP) and an increase in rainfall variability will render cultivation too risky and will result in a switch to more rangeland-based livestock production systems. There is also likely to be a switch to breeds and species (e.g., from sheep and goats to yaks) that are better adapted to more marginal conditions. Other changes include a greater frequency of loss of livestock assets, particularly through drought, a reduction in income, increased income inequalities, and a general reduction in livelihood security for people who derive their livelihoods primarily, or even in part, from rangelands. However, reducing all changes occurring on rangelands to "climate change" is a gross oversimplification and can be misleading.

Land cover changes cannot be only determined by changes in precipitation but also by complex rangeland management decisions. Climate (e.g., temperature and rainfall) is important, but the combined impact of grazing and stocking strategies, and other factors influencing decision making, are also key in shaping rangelands. Rather than singular stresses shaping and dominating the environment, a range of other factors also need to be understood, including the interaction of human settlements and changes in land use, as well as how various policies have an impact on land-use change (see below). Much more work also is required on how policy and understandings of rangeland changes are framed, reproduced, and mainstreamed into practice. These various interactions can act as critical drivers of rangeland bringing with them potential changes, including conflicts between different land-use sectors.

5.2.5 Current Trends and Future Growth Projections

Although the QTP has contributed little to China's global greenhouse gas emissions, it is considered to be the region most susceptible to climate change impacts because of (1) the large proportion of people who live in the region that is predicted to be affected by increased temperature and reduced precipitation; (2) the high dependence of people on natural resources, livestock, and agriculture for their livelihoods; (3) extreme poverty in many places, which makes it difficult for affected people to respond to an increased incidence of colder drought and floods; and (4) the degraded state of the natural resources in the Three Rivers Region which renders them less resilient to the impact of higher temperatures and lower and more variable precipitation.

In dry areas, including the QTP, it is necessary to get a balance between the functions (provision of goods and services[1]) supplied by natural resources and demands by societies exploiting these resources particularly in the context of predicted changes associated with climate change. This will necessitate sustained increases in the provision of goods and services that include food, forage, fuel, and fiber (Safriel and Adeel 2005) from existing production systems as lateral expansion of the agricultural sector is no longer tenable without significant losses in biodiversity and ecosystem function.

Considering extreme events of climate change and declining availability of appropriate quality water and/or highly productive soil resources for agriculture in dry-land regions and the need to produce more food, forage, fuel, and fiber will necessitate the effective utilization of marginal-quality soil and water resources.

---

[1] Environmental goods and services are of four kinds, provisioning (food, water, etc.) regulating (water supplies etc.), *supporting* crop pollination, nutrient cycling, and *cultural* spiritual and esthetic.

The efficient use of marginal-quality water and soil resources has the potential to improve livelihoods amid growing populations in dry areas while reversing the natural resource degradation trend. Although such natural resources are often viewed as representing major environmental and agricultural challenges in terms of providing goods and services, these resources are valuable and cannot be neglected, especially in areas where significant investments have already been made in infrastructure in general and irrigation systems development in particular.

Most of the people living in dry lands of the QTP eke out their livelihood from opportunistic agriculture and its allied activities in more favored areas (lower altitude, deeper soils, availability of supplementary irrigation) or from pastoralism over the vast expanses of rangeland. Dry-land farming faces daunting problems of soil erosion and drought/water stress due to untimely or inadequate rainfall and high evaporation exacerbated by the higher-than-usual levels of solar radiation experienced in these elevated regions.

All these factors, in addition to small land holdings, contribute to low and unstable agricultural yield which leads to food insecurity for the long cold winters that are common on the QTP. Viewed in the context of climate change and its effects, achieving food security is a very big challenge. It should be kept in mind that food security exists when all people, at all times, have nutritious food that meets their dietary needs and food preferences for an active and healthy life. Such a holistic view of food security depends on availability, affordability, and accessibility to food. Meeting this very challenge has become even more difficult in dry-land areas. The many impacts of changing climate on food security in dry-land areas have received consideration from a number of people, including government and academics, due to its close link with rising poverty and sagging economic conditions.

Vulnerability to climate change is comparatively high in the QTP where the occurrence of temperature increases is expected to be higher than global averages (Du et al. 2004). Resource-dependent communities are likely to be disproportionately affected by climate change. Yet, natural resource management policies continue to be developed and implemented without considering climate change adaptation (Klein et al. 2011). Rivers in QTP are predominantly fed by glaciers, which are reported to be receding quickly due to global warming (Hagg 2017). QTP's economy is almost entirely agrarian and hence highly sensitive to climate fluctuations including variability in precipitation, floods, and extended droughts that are likely to be experienced more often in future. Accounting for all these factors, food security in QTP is a serious threat. Vulnerability in the drier locations where agropastoralism is practiced is already high, and government policies should support safety nets and market and infrastructural development. Households in the slightly wetter areas need to manage risk and to increase crop productivity through better agronomic techniques, use of water-saving agriculture, practice of minimal tillage, etc. A critical requirement is knowledge transfer of appropriate techniques and skills.

We are already observing and experiencing the impacts of climate change. Environmental degradation exacerbated by climate change is threatening vital community infrastructure and is leading to forced displacement and relocation, especially in the "Three Rivers Region" that is the focus of this chapter. Reductions in precipitation and the continued experiences of prolonged drought affect soil quality and herding and agricultural practices. Drought impacts have been worsened by increasing evapotranspiration rates, reduced soil moisture, and intensified stress on vegetation and local water sources. The influx of invasive species and prolonged drought are disrupting subsistence practices. These impacts threaten traditional knowledge, food security, water availability, historical homelands, and territorial existence and may undermine traditional ways of life that have persisted and adapted for thousands of years. In some areas, permafrost melting and glacial retreat are making it more difficult for hunter/gatherers to access traditional habitats and are changing the migration patterns of certain species.

Insights about how these impacts are affecting ethnic minorities on the QTP can contribute to the development of policies, plans, and programs for adapting to climate change and reducing greenhouse gas (GHG) emissions A growing body of literature examines the vulnerability, risk, resilience, and adaptation of indigenous peoples to climate change (Adgers et al. 2007; Alexander et al. 2011; Whyte 2016). The knowledge and science of how climate change impacts are affecting peoples on the QTP contributes to the development of policies, plans, and programs for adapting to climate change and reducing greenhouse gas emissions.

Vulnerability is not characteristic of a community but the product of systems of inequality. Herders and agropastoralists differ in their vulnerability to climate change based on their distinct cultural practices and economies, and the vulnerability of a particular ethnic group's sociopolitical, economic, and eco-cultural systems may differ by geography and climate regime. Although groups may face similarities in terms of how climate change may affect their socioeconomic status and dependence on natural resources, distinct cultural practices influence how climate change vulnerability is experienced. Despite this variability, similarities among ethnic minority communities may exist in terms of the institutional barriers—including legal, administrative, and congressional policies—that affect adaptation and resilience among tribes. Government policies may have unintended consequences of limiting or removing climate adaptation options and in turn constraining, restricting, and undermining adaptation efforts within local ethnic minority communities. Vulnerability and resilience to climate change cannot be detached from the context of China's liberation of Tibet which created both the economic conditions for anthropogenic climate change and the social conditions that limit indigenous resistance and resilience capacity. *Resistance is another facet of the political, social, and cultural capacity to address and confront threats and stressors. Indigenous peoples have to confront the inadequacy of existing government policies and influence legislation to change or develop applicable policy. Indigenous social systems have to withstand external and internal challenges to maintain strong governance and traditional leadership structures to confront climate-related impacts.* As a result of China annexing Tibet, many of the traditional adaptation practices that allowed local communities to endure environmental changes are no longer possible. Social changes resulting from contact with Han Chinese and the administrative and educational systems they imposed, such as increased sedentarization (loss of mobility and contact) and compliance with the market economy and creation and enforcement of governmental subsistence harvesting regulations, shape the vulnerability to climate impacts by constraining efforts to respond to ecological change. Therefore, climate change adaptation is not only about responding to observable impacts of climate change—it is also about understanding and addressing the manner in which the broader political context can make communities more or less vulnerable to the impacts of climate change.

The challenge for achieving sustainable agriculture production systems and livelihoods on the Qinghai–Tibet Plateau lies with the establishment of planned and well-coordinated changes at the national as well as regional levels. Thus, appropriate supportive policies and functional institutions at the national level would be needed to capture the potential for improving agricultural productivity in general and water productivity in particular to meet the food demands and to generate employment and additional income opportunities. In addition, sustainability of soil rehabilitation and productivity enhancement would need to address capacity building of farmers, researchers, and regulatory institutions. The farmers alone cannot tackle the huge task of rehabilitating and managing degraded soils and marginal-quality water resources. In addition to researchers, the involvement of nongovernmental organizations and agropastoralists and herders would be important to address the long-term management and sustainability of water and land resources of only marginal quality. Community-based management of these resources can be used to help strengthen linkages among researchers, local bureau staff, agropastoralists, and herders to develop management options for marginal-quality water and land resources. The management options for marginal soil and water resources built on the accumulated wisdom of relevant stakeholders will assist in the adoption of conservation measures at the community level. Such participatory approaches would create a sense of ownership among the agropastoralists and help in strengthening linkages among land users, researchers, and policy makers.

5.2.6 Management and Policy Implications

Given that the QTP will be particularly negatively affected by climate change, what steps can be taken to mitigate and adapt to its impacts? *Firstly*, we need to know more about how QTP rangelands have responded to changing climates in the past and how land-use practices have combined with climate change to influence the production and composition of these systems. Climate change projections largely ignore the complex interplay that exists between people, their environment, and the linked urban–rural societies in which they live. The extent and rate of past environmental change and an understanding of the reasons for such changes are required. *Secondly*, and somewhat related, we also need to develop better monitoring systems, both of climate and of ecosystem response (Shang et al. 2016), and we need to expand the capacity of land users, administrators, and advisors through collaboration with international initiatives and agencies, to conceive, fund, implement, and manage such monitoring programs. Several major climate change adaptation-relevant initiatives and activities are already underway. A number of donor-driven activities as well as various institutions have also been established to address the role of such activities in shaping rangelands (e.g., the origins, outcomes, and processes of engagement of such activities). *Thirdly*, we need to intervene to lessen the impact of climate change on QTP's most vulnerable communities by adopting responses that include direct intervention to facilitate microlevel adaptations and technological developments, as well as market responses and institutional and policy changes. Although this broad general approach is an excellent starting point for any intervention strategy, knowing what approach could be used most effectively in which area is an important first step in knowing how to deal with the impact of climate change, and a greater understanding of the complex interactions in various areas is also required.

5.2.7 Coping and Adaptation: Can We Rely on Past Experiences?

Much is written about the value of local ecological knowledge (LEK) and about traditional knowledge (some of it folklore). And there is no doubt that LEK has a place in planning of mitigation and adaptation strategies. Some observers point out that the sorts of challenges being met with now bear little resemblance to those dealt with in the past and that it may be asking too much of traditional herders and even long-term farming households to come up with appropriate coping and adaptation strategies to deal with the challenges of global change (including climate change).

5.2.7.1 Traditional Knowledge

The term "traditional knowledge" refers to both individual pieces of information and the traditional "knowledge systems" embedded in indigenous ways of life. Such knowledge emerges from reciprocal relationships between indigenous peoples and place or the "nature-culture nexus." Therefore, traditional communities and knowledge holders have unique ways of knowing, experiencing, understanding, and practicing traditional knowledge. The dynamic and diverse body of knowledge and knowledge systems share common dimensions represented by the term traditional knowledge. Traditional knowledge can encompass culture, experiences, resources, environment, and animal knowledge. This knowledge is accumulated through experience, relationships, and upheld responsibilities toward other living beings and places and is passed down generationally from elder to youth through oral histories, stories, ceremonies, and land management practices. This traditional knowledge is considered by many to be a gift and come with certain responsibilities, such as determining when and with whom they should be shared.

5.2.7.2 Traditional Knowledge and Climate Change

Traditional knowledge is fundamental to proper understanding of climate change (Williams and Hardison 2013) for resilience against and adaptation to climate change, inspiring what Wildcat (2009) called "the ability to solve pressing life issues facing humankind now by situating our solutions in Earth-based local deep spatial knowledge." See Squires and Feng (this volume) for further discussion around this point. This raises the question to what extent past responses of agricultural households to climate variability and other stressors facilitate their adaptation to climate change in the future. There are limitations to drawing lessons in this way: past livelihood strategies and vulnerabilities emerged in a specific political–economic context which is different from the context within which people will adapt to climate change.

5.3 Herders' Perceptions and Attitudes to Land Degradation and to Climate Change

The people (pastoralists and agropastoralists) using the pastures are also concerned about the condition of the resource base (the land and water on which they depend). All the pasture users interviewed stated that vegetation has been getting poorer, referring to both vegetation quantity and quality. Moreover, other signs of land degradation such as wind and water erosion are widely seen on the pastures and threaten livestock keeping and the pasture infrastructure. These phenomena are especially visible in the appearance of large gullies formed by water erosion. A pasture user from the Three Rivers Region of Qinghai said:

> From 1973 I have been using the pasture and never in my life have I seen such gullies.... The problem is that livestock numbers are too high. They are put on the pasture before the grass is fully grown. Grass should root, but here sheep eat the grass before it can grow.

The majority of respondents acknowledge that not only is pasture vegetation decreasing in general, but the composition of plants is also changing. More unpalatable and poisonous plants decrease the pasture productivity. Pastoralists well understand the link between the decrease in pasture productivity and the decrease in their income. Many complained that their livestock are getting thinner or sick. As stated earlier, there is great dependency on animal husbandry with regard to households' livelihoods. This reliance on livestock keeping, alongside the long tradition of animal husbandry, is paramount:

> Listen, the main problem is that livestock are our main source of income. It does not make sense to live on the rangeland if you don't have livestock but if the pastures are bad, we will earn nothing.

Respondents state they have to invest additional financial resources in fodder throughout the winter, as pasture capacities are not sufficient for animals in the spring. A herder informs us:

> Degradation is bad for income as I have to buy extra fodder. That means I have to sell animals if I want to buy fodder.

During our interviews, pasture users usually blamed natural factors, such as rainfall and droughts. The high precipitation rate in the past years increased water erosion, while the droughts in previous years led to the spreading of weeds. Most pasture users named overgrazing as the second leading cause of land degradation, referring to large livestock numbers. We noted that pasture users do not necessarily make a connection between natural factors and overgrazing, although overgrazing affects the soil's ability to hold the vegetation. Only one user of the pastures mentioned mismanagement or inadequate herding practices as the main reason for land degradation and one which can be solved by enforcing pasture rotation. But more pastoralists from the community highlighted the lack of transhumant grazing patterns, leading to degradation, and the urgent need to coordinate pastoral migration. As summer pastures are located at higher altitudes, it is more difficult and costly to reach them. And even when pasture users move to summer pastures, they either ascend fairly late and/or descend very early in autumn. It seems to be a strategy and the dominating pasture use practice to increase livestock numbers and to occupy the most-accessible spring/autumn pastures near the villages and markets, without moving to summer pastures and practicing pasture rotation.

5.3.1 Mobility and Migration

Herd mobility and migration are a response tactic to environmental stressors used by families and clans to find sufficient fodder and crucial for nomadic pastoralism. Irrespective of the system used, however, livestock are central to the livelihoods of most people who rely on them for income from sales of milk, meat, and skins and for protein consumption, draft power, and ritual and spiritual needs, among other uses. Owning livestock is one way by which many people are able to diversify their risk, increase their assets, and improve their resilience to sudden changes in climate, disease outbreaks, and unfavorable market fluctuations.

As droughts increase in intensity and frequency and forage (especially winter feed) becomes insufficient for the expanding livestock inventory, herders are now forced to enlarge the distance of their migration patterns to lands outside their territories to compensate for diminishing resources. Grazing routes for migrating herds/flocks were once specific for territorial clans, and even now there are certain claims to unimpeded access to forage and water along the way. Conflict can erupt if "rights" are not respected by other pastoralist groups. The resulting conflict is a major reason why many women and young children no longer travel with the livestock and families are "semi-settling" within a few kilometers of a village for eight or more months at a time. Nomadic families used to rely on scouts to access water and feed, but new arrangements and government restrictions on transhumance (even transboundary to Pakistan) have made this activity less important.

Government policies, including implementation of the household contract responsibility system to pastoral lands beginning on the QTP the mid to later 1980s sought to allocate grazing user rights householders for specific tracts of rangeland in the expectation that this measure would promote stewardship on the part of the respective householders of the land so allocated. Under a major national project ("Returning Grazing Land to Grassland"), government policies, including implementation of the household contract responsibility system to pastoral lands beginning on the QTP the mid to later 1980s, sought to allocate grazing user rights householders for specific tracts of rangeland in the expectation that this measure would promote stewardship on the part of the respective householders of the land so allocated. Under a major national project ("Returning Grazing Land to Grassland" (*Tui mu huan cao* in Chinese), fencing of rangeland (in association with grazing bans) occurred in a bid to provide better control of livestock, facilitate rotational grazing, and fodder conservation and rangeland improvement. These policies have not had the outcomes that the government expected and may in fact have contributed to the present problems. These policies have not had the outcomes that the government expected and may in fact have contributed to the present problems (Hua et al. 2015; Squires and Hua 2015).

While many herders agree that overgrazing is one of the important causes of degradation, all interviewees would like to increase their herd size and plan to do so in the coming years. This highlights an important phenomenon: *awareness about overgrazing and its effects on pasture and livestock productivity has not been translated into action by pasture users.* The livestock populations had been increasing significantly since 1996, affecting pasture conditions. Winter pastures and the most-accessible spring/autumn pastures were overgrazed, as herders used them throughout the year without pasture rotation. Meanwhile, the most distant summer pastures went unused. Communities across the QTP face this problem of pasture use. The magnitude of overgrazing and deterioration of pastures poses a social dilemma in pasture use. Increasingly herders are putting individual benefit ahead of the good of the community. The situation in pasture use, where herders take individual and collective decisions, is a good illustration of Hardin's tragedy of the commons where "each herder is motivated to add more and more animals because he receives the direct benefit of his own animals and bears only a share of the costs resulting from overgrazing" (Ostrom 1990: 2).

5.4 Natural Resource Protection in the "Three Rivers Zone and the Role of 'Ecological Resettlement"

In recent years, the Chinese government placed high national priority on the QTP ecological problems and improvement of ecological protection. The government vigorously promoted ecological compensation, ecological restoration, and creation of national parks and reserves. Sustainable management of the alpine grassland was set as a major goal. However, the implementation of ecological initiatives faces serious problems, including the impact of past actions to control grazing, demarcate land, and so on. Large numbers of fences covered the rangeland, which fragmented the landscape and blocked migration routes taken by indigenous wildlife and transhumant herders. Fences had impact on biodiversity and were also a source of friction among herders who found that water points were no longer accessible. Ecological awareness has strengthened in recent years, and ecological education has enhanced protection consciousness of people. Many mining activities have stopped,

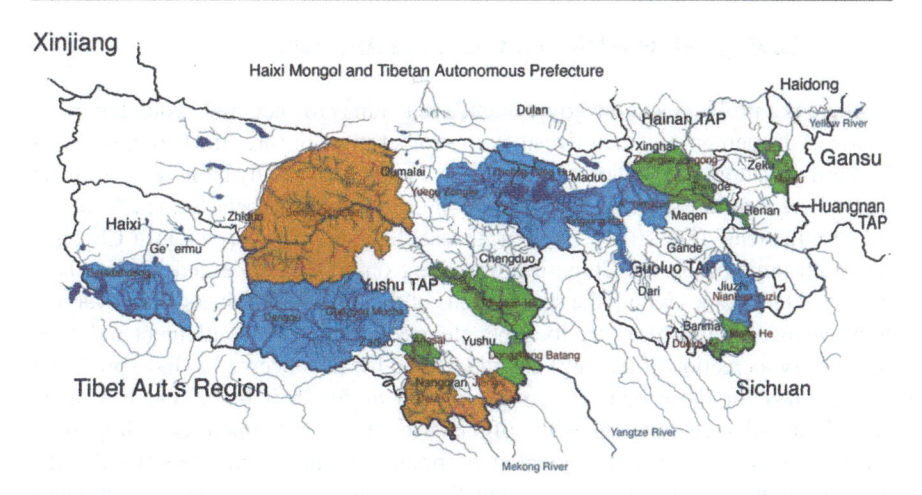

Fig. 5.4 The headwaters of the Three Rivers (see text) are in SW Qinghai, and the major parts of the source area are now part of the Sanjiangyuan National Nature Reserve

fences were removed in some reserves, and funds allocated for ecological restoration have gradually increased. After local people improved their livelihood, they attached more attention to their traditional culture. Until incomes had risen the major preoccupation was on subsistence. The role of traditional culture is very important in ecological protection. As a result, the development of culture and more secure livelihoods brought about an improvement in the ecological environment. All this is well and good but unfortunately there are serious ecological problems on the QTP that have impacts far from the plateau and which the national government is keen to remedy.

A key area is the so-called Three Rivers Region in Qinghai province where three major rise—the Yangtze (Changjiang), the Yellow (Huang He), and the Mekong (also called Lancan Jiang). The Three Rivers Nature Reserve, also referred to as the Sanjiangyuan National Nature Reserve (SNNR) is in Qinghai province. The SNNR was established to protect the headwaters of these three rivers (see map for origin of Yellow and Yangtze rivers). The reserve consists of 18 subareas, each containing three zones which are managed with differing degrees of strictness. The establishment of the nature reserve aims to protect the QTP ecosystem, with an emphasis on alpine swamp meadow and natural habitat of the unique wildlife in the region, and to promote sustainable economic development in the region (Fig. 5.4).

The source area of the Three Rivers has a population of approximately 200,000 people living within its 152,300 km^2—a land area larger than many European countries. Nearly 600 million people who live downstream depend on the proper functioning—and hence the long-term protection—of these rivers for their livelihood. People in the focus region face multiple stressors such as droughts, plant diseases, policy changes, and market fluctuations. Droughts, land-use rights (tenure), and climate variability in general are among the most important stressors (Squires 2012).

5.4.1 Ecological Resettlement: Is It the Answer?

So there is a problem and the government's reaction is to encourage voluntary relocation of the herders and their livestock to another place or force the herder to divest his herd and adopt a sedentarized lifestyle in a village especially built away from the Three Rivers zone. Such action on the part of government has been termed "ecological resettlement" (Du 2012). Ecological resettlement (*shengtai yimin* in Chinese) has been initiated by the Chinese government on a large scale and aims to help degraded landscapes within the source areas of the three major rivers to recover and improve living standards of the local people in SW Qinghai. Resettlement and grazing bans (via erection of fencing) are understood by all concerned to have profound implications for those being resettled, as well as for their home and host area. Overriding all such concerns is the official view that resettlement is an important means of conserving the ecological environment, improving people's livelihoods, and promoting urbanization. The last mentioned point is important from the point of view of provision of government services such as schooling and health and social service payments, pensions, etc. (Hao 2009; Du 2012).

The families of people relocated experienced drastic changes in the lives and livelihoods and identity. They found it hard to adapt. Loss of cultural continuity (Foggin 2011) is a major issue, especially for those who are moved to peri-urban locations. According to interview data, ecological resettlement to some extent improved housing, education, medical care, and transportation, but overall living standard actually fell following relocation (Du 2012). Partly this was because of the inability to continue to be self-sufficient and obtain subsistence from livestock and hunting and gathering and because of the lack of employment opportunities and the income stream so necessary in a market economy.

Some western scholars, scientists, and conservationists looking at ecological resettlement in western China interpret it as government-initiated "permanent resettlement" of seminomadic herders and agropastoralists from fragile ecological environments to new or existing settlements outside those ecologically vulnerable regions as justified (Dickinson and Webber 2007; West 2009) while acknowledging that Tibetans have sustained their livelihoods in these rangeland areas for hundreds of years (Yeh 2009; Qi 2011; Foggin 2011; Ptackova 2012). Others equate it to the resettlement of over 500,000 people from the Yangtze river valley to make way for the Three Gorges Dam—a major project in the national economic interest.

The relocation of rural people away from marginal or fragile lands is an increasingly common approach in China to achieve environmental and development objectives. But few studies have been published on the social impacts of such resettlement projects. Tashi and Foggins (2012) analyze and discuss several key social dimensions of a resettlement project in the Tibetan areas. According to Tashi and Foggin, the term "ecological resettlement" properly reflects and incorporates all three major elements of this fundamentally transformative development strategy: (i) the environmental rationale of the policy (cf. ecological conservation), (ii) the movement of rural people from marginally or ecologically fragile areas (cf. relocation), and (iii) a concomitant change in peoples' livelihoods (cf. sedentarization settlement). The term should be distinguished at the outset from the notion of "ecological refugees"—people who move in response to a change in the environment that can no longer support them or their families. They make a permanent move, i.e., "migrate." However, rural people who leave their homes (often coerced to do so) and move to new homes (and often new livelihoods) relocating in response to a development project have a different mindset. The motivation to relocate is not from the environmental situation per se. The people who live in urban or peri-urban areas where the relocated herders end up have expressed dissatisfaction even alarm at what they see as an "invasion" (Fischer 2008). For the people who are relocated, there is a sense of loss as social networks break down.

5.4.1.1 Social Networks

Social networks are friendship and family ties that provide people with social capital and resource security during stressful times and hardships. This approach is necessary because of increases in population density and close proximity to rival tribes, as well as the rising encroachment of private lands and devastation of livestock numbers. Essentially social networks play a key role in maintaining herd wealth and pastoral livelihoods. Networks may be broad and it was discovered that, although all men and women do not directly know one another, they indirectly connect to each other through a friend, relative, in-law, or herding partnership, which offers friendship, shared labor, protection, and help in time of need.

5.5 Conclusions and Summing Up

Climate change is affecting the culture, health, economies, and lifestyles of ethnic minority peoples on the QTP. QTP's rangelands are dominated by extensive livestock production systems (Miller 2002, 2005; Bedunah et al. 2006), and securing these assets, particularly for poorer households in the face of climate change, is a major challenge. Some of the most important suggestions for how to do this focus on enabling herd mobility through securing better access to water resources and increasing access to more land, particularly when it is marginal for crop production (West,). Other suggestions include the improvement of early warning systems and dissemination of this information, enabling pastoral groups to better engage with policy debates, building stronger conflict management institutions, and supporting a diversification of livelihoods, perhaps through tourism and conservation. The non-equilibrium perspective (Behhke et al. 1993(that depicts ecosystems as existing in multiple stable states (Westoby et al. 1989) depending on both external and internal drivers is a useful paradigm to follow in this regard.

It should be remembered, however, that the impact of climate change is not the only factor that will affect livestock production on the QTP and the rangelands that sustain these activities (Klein et al. 2005). Although these impacts should not be ignored, they should also not be exaggerated. Population growth, rapid rates of urbanization (a major objective in China's 12th and 13th 5-year plans), and land reform initiatives are only a few of many other factors that will influence QTP's rural environments, including the rangelands, in the twenty-first century. Climate change impacts might not all be negative, and a better understanding of where and why responses have been successful is also required to ensure that policy interventions are properly directed.

References

W.N. Adger, S. Agrawala, M.M.Q. Mirza, C. Conde, K. O'Brien, J. Pulhin, R. Pulwarty, B. Smit, K. Takahashi, Assessment of adaptation practices, options, constraints and capacity, in *Contribution of Working Group II to the Fourth Assessment Report of the Intergovernmental Panel on Climate Change*, ed. By M. L. Parry, O. F. Canziani, J. P. Palutikof, P. J. van der Linden, C. E. Hanson, (Cambridge University Press, Cambridge, UK, 2007), pp. 717–743. Chapter 17

C. Alexander, N. Bynum, E. Johnson, U. King, T. Mustonen, P. Neofotis, N. Oettlé, C. Rosenzweig, C. Sakakibara, V. Shadrin, M. Vicarelli, J. Waterhouse, B. Weeks, Linking indigenous and scientific knowledge of climate change. Bioscience **61**(6), 477–484 (2011)

D.J. Bedunah, J. McArthur, E. Durant, *Rangelands of Central Asia: Proceedings of the Conference on Transformations, Issues, and Future Challenges, Proceeding RMRS-P-39* (U.S. Department of Agriculture, Forest Service, Rocky Mountain Research Station, Fort Collins, 2006)

R. Benhke, I. Scoones, C. Kerven (eds.), *Range Ecology at Disequilibrium: New Models of Natural Variability and Pastoral Adaptation in African Savannas* (Overseas Development Institute, London, 1993)

D. Dickinson, M. Webber, Environmental resettlement and development, on the steppes of Inner Mongolia, PRC Journal of Development Studies **43**(3), 537–561 (2007)

F. Du, Ecological resettlement of Tibetan herders in the Sanjiangyuan: a case study in Madoi county of Qinghai. Nomadic Peoples **16**(12), 116–133 (2012)

M.Y. Du, S. Kawashima, S. Yonemura, X.Z. Zhang, S.B. Chen, Mutual influence between human activities and climate change in the Tibetan Plateau during recent years. Glob. Planet. Chang. **41**, 241–249 (2004)

F. Ellis, Household strategies and rural livelihood diversification. J. Dev. Stud. **35**(1), 1–38 (1998)

A.M. Fischer, "Population invasion" versus urban exclusion in the Tibetan areas of western China. Popul. Dev. Rev. **34**, 631–662 (2008)

J.M. Foggin, Rethinking "Ecological migration" and the value of cultural continuity: a response to Wang. J. Hum. Environ. **40**(1), 100–101 (2011)

W. Hagg, Water from the mountains of Greater Central Asia: a resource under threat, in *Sustainable Land Management in Greater Central Asia: A Regional and Integrated Approach*, eds. V. Squires, Lu Qi (Routledge, UK. 2017)

R. Hao, Qinghai province in the perspective of ecological anthropology: ecological crisis, economic development and preservation of cultural identity, in *Disasters, Cultures, Politics: Chinese Bulgarian Anthropological Contribution to the Study of Critical Situations*, ed. By F. Tzaneva, S. Fang, M. Liu, (Cambridge Scholars Publishing, Newcastle, 2009), pp. 12–13

L.M. Hua, S.W. Yang, V. Squires, G.Z. Wang, An alternative rangeland management strategy in an agro-pastoral area in western China. Rangel. Ecol. Manag. **37**(2), 95–97 (2015)

J.A. Klein, J. Harte, X.Q. Zhao, Dynamic and complex microclimate responses to warming and grazing manipulations. Glob. Chang. Biol. **11**(9), 440–1451 (2005)

J.A. Klein, E. Yeh, J. Bump, Y. Nyima, K. Hopping, in *Climate Change Adaptation in Developed Nations: From Theory to Practice*, ed. by J.D. Ford, L. Berring-Ford. Coordinating Environmental Protection and Climate Change Adaptation Policy in Resource-Dependent Communities: A Case Study from the Tibetan Plateau, (Springer, Dordrecht, 2011), pp. 423–438

L.K. Limbu, in *Rangeland and grassland in Nepal's alpine regions: problems and prospects in Rangelands along the Silk road: Transformative Adaptations to Climate and Global change*, ed. by V. R. Squires, Z.H. Shang, A. Ariapour, (Nova Science Publishers, New York, 2017)

R.J. Long, L.M. Ding, Z.H. Shang, X.H. Guo, The yak grazing system on the Qinghai-Tibetan Plateau and its status. Rangel. J. **30**, 241–246 (2008)

S.Z. Ma, M. Peng, G.C. Chen, G.Y. Zhou, Q. Sun, Feature analysis of vegetation degradation on alpine grasslands in the Yellow River source regions. Reg. Pratacultural Sci. **10**, 19–23 (2004.) (in Chinese)

D.J. Miller, The importance of China's nomads. Rangelands **24**, 22–24 (2002)

D.J. Miller, The Tibetan steppe. Plant production and protection series no.34, in *Grasslands of the World*, ed. By J. M. Suttie, S. G. Reynolds, C. Batello, (FAO (UN Food and Agriculture Organization), Rome, 2005), pp. 305–342

E. Ostrom, *Governing the Commons: The Evolution of Institutions for Collective Action*, vol 1990 (Cambridge University Press, Cambridge, 1990), 280 p

J. Ptackova, in *Implementation of Resettlement Programs Among Pastoralist Communities in Eastern Tibet, in Pastoral Practices in High Asia*, ed. by H. Kreutzmann (Springer, Berlin, 2012), p. 217–235

J.Y. Qi, Research on grassland ecological Emigration and cultural adaptability: a case study of Applied Anthropology for the source of the Yellow River. National. Res. Qinghai (Qinghai Minzu Yanjiu), **22** (1) [In Chinese] (2011)

U. Safriel, Z. Adeel, Dryland systems, in *Ecosystems and Human Well-Being, Current State and Trends*, ed. By R. Hassan, R. Scholes, N. Ash, vol. 1, (Island Press, Washington, 2005), pp. 625–658

Z. H. Shang, Q. Dong, A. Degen, R. Long, in *Ecological Restoration: Global challenges, Social Aspects and Environmental Benefits*, ed. by V. R. Squires. Ecological restoration on Qinghai-Tibetan Plateau: Problems, strategies and prospects, (NOVA Science Publishers, New York, 2016), pp. 151–176

V.R. Squires, *Rangeland Stewardship in Central Asia: Balancing Livelihoods, Biodiversity Conservation and Land Protection* (Springer, Dordrecht, 2012.) 480 p

V.R. Squires, L M. Hua, On the failure to control overgrazing and land degradation in China's pastoral lands: Implications for policy and the research agenda, in *Rangeland Ecology, Management and Conservation Benefits* (Nova Press, N.Y. 2015), pp. 19–42

H. Sun, D. Zheng, T. Yao and Y Zhang, Protection and construction of the national ecological security shelter zone on Tibetan plateau. Acta. Geographica. Sinica. (1): 3–12 (2012)

B. Tache, G. Oba, Is poverty driving Borana herders in southern Ethiopia to crop cultivation? Hum. Ecol. **38**, 639–649 (2010)

X. Wang, On the formation reasons of Eco-refugee from a geography perspective in the source of yellow river, lancang river and yangtze river. Ecol. Environ. **9**, 131 (2006b.) [In Chinese with English Abstract]

P. Wang, J.P. Lassoie, S.L. Morreale, S.K. Dong, A critical review of socioeconomic and natural factors in ecological degradation in the Qinghai-Tibetan Plateau, China. Range. J. **37**, 1–9 (2015)

J.J. West, *Perceptions of Ecological Migration in Inner Mongolia, China: Summary of the Work and Relevance to Climate Adaptation*. CICER (Center for International climate and Environmental research) Report. 2009: 04 (2009)

M. Westoby, B. Walker, I. Noy-Meir, Opportunistic range management for rangelands not at equilibrium. J. Range Manag. **42**, 265–273 (1989)

K.P. Whyte, Indigenous Peoples, Climate Change Loss and Damage, and the Responsibility of Settler States (Michigan State University, East Lansing, 2016) 20p. https://www.academia.edu/7485990/Indigenous_Peoples_Climate_Change_Loss_and_Damage_and_the_Responsibility_of_Settler_States. Accessed 30 July 2016.

D.R. Wildcat, *Red Alert! Saving the Planet with Indigenous Knowledge* (Fulcrum Publishing, Golden, 2009.) 128 p

T. Williams, P. Hardison, Culture, law, risk and governance: contexts of traditional knowledge in climate change adaptation. Clim. Chang. **120**, 531–544 (2013)

T. Yan, W.Y. Qian, Environmental migration and sustainable development in the upper reaches of the Yangtze River. Popul. Environ. **25**(6), 6 (2004)

T. Yao, J. Pu, A. Lu, Y. Wang, W. Yu, Recent Glacial Retreat and Its Impact on Hydrological Processes on the Tibetan Plateau, China, and Surrounding Regions. Arctic, Antarctic, and Alpine Research **39**(4), 642–650 (2007)

E.T. Yeh, Greening western China: a critical view. Geoforum 40, 884–889 (2009)

G. Yu, C. Lu, G. Xie, Z. Luo, L. Yang, Grassland ecosystem services and their economic evaluation in Qinghai-Tibetan plateau based on RS and GIS. PlosOne **8**(3), 56–58 (2013)

G. Zhang, T. Yao, H. Xie, S. Kang Y. Lei, Increased Mass over the Tibetan Plateau: From Lakes or Glaciers? *Geophysical Research Letters* 40: 2125–2130 (2013)

Further Readings

W.N. Adger, S. Huq, K. Brown, D. Conway, M. Hulme, Adaptation to climate change in the developing world. Prog. Dev. Stud. **3**(3), 179–195 (2003)

W.N. Adger, P.M. Kelly, Social vulnerability to climate change and the architecture of entitlements. Mitig. Adapt. Strateg. Glob. Chang. **4**, 253–266 (1999)

W.N. Adger, K. Vincent, Uncertainty in adaptive capacity and the importance of scale. Compt. Rendus Geosci. **337**, 399–410 (2005)

A. Byg, J. Salick, Local perspectives on a global phenomenon: climate change in eastern Tibetan villages. Glob. Environ. Chang. **19**, 156–166 (2009)

Y. Cao, A Study of the Impact Herdsman Economic Activities on the Ecological Environment in Sanjiangyuan Area, Qinghai University, Master Dissertation [In Chinese with English Abstract], (2016)

S.B. Chen, Y.F. Liu, A. Thomas, Climatic change on the Tibetan Plateau: potential evapo – transpiration trends, 1961–2000. Clim. Chang. **76**, 291–319 (2006)

G.D. Cheng, T.H. Wu, Responses of permafrost to climate change and their environmental significance, Qinghai-Tibet Plateau. J. Geophys. Res. Earth Surf. **112**, F02S03, doi:10.1029/2006JF000631. (2007)

R.P. Cincotta, Y.Q. Zhang, X.M. Zhou, Transhumant alpine pastoralism in north eastern Qinghai province: an evaluation of livestock population response during China's agrarian economic reform. Nomadic Peoples **30**, 3–25 (1992)

G.E. Clarke, *China's Reforms of Tibet and Their Effect on Pastoralism*, IDS discussion. Paper no. 237, (Brighton, 1987)

G.E. Clarke, *Development, society and environment in Tibet, and socio-economic change and the environment in a pastoral area of Lhasa municipality*, in Bob Clarke (ed) Development, Society, and Environment in Tibet (Graham Publishing, Denver, 1998)

Climate and Traditional Knowledges Workgroup [CTKW], Guidelines for considering traditional knowledges in climate change initiatives (2014), http://climatetkw.wordpress.com/. (29 April 2016)

G. Davidson, R.H. Behnke, C. Kerven, Implications of rangeland enclosure policy on the Tibetan plateau. UPDATEMagazineNo.2,October2008. International Human Dimensions Programme on Global Environmental Change (2008), pp. 59–62

M. Domros, D. Schafer, *Recent climate change in China–statistical analyses of temperature and rainfall records, TerraNostra*, 2003/6. (Klimavariabilitat, Schriftender Alfred-Wegener-Stiftung, Berlin, 2003). pp. 371–374

S. Dong, Characteristics and problems of western China's eco-economic regions. Resour. Sci. **27-6**, 104 (2005.) [In Chinese with English Abstract]

J.M. Foggin, Depopulating the Tibetan grasslands: national policies and perspectives for the future of Tibetan herders in Qinghai Province, China. Mt. Res. Dev. **28**(1), 26–31 (2008)

O.W. Frauenfeld, T.J. Zhang, M.C. Serreze, Climate change and variability using European Centre for Medium-Range Weather Forecasts reanalysis (ERA-40) temperatures on the Tibetan Plateau. J. Geophys. Res. **110**, –D02101 (2005)

X.Q. Gao, Preliminary analysis of climate changes in some regions of the Northern part of the Qinghai Xizang Plateau on a scale of decades, in *Evolution and Changes in the Geology, Environment and Ecosystems of the Tibetan Plateau* (Science Press, Beijing, (in Chinese), 1995), p. 297–303

Q.Z. Gao, Y. Li, H. Xu, Y. Wan, S.K. Dong, Y.L. Wang, Adaptation strategies of climate variability impacts on alpine grassland ecosystems in Tibetan plateau. Mitig. Adapt. Strateg. Glob. Chang. **19**, 199–309 (2010)

M.C. Goldstein, C.M. Beall, Changing patterns of Tibetan nomadic pastoralism, in *Human Biology of Pastoral Populations*, ed. By W. R. Leonard, M. H. Crawford, (Cambridge University Press, Cambridge, 2002), pp. 131–150

M.C. Goldstein, C.M. Beall, R.P. Cincotta, Traditional nomadic pastoralism and ecological conservation on Tibet's Northern Plateau. Nat. Geogr. Res. Rep. **6**, 139–156 (1990)

A.L. Gu, Biodiversity of Tibet's rangeland resources and their protection, in *Tibet's Biodiversity: Conservation and Management*, ed. By N. Wu, D. J. Miller, Z. Lu, J. Springer, (China Forestry Press Publishing House, Beijing, 2000), pp. 94–100

S.Z. Gu, Approaches for sustainable development of agriculture and animal husbandry in Tibet. Resources Science **22**, 44–49 (2000b.) (in Chinese)

R.B. Harris, Rangeland degradation on the Qinghai-Tibetan plateau: a review of the evidence of its magnitude and causes. J. Arid Environ. **74**, 1–12 (2010)

P. Ho, Ownership and control in Chinese rangeland management since Mao: a case study of the free-rider problem in pastoral areas in Ningxia, in*Cooperatives and Collective in China's Rural Development: Between State and Private Interests*, ed. By E. B. Vermeer, P. FN, W. L. Chong, (M.E. Sharpe, Armonk, 1998), pp. 196–235

X.Y. Hou, J.Z. Shi, *Pastoral Grasses of Western China* (Chemical Industry Press, Beijing, 2002.) (in Chinese)

Z.Z. Hu, D.G. Zhang, China's pasture resources, in *Transhumant Grazing Systems in Temperate Asia. Plant Production and Protection Series 31*, ed. By J. M. Suttie, S. B. Reynolds, (Food and Agriculture Organization of the United Nations (FAO), Rome, 2003), pp. 81–133

Intergovernmental Panel on Climate Change [IPCC], *Climate Change 2007: Impacts, Adaptation and Vulnerability. Contribution of Working Group II to the Fourth Assessment Report of the Intergovernmental Panel on Climate Change* (Cambridge University Press, New York, 2007.) 976 p

J. Jin, S.X. Li, G.D. Cheng, S.L. Wang, X. Li, Permafrost and climatic change in China. Glob. Planet. Chang. **26**, 387–404 (2000)

X.C. Kang, L.J. Graumlich, P.R. Sheppard, Construction and preliminary analysis of a 1835 year sequence of annuli in the Dulan area of Qinghai. Chin. Sci. Bull. **42**, 1089–1091 (1997.) (in Chinese)

X. Liu, B. Chen, Climatic warming in the Tibetan Plateau during recent decades. Int. J. Climatol. **20**(14), 1729–1742 (2000)

W. Liu, Q. Gui, Y. Wang, Temporal-spatial climate change in the last 35 years in Tibet and its geo-environmental consequences. Environ. Geol. **54**, 1747–1754 (2008)

H. Ma, Explanation about the causes of ecological degradation in the "Sanjiangyuan" district under the perspectives of new institutional economics. Tibetan Stud. **3**, 88 (2007)

D.J. Miller, Nomads of the Tibetan plateau rangelands in western China, part three: pastoral development and future challenges. Rangelands **21**(2), 17–20 (1999)

D.J. Miller, Tough times for Tibetan nomads in western China: snowstorms, settling down, fences and the demise of traditional nomadic pastoralism. Nomadic Peoples **4**(1), 83–109 (2000)

D.J. Miller, D.J. Bedunah, D.H. Pletscher, R.M. Jackson, From open range to fences: changes in the range-livestock industry on the Tibetan Plateau and implications for development planning and wildlife conservation, in *Proceedings of the 1992 International Rangeland Development Symposium, February 11–12, 1992*, ed. By G. K. Perrier, C. W. Gay, (Spokane, Washington, 1992), pp. 95–109

U. Safriel, Z. Adeel, Development paths of drylands: thresholds and sustainability. Sustain. Sci. **3**, 117–123 (2008). doi:10.1007/s11625-007-0038-5

V.R. Squires, H Y. Feng, *Humans as Agents of Change in Arid Lands: With Special Reference to Qinghai-Tibet Plateau, China.* this volume (2017), p. xx

G. Tashi, M. Foggin, Resettlement as Development and Progress? Eight years on: Review of emerging Social and Development Impacts on an "Ecological Settlement" project in Project in Tibet Autonomous Region, China. Nomadic Peoples **16**(10), 134–151 (2012)

D. Tsering, The theoretical relations between the poverty oriented resettlement and environmental migration. Tibet. Stud. **5**, 45 (2014.) [In Chinese with English Abstract]

X. Wang, Studies on ecological refugee in the source of Yellow River, Lancang River and Yangtze River. J. Qinghai Nat. Univ **17**, 23–31 (2006a.) [In Chinese]

K.P. Whyte, Justice forward: tribes, climate adaptation and responsibility. Clim. Chang. **3**, 517–530 (2013)

X.U. Xiaoling, Research on ecological vulnerability change and interaction development model of economy and ecology in Sanjiangyuan Region, Shanxi Normal University, PhD dissertation, (2007)

X.J. Xun, H Q. Zhang, in *An Overview of the General Plan for the Ecological Protection and Construction of Sanjiangyuan Nature Reserve,* ed. by Editing Committee of Ecological Protection and Construction of Sanjiangyuan Nature Reserve). [In Chinese], (2007)

E.T. Yeh, Tibetan range wars: Spatial politics and authority on the grasslands of Amdo. Dev. Chang. **34**(3), 1–25 (2003)

E.T. Yeh, Green governmentality and pastoralism in Western China: 'Converting pastures to grasslands'. Nomadic Peoples **9**(1), 9–29 (2005)

E.T. Yeh, *Taming Tibet: Landscape Transformation and the Gift of Chinese Development* (Cornell University Press, Ithaca, (2013)

Integrated Watershed Management Approach in Drylands of Iran

Hamid R. Solaymani and Hossein Badripour

6.1 Introduction: The Context

Iran is the second-largest country in the Middle East. It is bordered by Iraq, Turkey, Armenia, Azerbaijan, Turkmenistan, Afghanistan and Pakistan. Total land area is about 1,648,000 km^2. More than 90% of the land area in Iran is classified as arid or semiarid. The mean annual precipitation of the 87 million ha of the mountainous regions and 77.8 million ha of the plains areas are 365 mm and 115 mm, respectively. Approximately one-half of Iran's water supply comes from surface waters, with most of the remainder coming from groundwater aquifers which are significantly overdrawn. Therefore, water shortage is an ever present threat to most of Iran.

Despite having less than 1% of the world population, Iran has at its disposal less than 0.36% of the world's fresh water, and it suffers from an uneven distribution. The per capita water 40 years ago was 4000 cm/year (cubic metre per year), this has currently been reduced to 1700 cm/year and will further reduce to less than 900 cm/year in upcoming next 40 years (Madani 2014). The Iranian water shortage is not only limited to surface water but also the subsurface waters are experiencing overdraft (Döll et al. 2014; Gleeson et al. 2012). Many researches have reported the groundwater to be extremely critical in some parts of the country (Bagheri and Hosseini 2011; Hojjati and Boustani 2010; Izady et al. 2012; Soltani and Saboohi 2009).

Soil and water conservation is the cornerstone and present crucial factors to sustain the natural resources and food production in the world (Molden, 2007). The exploitation and management of these factors follow the various patterns across the globe. Whereas in some states of the world proper management practices are employed, in others these vital resources are facing collapse (Solaymani and Haghighi 2010). The emerging crises in most parts of the world as a result of serious water scarcity are the immediate results of increased population growth (Statistical

---

H.R. Solaymani (✉) • H. Badripour
Forest, Range and Watershed Management Organization, Tehran, Iran
e-mail: hrsolaymani@yahoo.com; badripour@yahoo.com

© Springer International Publishing AG 2018
M.K. Gaur, V.R. Squires (eds.), *Climate Variability Impacts on Land Use and Livelihoods in Drylands*, DOI 10.1007/978-3-319-56681-8_6

Yearbook 1986, 1997; Giam Pietro and Pimental 1993), negative impacts of the climate change and decline of the water quality because of the water mismanagement. Compounding water and land mismanagement make the people more vulnerable to extreme climate conditions and leads to widespread over exploitation of natural resources.

Soil surface and gully erosion, because of the land-use change, overgrazing and cropping on hill slopes, can threaten the future of the land productivity. Water quality becomes impaired when coupled with high sediment levels and contamination. The capability of water resource management agencies constrains development of sustainable land use in some regions. Clearly, actions must be taken on many fronts to develop sustainable solutions and improved management of land, water and all the detail relevant in dryland environments.

Formulation of region-specific practices at watershed scale gains increasing importance in the face of the global variations in climatic and demographic conditions (Brooks et al. 1992, 1994, 1997; Gregersen et al. 1987, 1996; National Research Council 1999). According to a report prepared by the Dialogue on Water, Food and Environment, a consortium of international organizations concerned with the status of world's water resources, about 2.7 billion people, or nearly one-third of the world's population, will live in regions of acute water scarcity by 2025 (Brooks et al. 1997).

6.2   Water: A Resource Under Threat in Iran

Under Iran's drylands climatic conditions, water is one of the most scarce and most vital resources that have witnessed drastic quantitative fluctuations in recent years. Natural resources in the arid regions in particular have been experiencing wide variations in the timing and amount of stream flows. Water flows under gravity and easily escapes access. This situation calls for a firm planning for prudent water consumption in the years to come. Integrated watershed management has a major role to play (see Sect. 6.3 below).

Water inflicts great damage to human beings and property during floods and disrupts and endangers life during droughts. It is, therefore, essential to establish a balance between supply and demand prior to, and as a preparation for, the emergence of water crises. Human intervention in this water cycle leads either to production or to regression and deterioration of these resources. It is evident that human intervention in natural cycles can have both negative and positive effects, Iran experienced a high birth rate in the decade from 1976 to 1986. The dire consequences of this rather high population explosion and the war from 1980 to 1988 are currently unfolding in all social and environmental aspects (Solaymani and Haghighi 2010). Planners, researchers and managers in their concerted efforts to preserve the natural environment and its resources for future generations are busily engaged in finding sustainable ways to utilize soil, water and natural resources. The farmlands and natural resources areas are the hotspots of poverty, malnutrition and food insecurity and are prone to severe land degradation and water insecurity (Rockström et al.

2007; Wani et al. 2007). Poor social and institutional infrastructures are characteristic of such places.

Just as population increases, overexploitation and the employment of heavy machinery can destroy natural resources, well-defined and thought-out management and resource utilization practices can bring with them higher yields, more productivity and sustainability. It is the coexistence of soil, water and human resources that requires a sound management to create a balance in life, production, vegetation cover and forest; without the human element, nature will find a balance of its own.

6.3 Integrated Watershed Management (IWM)

Water resources management is of special importance because of its vast temporal and spatial dimensions, its relationships with both natural and social laws, its interactions with governmental and non-governmental organizations and its role in food security, social services and community infrastructures. Water resources management involves the management of the geo-system, the bio-system and the human community system (Squires, Milner and Daniell, 2014). That is why water resources management must of necessity include the management of the whole water cycle. Such a management system will involve coordination and cooperation in the management of the soil and water to achieve specific objectives. For example, harvesting reliable and pollution-free water; allocating timely delivery of water for irrigation purposes; supplying adequate drinking water; supplying for industrial, energy and environmental water demands; soil erosion control in catchments; and flood and sediment control and increasing the effective life of dam structures. There are different aspects of this management that involve not only the various sectors but also the community and its culture. This vast and inclusive phenomenon will necessarily call for a national and comprehensive approach.

Watershed is defined as a natural unit for the management and preparation of land. The management of this natural unit involves the management of environmental resources aimed at maximum productivity at the least social, environmental and economic cost. Integrated watershed management is coordinated management of economic, social, biological and physical systems with the least negative effects on the resources while also securing and supplying for the benefits of the community (Powell 1986; Burton 1988; Solaymani and Haghighi 2010) (Fig. 6.1).

Management normally deals with establishing coordination among disparate entities and/or components with casual reference to the role of the manager. This approach, however, fails to distinguish the role played by the society and its culture in management. IWM entails coordination and cooperation as well as the management of soil and water for the attainment of several objectives. There are numerous examples and regulations to be cited from countries like the USA, India and Australia that clearly depict measures and activities directed at the management of soil, water and other environmental resources.

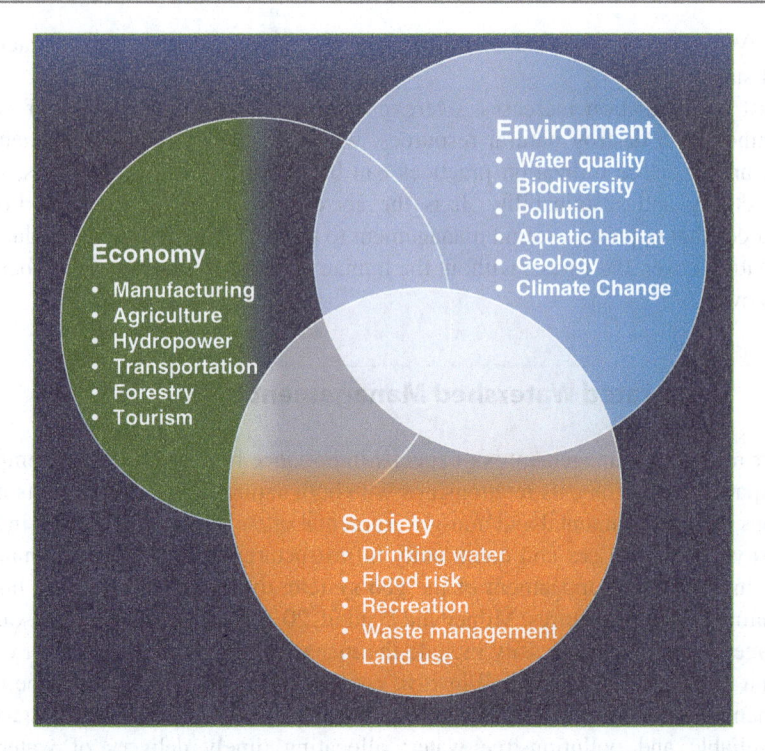

Fig. 6.1 The details of the IWM (http://conservationontario.ca/what-we-do/what-is-watershed-management/integrated-watershed-management)

Box 6.1 MENARID in Iran

MENARID Integrated Watershed Project

MENARID's (Middle East and North Africa Regional Programme for Integrated Sustainable Development) international project

In the framework of the MENARID project, integrated watershed management practices have been implemented in four demonstration sites located in four provinces: Semnan, Kermanshah, Yazd and Sistan-Baluchistan. Those measures include soil and water conservation practices, improved farming practices (organic, pest management, irrigation system) and improved livestock management (permitting systems, cooperative grazing, rangeland conservation).

In order to monitor the project, to justify its utility and to implement payment for environmental services (PES) mechanisms, it is essential to find a method to assess environmental benefits from IWM practices carried out or to be carried out in the demonstration sites. It is not simple to assess the performance of integrated natural resource management practices given that such systems generally have multiple scales of interaction and response: high frequency of nonlinearity, uncertainty, time lags, multiple stakeholders with contrasting objectives and a high degree of context specificity.

Integrated watershed management, as a new concept, calls for the collaborative activities of an assortment of expert groups including hydrologists, hydraulic engineers, soil conservationists, natural resources experts, geologists and a multitude of other engineers, experts and social scientists in a concerted effort assisted by local communities to seek ways of preventing further losses and degradation of natural resources. The MENARID project in Iran (Box 6.1) is an example of such an attempt to engage experts, land users and water managers in an integrated approach to deal with several objectives.

There is universal consensus among all those involved that, on a large scale, it is essential to exercise governmental authority in watersheds so as to regulate and define land use in such a way to observe the overall capacity of the watershed, to control water quality and quantity and to minimize soil erosion and sediment deposition. However, most experts and practitioners maintain that, on a smaller scale, operations such as erosion and sedimentation control in smaller watersheds must be implemented as executive and corrective measures in those regions. Salinity control and supervision for optimal use of the resources may also be included under the rubric of corrective measures.

Recognizing the watershed as the best unit for planning, development and resource management encourages appropriate and proper institutional arrangements in the watershed for a comprehensive management system. The different stages of management in the watershed require different institutional arrangements. The different stages include production, balance and sustainability of the resources, harvesting, distribution, utilization and consumption. There are numerous factors involved in resource management that make watershed management a quite complicated task during the production and sustainability stages, requiring the coordination and collaboration among many organizations and sectors. Although initially managed easily with the erection of control infrastructures and water transfer systems, the stage of development, harvesting and distribution faces more and more management complexities with increasing control on resources, increasing prices, emergence of environmental problems, pollution of resources and overexploitation of groundwater resources. When these issues are integrated with consumption and reuse management into a unified whole, then the need for more coordinated arrangements among the sectors and executive bodies involved becomes evident while the differences and conflicts for meeting the dynamic demands are also disclosed.

6.4 Characteristics and Future Vision of Drylands in Iran

Drylands of Iran face numerous challenges linked to desertification, population growth, climate change and groundwater overharvesting and mismanagement. Changing land uses and practices such as the transformation of rangelands and forests systems to cultivated croplands and residual area, wasteful and unsustainable water use, inappropriate cultivation and overgrazing practices and the overharvesting of wood fuel are leading to land degradation and desertification, water shortages and major losses of environmental services. But Iran is not the only country that is affected by water management challenges (Fig. 6.2).

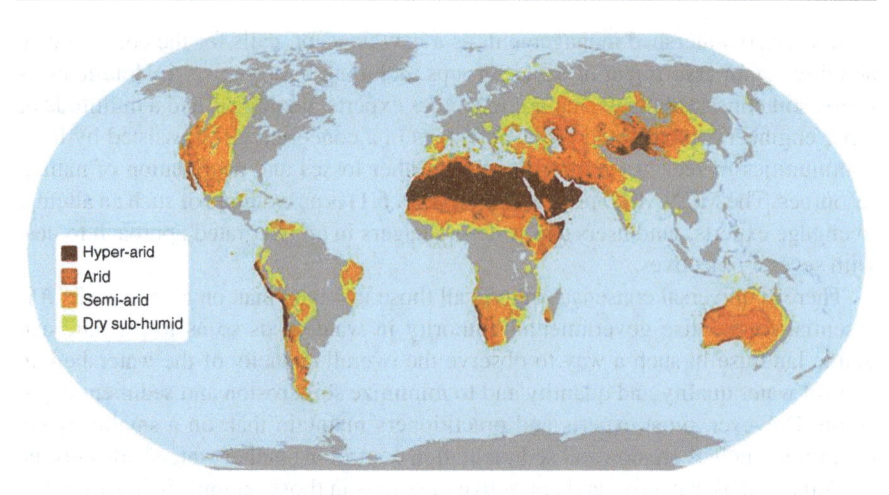

Fig. 6.2 World's drylands and subtypes (Adapt: https://www.unep-wcmc.org/)

Iran with over 1.4 million km^2 is 90% dryland that can be divided into three aridity zones: arid, semiarid and dry subhumid according to criteria set out by UNEP (1992). Many researchers have reported the groundwater to be extremely critical in some parts of the country (Bagheri and Hosseini 2011; Hojjati and Boustani 2010; Izady et al. 2012; Soltani and Saboohi 2009).

The next limiting factor is *soil*, whose interactions with water and plant life are essential. Soil is being lost at an alarming rate. Soil erosion is the one major problem in Iran. Soil washed into rivers can cause flooding and further erosion. This disturbs the river habitats, and the sedimentation downstream affects the storage capacity of streams, lakes and reservoirs. Sediment in the river and reservoir causes environmental problems that have undesired effects on water quality and create large losses and dangers. On the basis of measured data and monitoring in a number of reservoirs, the country will annually be facing 250 MCM (million cubic metre) of sedimentation in the reservoirs and 400 MCM of sedimentation downstream of dam reservoirs and irrigation networks (Sharifi and Haydarian 1999). The analysis of erosion values in various climates showed that the maximum erosion rate occurs in the semiarid and dry subhumid of Iran (Sharifi and Haydarian 1999). A possible explanation for this might be because of the major agricultural activities and land-use changes in dry subhumid area (Fig. 6.3).

The sedimentation production from dryland watersheds in Iran has increased over the three or four decades (ending 2010) because of the intensified soil erosion and flood events. The sediments have not only inflicted destructive damages on different water storage and transfer facilities but also have caused adverse geomorphologic changes in rivers as well. This will naturally result in flooding risks in agricultural lands and industrial and municipal facilities in the neighbouring areas due to sediment deposition and reduced flow capacity. The environmental impacts due to sedimentation such as adverse effects on water quality and aquatic habitats as well as raising of wetland beds and natural lakes are alarming signals of enormous

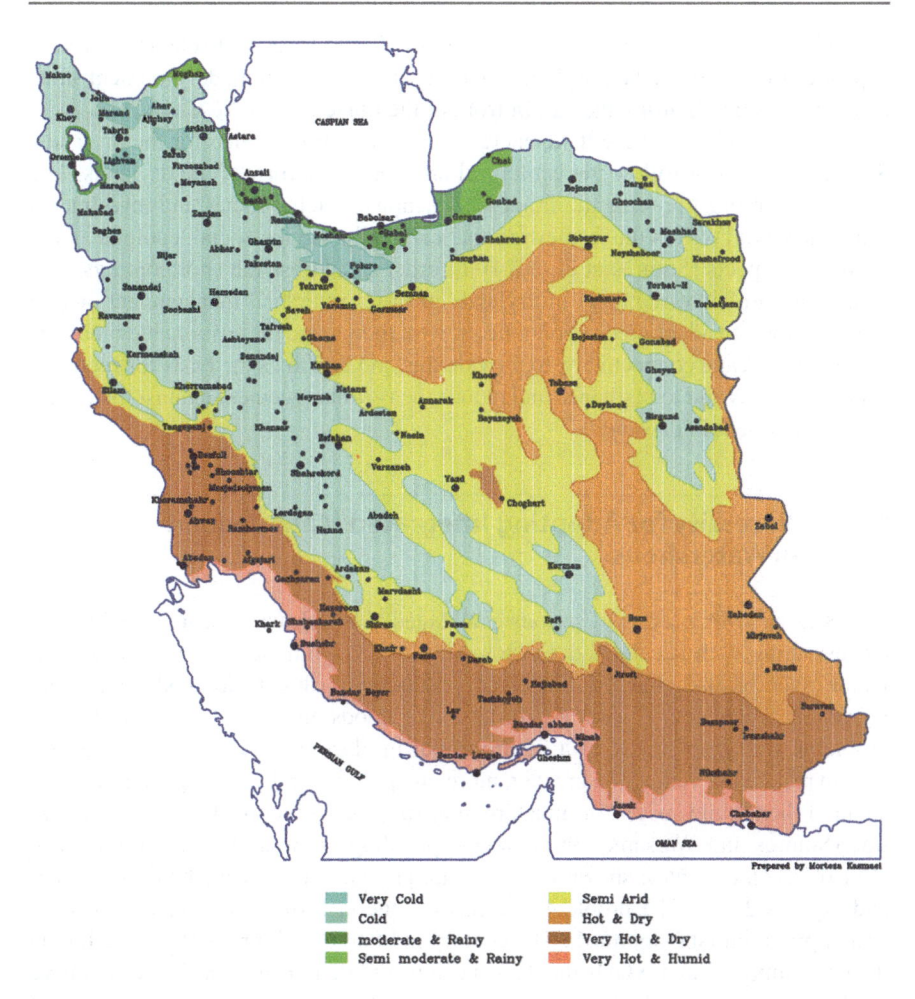

Fig. 6.3 Iran's drylands distribution based on the climate zone (Source: http://climaticdesign.net/?page_id=312)

nationwide economic havoc. During only one flood event, approximately 10% of the storage capacity of Golestan dam reservoir was lost due to sediment deposition. The immediate economic damage being estimated at around $150 million (Sharifi et al. 2002) but the reduction in storage capacity continues.

The drylands are different in a number of important ways from humid lands. In many cases, development pathways for the drylands are driven by a distorted idea of how drylands should or could exist. Concepts of greenness and 'greening the desert' often expose a deep misunderstanding of dryland environments. There are also many examples of efforts to 'green' drylands that have been environmentally damaging (Keck 2012) diverting attention away from sustainable and adaptive management of limited resources. The consequences of these diversions are that rather than adapting development strategies to sustain the drylands, considerable effort is expended on trying to adapt drylands to be capable of sustaining development strategies. This is like 'putting the cart in front of the horse'.

In order to address these deficiencies, a clear vision is needed for the drylands. Such a vision must follow the ecological and social realities of the drylands. The agreed vision must provide the framework against which policies and investments can be judged. IWM is the framework that can equipoise the social, ecological and economic aspects (Squires et al. 2014). In order to address the shortcomings (e.g. water and rainfall variability) in drylands in Iran, watershed management can be used as a tool, to adapt green economic growth, to manage the sustainable biodiversity and landscape connectively, to provide the land health as the basis for food and water securities and to resilience and reduce the risk management against the uncertain environment in dryland.

6.5 Necessity for Adopting Integrated Management in Watersheds

Iran is faced with a series of serious challenges of which water scarcity is high on the list along with accelerated soil loss from the catchment areas (watersheds of major river basins), loss of rangeland productivity, threats to biodiversity and increasing frequency of natural disasters, like floods and landslides. All of these problems are symptomatic of a failure to put in place a workable national action plan to realize the vision of a greener, more prosperous rural landscape. Past efforts, some of them quite successful in addressing specific problems like sand dune fixation (Squires and Warsame 1998), water spreading (Kowsar 1991), gully control (Nejad 2013), etc., have shown that there are proven practices available (Heshmati and Squires 2013). The problem Iran faces is the lack of a coordinated and integrated programmatic approach. The country is littered with projects funded by the donor community and NGOs that have been useful demonstrations but which have not been replicated and scaled up. Attention needs to be given to identify the root causes of land degradation (rather than just the symptoms). There are direct drivers and indirect drivers; both categories may be amenable to manipulation via a coordinated approach such as IWM. There are many examples of successful application of IWM (Squires et al. 2014; Brooks et al. 1994, 1997).

Although agriculture and livestock production have a history of over 5000 years in Iran, evidence shows that the major trend in soil erosion and flood flows is of a recent history of no more than 100 years. Like in other parts of the world, there have been great efforts in Iran being put to the development and proper utilization of soil and water resources. Introduction of mechanization and machinery in these areas and easy access to new equipment have enabled individuals and local communities to exercise a more rapid influence on nature. However, some have brutally invaded nature either for their base living or merely for more profits. Overgrazing of rangelands became more serious after the rangelands were nationalized because a situation was created where privately owned livestock are raised on public land. Local householders increased their flock/herd size as one of the few ways available to them to increase wealth or even to maintain a subsistence level. The regulatory framework has never been tailored to the problem faced.

Official statistics indicates that Iran once had an estimated forest area of 19.5 million ha. Today, this has dropped to about 12.4 million ha, while there has also been a drastic qualitative decline in forest areas. During the years 1985–1995, the poor, average and good quality forest ranges in Iran decreased from 14, 60 and 16 million ha to 9.3, 37.3 and 43.3 million ha, respectively (Sharifi and Haydarian 1999). It can thus be claimed that forests and ranges in Iran have paved a regressive path. This is why Iran is more fragile and susceptible than so many parts of the world.

Iran has a long tradition of water management throughout its history of civilization. More than 80,000 operational and non-operational qanats (underground water galleries), dams and weirs and a lot more of historical water structures have been reported by historians to exist along with a multitude of forests (Ahmadi et al. 2010). But sadly today, we are witnessing an imbalance in our environment. The British Consulate in Iran reports in his travel accounts the names of the then thick forests in Sarakhs region, a region that is currently barren and dry (cited by Sharifi and Haydarian 1999). The elders in many parts of the country such as Fars, Kurdistan and Azerbaijan have also reported of forests and plant covers so thick that they were impassable.

Population growth and need for food from the main sources of income in watershed area (agriculture and animal husbandry) overexploit the land. A decrease of vegetation cover and deforestation due to overgrazing and cutting the trees for fuels or cash (or to convert to arable land) results in more flooding and landslides and destroys infrastructures. The climate change because of the human activities has also changed rainfall pattern and has increased the risk. Based on the impact assessment research, projections are for increased pressure on water and food securities in Iran (Solaymani and Gosain 2014; Abbaspour et al. 2009; Evans 2009; Jamali et al. 2012; Zarghami et al. 2011).

Implementation of watershed management practices and rehabilitation of vegetation cover will provide the conditions for a proper resource exploitation led by scientific management. It is essential to know that rehabilitation is not the final job and subsequent exploitation must be guided by technical and scientific considerations. Otherwise, degradation and desertification will occur again and the investments will be wasted. It is even more essential to prevent rather than cure, i.e. technical consultation must be sought in the maintenance and conservation of catchments prior to their degradation.

The relationship between environment change, land degradation, natural disasters, poverty and climate change is evolving and gradually being better understood. This chain of events is a major threat to the development of the poorest and most marginalized people. This chain, if not controlled, creates a moving vicious cycle, which will be growing and expanding, and more people will be affected by it. The disruption in the natural environment not only leads to desertification and drying of the land but influences the constant and variable features of the strata in the biosphere as well. The air, soil and subsoil strata, plant and animal communities and even the human legacy on the earth are all endangered by this disruption. The only way to reverse or to slow down this dangerous trend is a well-defined management system and a set of proper activities (Solaymani and Haghighi 2010).

The establishment of Ministry of Jihad Construction in 1979 initiated the establishment of large-scale rural development activities. Parallel to the infrastructure development, the shortcomings and constraints in water resources development in the country have led the decision-makers to select agricultural developments as the key and pivotal indicator of development in general. Thus, a large-scale programme of dam construction and water resource development has been initiated to meet the demands of agricultural development in drylands. Unfortunately, the activities conducted in some area could not stop land degradation and migration and in some cases accelerated the process. Ignorance of proper methods of watershed management and overexploitation by people in some marginal areas causes these watersheds to soon change into dry deserts (Solaymani and Haghighi 2010).

Lack of a comprehensive or holistic approach, improper interventions and inadequate legislation with varied conflicting regulations are among the major causes for the natural resource degradation occurring in Iran. As an example of this devastating trend, rather than developing plans proportionate to watershed capacity, a plan was developed to secure self-sufficiency in meat production that inevitably increased livestock numbers from three to seven times the carrying capacity of ranges (Sharifi and Haydarian 1999). This was while another plan was developed for self-sufficiency in agricultural products that changed more rangeland into rain-fed cropland in drylands in Iran (Lal 1984; IECA 2000; FRWO 2001). The two conflicting plans had drastic effects on the degradation of rangelands in Iran (but fortunately, a Livestock Balancing Act has been recently developed and approved that will hopefully reverse the trend). Other destructive activities include the application of excessive machinery on land, eradication of bushes, deforestation and improper land-use changes. These have cumulatively resulted in more flood occurrences (a tenfold increase) with their subsequent droughts and groundwater level decline in most areas (FRWO 2001).

The gradual salinization of soil, the devastation of vegetation cover, reduced soil productivity, increased erosion and sedimentation and increasing chemical pollution were first noticed in Australia where, in 1985, the integrated watershed management was adopted and watershed was recognized as a management unit. In the USA, the change in the course of the Kimi River located in the Mississippi drainage basin and the reduction of its length by half resulted in the devastation of 18,000 ha of wetland with the subsequent extinction of valuable fish species, increased environmental pollution and loss of beautiful natural sites. This fatal event led to Food Security Act of 1985. In 1987, a Resource Conservation and Development Program (RC&D) was also passed by the legislature, which enabled the government to control not only soil erosion but also farm production. According to this programme, about eight million hectares of land that had been retired from farming due to excessive erosion was brought under tree and forage cultivation within a 10-year plan (Myers 1988).

Decades have passed now since the idea of watershed management was first put forth in countries like the USA, Australia and New Zealand. In others such as Canada and the UK and also India, it has been found that drinking and irrigation water supply must be left to the private sector, while watershed management must remain within the hands of the government assisted by public participation. There are still other countries that are in the process of reviewing their policies with regard to the management of their strategic resources.

The big battles are between the traditional sustainable groundwater withdrawal through qanats and the drillers of wells. Most qanats have lost the hydraulic head battle and dried up (also many springs). The groundwater pumping, changing the socioeconomic habits of the people and the land reforms in the White Revolution in 1963 are the main reasons for a drastic lowering of groundwater tables (Madani 2008). Declining groundwater tables have also caused significant land subsidence throughout the country's drylands, e.g. in plain of Tehran (Dehghani et al. 2009, 2013; Motagh et al. 2007; Mousavi et al. 2001).

Salinization is a looming threat associated with improper management of dryland watersheds. Increased salinity in both surface and groundwater resources can be ascribed to overexploitation, depletion of the vegetation cover, deforestation, range degradation and inefficient methods of irrigation. All in all, these factors have led to more runoff and to rising water level in low-lying areas. Salinity is a threat to soil productivity and public health (e.g. mosquito borne diseases) in rural and urban communities. Rural farms, urban development, infrastructures (bridges and roads), water users and the environment are adversely affected by salinity. Although salinity may be a natural phenomenon in some areas, increased salinity in other parts is the direct result of a rising water table which is itself caused by changes in land use such as depletion of the vegetation cover, urban development, river flow regulation, irrigation and cropping. Such factors result in what is commonly called 'secondary salinization'. If left unattended to, not only will salinity lead to decreasing harvests, as its first and foremost consequence, but sustained salinization will also result in abandonment of the land and its ultimate desertification. When this happens, it will be too late to take any corrective measures, and we should only witness the expansion of deserts, immigrations, rapid livestock mortality and loss of our national legacy.

6.6 Watershed Management in Iran

Among the first steps to be taken in implementing a watershed management system are strengthening the intersectional coordinating institutions, identifying critical and strategic watersheds (e.g. in terms of their importance for drinking water supply), defining types of land uses according to definite standards and supervising the general trend of activities in the watershed. Any sectoral approach must be abandoned in favour of an integrated, inter-sectoral approach. In sectoral approaches, it is only the economics of the cubic metre of harvested water that is given first priority in investments at the expense of environmental and community relocation costs. It must be borne in mind that local communities are the backbone to all development.

In profit-seeking development initiatives, limited approaches, even the costs for implementing upstream watershed management operations and the costs of constructing water transfer and distribution facilities, are not taken into account. Integrated Management prioirity is given to supplying water for communities and alleviating economic stresses on these communities. It is characterized by decentralized rather than centralized investments in environmental plans and artificial recharge schemes, utilization of precipitation and increasing soil moisture to assist rangeland and forest reclamation.

In dryland states, about 90% of water resources are often used in irrigation (Seyf 2006, 2009). This is while precipitation in these areas undergoes considerable variations through the year and from year to year. The amount of precipitation in different areas is directly related to the altitude of the watershed. In drylands, evaporation is also rather high, usually several times the quantity of precipitation. This combination of climatic and demographic conditions have led to the development of decentralized water utilization organizations through which participatory systems such as the qanat, local water storage systems (Abanbar), Khooshab and ice water systems have evolved through the centuries.

The population density varies between upstream and downstream within a watershed that may not exceed 600 people per km^2 in some areas in Iran. Farm size in these areas are normally small, and water consumption has naturally been organized in participatory systems. Single cropping and centralized water use is rarely ever found in drylands and is associated with numerous problems in cases where such instances exist. For instance, in situations where the population density is low and where population is mainly located along coastal areas or river banks, there is usually no concern for water evaporation and/or wastage so that no human intervention is generally needed to optimize consumption or to increase the water available. The only concern in such areas would be transfer of water to population centres.

In contrast, in areas with high population density and high evapotranspiration rates, watershed management is focused on water transfer and, even more important, on reducing evaporation and on salvaging a percentage of the precipitation before it is lost to evaporation, e.g. subsurface dams (Fig. 6.4). In most rural watershed areas, the main difficulty lies in supplying drinking water, combating dry spells and providing water for supplementary irrigation for dryland farming. Centralized investments by the government in constructing large dams to procure drinking water for downstream large cities or establishing centralized agricultural sites may not be considered as solutions to rural community problems. Instead, these communities always look for ways to mitigate dry spells and periodical droughts through water storage in underground reservoirs, through preservation of soil moisture or through storage of water in places as near to their residence as possible.

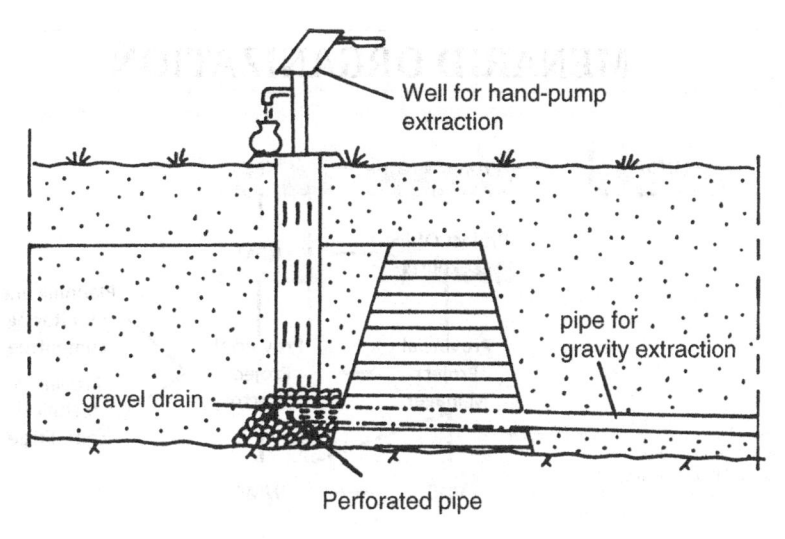

Fig. 6.4 Schematic diagram of a subsurface dam (Nilsson 1988)

6.6.1 Water Council Act of 2000 and Non-sectoral IWM

There is no doubt that the Water Council Act of 2000 passed by the Iranian Parliament has triggered an important movement in non-sectoral integrated watershed management. The legislation has as its objective the formation of a supreme watershed management body to serve as the highest decision-making body in defining national policies and guidelines in this area. The need for the formation of a national water council had already been emphasized and reiterated by various authorities over the years. It is certain that the council will establish different subcommissions to address its agenda items including the major aspects of defining policies, major allocations, inspection and information dissemination, as well as the financial aspects of water concerning water supply and use, and priorities in investments. It is expected that the duties and liabilities of each subsector will be defined after the subcommissions are set up. The National Water Council will be expected to establish, in a later stage, regional watershed management councils as management-executive bodies as well as local watershed management councils, which will be in charge of daily affairs and activities within the watershed. These lower councils will consist of political managers, executive managers and users' representatives.

A next step by the council will be to define an upper uptake limit based on the base water year, water rights and environmental considerations. Independent mechanisms will have to be developed for water allocations to environmental projects (watershed management, desertification control, range and forest reclamation). The water pricing must become realistic in order to economize on water use

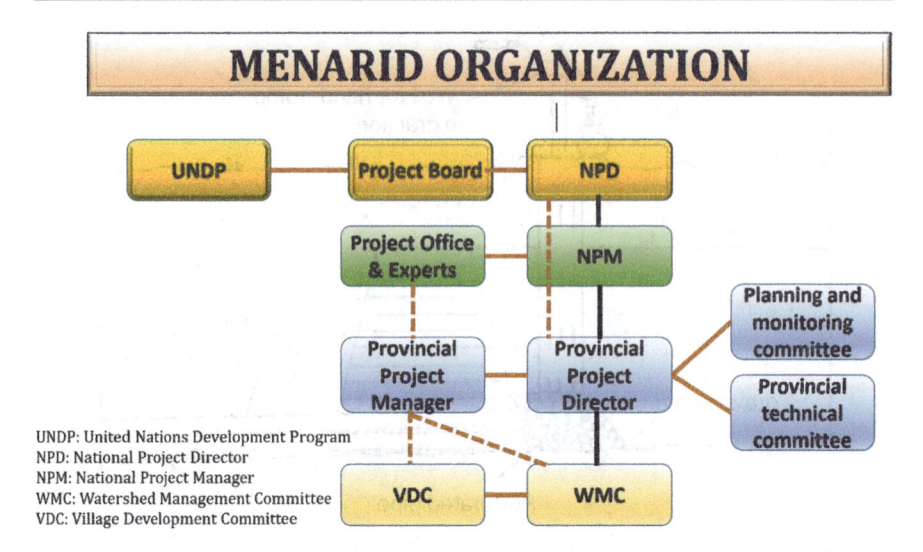

Fig. 6.5 MENARID coordinates several levels of government and non-government groups, and local land users are represented on key committees (http://www.menarid.ir/en)

and make water supply economical. Special exemptions or subsidies need to be given to small landowners and environmental projects. For agricultural and industrial establishments, the pricing system should be based on water consumption rates and total production rates for every cubic metre of water used. Volumetric measurement installations will allow exchange and trading on a larger scale between the downstream and upstream communities. In order to conserve the physical, chemical and biological quality of water, a quantified system of quality indices must be defined so that every salinity unit can be commercialized for industrial and agricultural users according to a legal regime. Privatization and services provided in irrigation management and watershed management must be organized within a governmental structure.

6.6.2 Implementing MENARID in Iran

Another initiative is the trialling of MENARID (Middle East and North Africa Regional Program for Integrated Sustainable Development). This international project follows the IWM approach that is executed based on its working chart (Fig. 6.5). The MENARID project in Iran (Box 6.1) is an example of such an attempt to engage experts, land users and water managers in an integrated approach to deal with several objectives.

It involves several provinces (Semnan, Kermanshah, Yazd and Sistan-Baluchistan) and four demonstration sites.

6.7 Concluding Remarks

A new era has definitely started in Iran in the integrated watershed management that so far pursued disparate and independent objectives such as supplying irrigation or drinking water and neglecting issues such as flooding, erosion, sedimentation or the various environmental changes. It is certain that this trend has become a thing of the past, as watershed areas cannot be maintained intact. They must be managed for various objectives and beneficial uses. It is also definite that industrial, mining, recreational, agricultural and range uses of watersheds will continue even at a more intensified level due to the growing demands.

The development operations of the recent past concentrated on single objectives such as water harvesting or its use, thus pooling all investments on a single objective. However, while water economy must be regarded as a major concern, it is necessary to define new policies for such areas as natural resources and livestock breeding, the environment, erosion, sedimentation, flooding, etc. Thus, creating coordination among various bodies and agreement on the policies and planning must form an essential objective of the integrated watershed management.

An integrated and demand-driven strategy for sustainable development of watersheds with empowerment of people and active participation and partnership of local communities is necessary to ensure productivity and sustainability in the watersheds. Any successful approach should closely involve community sectors at grass-root levels, including NGOs, women and youth, in formulation, planning and implementation as well as launching awareness to sensitize all stakeholders to understand the impacts and to identify their roles.

In most regions, comparing with other natural resources, water is the scarcest substance. As a result, production relies heavily on underground and surface water, which become increasingly scarcer. Water will remain a barrier to the achievement of poverty alleviation and food security. Most of renewable water resources have already been committed by conventional method of dam construction, but the demand for water is exceeded renewable water supplies. Therefore the future emphasis must be directed towards increasing the efficiency of water management and increasing water productivity and producing more crops per cubic metre. Unless there is an increase in watershed investment to generate higher employment, income, productivity and production opportunity for local inhabitants, the trend of watershed degradation will continue to exist.

A second question to be addressed is whether or not the existing legislation is adequate for our long-term objectives. The answer is clearly negative. Currently, a major problem in the way of integrated management is the lack of a set of comprehensive laws that can serve as the foundation and supporter of an integrated watershed management.

The next issue of concern is the planning and financing system. Most of the organizations involved are entangled and choked by the national bureaucracy inherent in the planning and financing system. They normally face resistance against reforms and improvements even in cases where the sources of the problem are identified and/or disclosed. Lack of financial and human resources most often discourage them. Measures must be taken to lift these obstacles.

Public participation by the local communities in watersheds must be taken more seriously than ever. Mechanisms must be developed to remove executive undertakings by the government in favour of a privatized system of watershed executive management so that the public sector will be in a position to take its more fundamental role of steering and mentoring the implementation of projects. Along these lines, it is essential for banks and the related systems to be empowered in order to offer financial facilities.

Decades have passed now since the idea of integrated watershed management was first put forth in countries like the USA, Australia and New Zealand. In others such as Canada and the UK and also India, it has been found that drinking and irrigation water supply must be left to the private sector, while watershed management must remain within the hands of the government assisted by public participation. There are still other countries (Iran is one of them) that are in the process of reviewing their policies with regards to the management of their strategic resources.

References

K.C. Abbaspour, M. Faramarzi, S.S. Ghasemi, H. Yang, Assessing the impact of climate change on water resources in Iran. Water Resour. Res. **45**(10), W10434 (2009). doi:10.1029/200 8WR007615

H. Ahmadi, A. Nazari Samani, A. Malekian, The qanat: a living history in Iran, in *Water and Sustainability in Arid Regions*, ed. By G. Schneier-Madanes, M.-F. Courel, (Springer, Netherlands, 2010), pp. 125–138. doi:10.1007/978-90-481-2776-4_8

A. Bagheri, S.A. Hosseini, *A System Dynamics Approach to Assess Water Resources Development Scheme in the Mashhad Plain, Iran, Versus Sustainability*. Proceedings of the 4th International Perspective on Water Resources & the Environment (IPWE), Jan 2011, (Singapore 2011)

K.N. Brooks, P.F. Ffolliott, H.M. Gregersen, L.F. DeBano, *Hydrology and the Management of Watersheds* (Iowa State University Press, Ames, 1997)

K.N. Brooks, P.F. Ffolliott, H.M. Gregersen, K.W. Easter, *Policies for Sustainable Development: The Role of Watershed Management* (EPAT/MUCIA Policy Brief No. 6, Department of State, Washington, DC, 1994)

K.N. Brooks, H.H. Gregersen, P.F. Ffolliott, K.G. Tejwani, Watershed management: a key to sustainability, in *Managing the World's Forests: Looking for Balance Between Conservation and Development*, ed. By N. P. Sharma, (Kendall/Hunt Publishing Company, Dubuque, 1992), pp. 455–487

J.R. Burton, Catchment management in Australia. Civ. Eng. Trans. 30(4), 145–152 (1988).

M. Dehghani, M.J. Valadan Zoej, I. Entezam, A. Mansourian, S. Saatchi, InSAR monitoring of progressive land subsidence in Neyshabour, Northeast Iran. Geophys. J. Int. **178**, 47–56 (2009). doi:10.1111/j.1365-246X.2009.04135.x

M. Dehghani, M.J. Valadan Zoej, A. Hooper, R.F. Hanssen, I. Entezam, S. Saatchi, Hybrid conventional and persistent scatterer SAR interferometry for land subsidence monitoring in the Tehran Basin, Iran. ISPRS J. Photogramm. Remote Sens. **79**, 157–170 (2013). doi:10.1016/j.isprsjprs.2013.02.012

P. Döll, H. Müller Schmied, C. Schuh, F.T. Portmann, A. Eicker, Global-scale assessment of groundwater depletion and related groundwater abstractions: combining hydrological modeling with information from well observations and GRACE satellites. Water Resour. Res. (2014). doi:10.1002/2014WR015595

EMG, *Global Drylands: A UN System-Wide Response*. United Nations Environmental Management Group, United Nation (2011)

J. Evans, 21st century climate change in the Middle East. Clim. Chang. **92**, 417–432 (2009)

Forest, Range and Watershed Management Organization (FRWO), *Vision of the Iranian Watershed* (FRWO Publishing, Persian, 2001)

M. Giam Pietro, D. Pimental, *The NPG Forum* (Negative Population Growth, Teaneck, 1993)

T. Gleeson, Y. Wada, B. MFP, v.B. LPH, Water balance of global aquifers revealed by groundwater footprint. Nature **488**, 197–200 (2012)

H.M. Gregersen, K.N. Brooks, J.A. Dixon, L.S. Hamilton, *Guidelines for Economic Appraisal of Watershed Management Practices* (FAO Conservation Guide No. 16, Rome, 1987)

H.M. Gregersen, K.N. Brooks, P.F. Ffolliott, T. Henzell, A. Kassam, K.G. Tejwani, Watershed management as a unifying framework for researching land and water conservation issues. Land Husb. **2**, 23–32 (1996)

G.A. Heshmati, V.R. Squires, *Combating Desertification in Asia, Africa and the Middle East: Proven Practices* (Springer, Dordrecht, 2013)

M.H. Hojjati, F. Boustani, An assessment of groundwater crisis in Iran, case study: fars province world academy of science. Eng. Technol. **4**, 427–431 (2010)

IECA, Journal of Erosion Control (International Erosion Control Association, Santa Barbara, United Nation 2000), p. 104, pp. 68–75

A. Izady, K. Davary, A. Alizadeh, B. Ghahraman, M. Sadeghi, A. Moghaddamnia, Application of "panel-data" modeling to predict groundwater levels in the Neishaboor Plain, Iran. Hydrogeol. J. **20**, 435–447 (2012). doi:10.1007/s10040-011-0814-2

S. Jamali, A. Abrishamchi, M.A. Marino, A. Abbasnia, Climate change impact assessment on hydrology of Karkheh Basin, Iran. Proc. ICE-Water. Manag. **166**, 93–104 (2012)

Keck, A., NASA sees fields of green spring up in Saudi Arabia 2012. Available at: http://www.nasa.gov/topics/earth/features/saudi-green.html. Accessed 25 June 2012

A. Kowsar, Floodwater spreading for desertification control: an integrated approach. Des. Con. Bull. (UNEP) **19**, 3–18 (1991)

Lal, R. 1984. Productivity Assessment of Tropical Soils and the Effects of Erosion.

K. Madani, 2008. Reasons Behind Failure of Qanats in the 20th Century. Paper presented at the World Environmental and Water Resources Congress 2008, Honolulu, ASCE, 10.1061/40976(316)77

K. Madani, Water management in Iran: what is causing the looming crisis? J. Environ. Stud. Sci. **4**, 315 (2014). doi:10.1007/s13412-014-0182-z

D. Molden, Water for food, water for life: a comprehensive assessment of water management in agriculture, in *International Water Management Institute IWMI*, ed. By D. Molden, (Earthscan, London, 2007)

M. Motagh, Y. Djamour, T.R. Walter, H.-U. Wetzel, J. Zschau, S. Arabi, Land subsidence in Mashhad Valley, northeast Iran: results from InSAR, levelling and GPS. Geophys. J. Int. **168**, 518–526 (2007)

S.M. Mousavi, A. Shamsai, M.H.E. Naggar, M. Khamehchian, A GPS- based monitoring program of land subsidence due to groundwater withdrawal in Iran. Can. J. Civ. Eng. **28**, 452–464 (2001). doi:10.1139/l01-013

P.C. Myers, Conservation at the crossroads. J. Soil Water Conserv. **43**(1), 10–13 (1988)

A.N. Nejad, Soil and water conservation for desertification control in Iran, in *Combating Desertification in Asia, Africa and the Middle East: Proven Practices*, ed. By G. A. Heshmati, V. R. Squires, (Springer, Dordrecht, 2013), pp. 377–400

National Research Council, *National Water Master Plan- Tehran, Water Resources Organization of Iran* (JAMAB Conculting Consulting Company Pub, 1999) (In Persian).

A. Nilsson, *Ground Water Dam for Small-Scale Water Supply* (IT Pub, London, 1988)

F.J. Pierce, W.E. Larson, R.H. Dowdy, W.P. Graham, Productivity of soils: assessing long-term changes due to erosion. J Soil Water Conser. **38**, 39–44 (1983)

J.W. Powell, Seven Venerable Ghosts. J. Am. Anthropol. **A9**, 67–92 (1896). doi:10.1525/aa.1896.9.3.02a00000

M. Rahbari, G.H. Roshani (Translators), Khorasan and Sistan (by Yate, E., and Sir Charles Edward) (Yazdan Publisher, 1986) (In Persian)

J. Rockström, N. Hatibu, T. Oweis, S. Wani, J. Barron, A. Bruggeman, Z. Qiang, J. Farahani, L. Karlberg, Managing water in rain-fed agriculture, in *Water for Food, A Comprehensive Assessment of Water Management in Agriculture, International Water Management Institute*, ed. By D. Molden, (Earthscan, London, 2007)

A. Seyf, On the importance of irrigation in Iranian agriculture. Middle East Stud. **42**, 659–673 (2006). doi:10.1080/00263200600642399

A. Seyf, Population and agricultural development in Iran, 1800–1906. Middle East Stud. **45**, 447–460 (2009). doi:10.1080/00263200902853439

F. Sharifi, A. Haydarian, *Natural Resources Management Strategy in Iran, Proceedings of the Regional Workshop on Traditional Water Harvesting Systems* (UNESCO, Tehran, 1999), p. 345

B. Sharifi, Saghafian, A Telvari, *The Great 2001 Flood In Golestan Province, Iran: Causes and Consequences* (International Conference on Flood Estimation, Berne, 2002)

H.R. Solaymani, A.K. Gosain, Assessment of climate change impacts in a semi-arid watershed in Iran using regional climate models. J. Water Clim. Chang. **6**(1), 161–180 (2014). doi:10.2166/wcc.2014.076

H.R. Solaymani, M.B. Haghighi, *Relationships Between Land Degradation and Natural Disasters and Their Impacts on Integrated Watershed Management in Iran* (2010) p. 285–296. doi: 10.1007/978-90-481-8657-0_21

G Soltani, M Saboohi, *Economic and Social Impacts of Ground- Water Overdraft: The Case of Iran*. Paper presented at the Proceedings of the 15th Economic Research Forum (ERF) annual conference (2009)

V. R. Squires, H. Milner, K. A. Daniell (eds.), *River Basin Management in the Twenty-First Century; Understanding People and Place* (CRC Press, Boca Raton, 2014)

V.R. Squires, A.K. Warsame, Sand dune fixation – lessons learned from China, Iran and Somalia pp. 241–250, in *Sustainable Development in Arid Zones: Assessment and Monitoring of Desert Ecosystems*, ed. By S. Omar, R. Misak, D. Al-Ajmi, (A.A. Balkema, Rotterdam, 1998), pp. 19–30

Statistical Yearbook, Iran Statistics Center, Ministry of Budget and Planning of I.R. of Iran, (1986) p. 391.

Statistical Yearbook, Ministry of Budget and Planning of I.R. of Iran, (1997) p. 174.

UNEP, *World Atlas of Desertification* (Nairobi 1992).

S.P. Wani, P.K. Joshi, Y.S. Ramakrishna, T.K. Sreedevi, P. Singh, P. Pathak, A new paradigm in watershed management: a must for development of rain-fed areas for inclusive growth, in *Conservation Farming: Enhancing Productivity and Profitability of Rain-Fed Areas*, ed. By A. Swarup, S. Bhan, J. S. Bali, (Soil Conservation Society of India, New Delhi, 2007), pp. 163–178

R.P. White, J. Nackoney, *Drylands, People, and Ecosystem Goods and Services: A Web-Based Geospatial Analysis* (World Resources Institute, Washington, DC, 2003)

M. Zarghami, A. Abdi, I. Babaeian, Y. Hassanzadeh, R. Kanani, Impacts of climate change on runoffs in East Azerbaijan, Iran. Glob. Planet. Chang. **78**, 137–146 (2011). doi:10.1016/j.gloplacha.2011.06.003

Thar Desert: Its Land Management, Livelihoods and Prospects in a Global Warming Scenario

7

R.P. Dhir

7.1 Introduction

The northwest arid zone of India encompasses the popularly known Thar Desert which marks the eastern end of the mid-latitude desert belt. It is characterized by a monsoonal pattern of rainfall distribution. The mean annual rainfall ranges from ~450 mm in the wettest eastern margin to less than 150 mm in the driest part. However, this constitutes only a small fraction of the mean annual potential evapotranspiration that has a spread of 1,400 mm to more than 2,200 mm. Major landforms comprise sandy plains and dunes, alluvial plains and rocky-gravelly pediments with scattered hills. The region has more than 700 species of vegetation. Most of these are well-adapted to harsh and uncertain environment and useful as a feed, fuel wood, or thatch making. Despite its high aridity, the region boasts of a world-renowned civilization, namely, the Indus/Harappan/Saraswati, that flourished during 3300–1600/1300 CE in the then riverine tract at the western margin of the Thar Desert. However, this civilization did not extend from its niche to the arid zone proper. That region instead was occupied by hunters and food gathering tribes.

The real influx of people into the region began about 2,000 or so years ago. However, population all through was marked by a sluggish growth with large variation in numbers due to drought-related famines and associated disease epidemics. Major livelihood source during mediaeval times was animal husbandry but since then arable farming has taken over. The twentieth century has been marked by a huge population expansion and intensification of agriculture. This has meant an overexploitation and degradation of land resources. For many a centuries, livelihoods have been at the subsistence level in a conservative, traditional environment, but in the past few decades, there have been profound improvements in lifestyle and

---

R.P. Dhir (✉)
Central Arid Zone Research Institute, 498, Defense Colony, Jodhpur 342009, India
e-mail: dhirrp08@gmail.com

social milieu. Scientific research has produced a wealth of knowledge and new technologies. With greenhouse-induced warming becoming a global agenda, studies have been made on future climate, particularly the predicted rainfall regimen in the Thar. Hot topics for discussion and debate include land use, socioeconomics, livelihoods and global warming. Several of these have a long history of research and study behind them.

7.2 The Indus/Harappan/Saraswati Civilization and the Climate

The archaeological sites of this grand culture, when discovered in year 1931, lay along Indus River and its tributaries, but investigations since then have shown far greater concentration of these along the Ghaggar-Hakra River, the course of which mostly ran parallel to and a few kilometre to the east of Indus River and had an independent exit to Arabian Sea. However, presently this river is lying in a dysfunctional and disorganized form with no flow to the sea. This culture is subdivided into three stages, namely, Early Harappan (3300–2600 CE), Mature Harappan (2600–1900 CE) and Late or Degenerate Harappan (1900–1600 CE). This civilization distinguished by its well-developed and diversified agricultural economy, societal organization and development of a unique script (Fig. 7.1). At

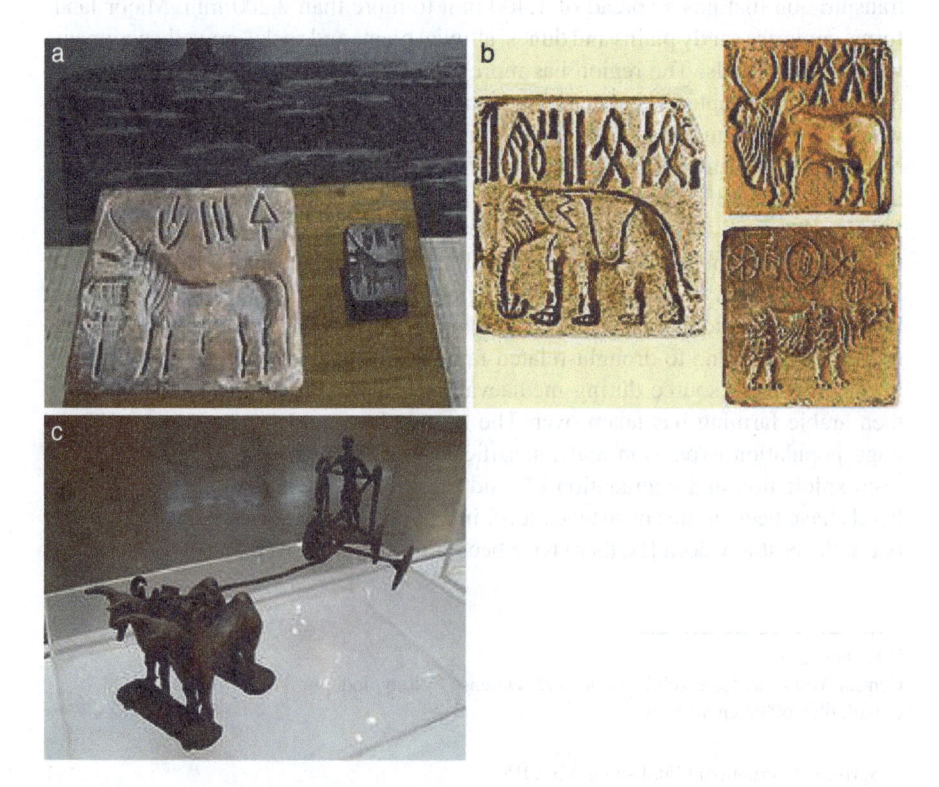

Fig. 7.1 Some artefacts of Harappan civilization: (**a, b**) seals; (**c**) a terracotta (*Source*: Wikipedia)

its peak, i.e. Mature Harappan period, it enjoyed a high degree of affluence, agricultural surpluses and outstanding urbanization that permitted blossoming of art and culture and long-distance trade. A variety of household ceramics and exquisite jewellery made from semi-precious and precious stones, bone and seashells are evidences of the same.

For our present purpose, this grand civilization is discussed from the view point of climate change. Regarding the decay and end of this culture, several arguments have been put forward, namely, a devastating flood, climate deterioration, soil exhaustion, land degradation, Aryan invasion, shift of river courses, earthquake, etc. (Wright 2010). Hypothesis favouring the drainage desiccation arising from shift in river courses, not an uncommon feature for the area, were put forward by Agrawal and Sood (1982) and by Misra (1984) on the ground that the existing catchment of river Ghaggar/Hakra, assuming even a higher rainfall of the Mid-Holocene, could not have generated a flow in terms of durability and volume needed to sustain this exalted civilization at the tail end, some 800 km away from the source. Prior to this, Yashpal et al. (1980) have shown from remote sensing the existence of several paleochannels, both of the present-day Sutlej and of Yamuna, that flowed to the Ghaggar in the past. Valdiya (2013), based on geological and *Vedic* literature, also favoured this possibility. However, studies of Giosan et al. (2012) based on incision pattern in upstream region, and of Clift et al. (2012) based on sediment dating and the mineralogy, suggest that the Yamuna River had abandoned its southwestern course to Ghaggar about 50,000 years in the past. Clift et al. (2012) further showed that Sutlej continued to flow into Ghaggar for a long time thereafter but shifted its course to become a tributary of Indus system well before 10,000 years BP.

Climate deterioration has been suggested as another possibility for decay of this grand civilization. In a comprehensive review, Fuller and Madella (2001) show dominance of wheat and barley in cropping pattern as also presence of peas and perennial horticultural crops during the Mature Harappan period. The summer crops, particularly low water requiring millets, constituted a small fraction only of the cropping. However, during Late or Degenerate Harappan period, summer crops dominated the scene (Madella and Fuller 2006). Similarly, there was a large change in distribution and size of settlements also. The settlements during Late Harappan period not only got smaller but like the settlements of the succeeding/contemporary Painted Grey Ware Culture, they spread to the north in plains of present-day Punjab, Haryana and western Uttar Pradesh as also to south in northwestern Gujarat (Suraj Bhan 1973; Mughal 1982). Further, an independent support of the climate deterioration theory comes from the Thar Desert proper. Singh et al. (1990) showed from palynological, salt composition and distribution and lake-level studies on Didwana Salt Lake that during early and middle part of Holocene, i.e. the period from 10 to12,000 years BP, and particularly from 7,500 to 6,200 years BP, was largely wetter than the present, though with some fluctuations of lake level. Thereafter, arid conditions set in leading to complete drying up of the lake. Another study (Enzel et al. 1999) from Lunkaransar Lake, some 150 km to the northwest of Didwana Lake, while supporting mid-Holocene wet regime, showed that lake-full stage started declining from 4,800 BP and that lake completely dried by 4,000 years BP. The lake continues to exist in its dry state ever since. Importantly, they, like Singh et al. (1990), showed also that besides the summer season monsoon, winter rainfall made a significant contribution also to lake hydrology during the wetter past. Thus, presently amongst the various hypotheses, climate deterioration remains the most plausible reason for the demise of this grand culture, although the scope is not exhausted for more investigations.

7.3 History of Human Settlement, Population Growth, Land Use and the Recent Phenomenal Developments

Though the above-described Saraswati and the subsequent Chalcolith civilizations had flourished at its western and northern fringes, the Thar proper was devoid of any settlements of that period. Historians believe that influx of people into this region started at the beginning of the first millennium. This entry happened from the settlements up in the north, where apprehending invasions from the west, the people moved into this environmentally harsh but otherwise secure region. Further, there was a movement from the present-day Gujarat region in the south, where a well-established population was in existence also. This is evidenced by the presence of ancient towns like *Bhatnair* (present-day Hanumangarh) in the north and the Bhinmal in the south (Reu 1935). From then on, the population spread such that by the middle of first millennium, the central part comprising present-day Nagaur, then known as *Itchipur*, was well populated and also socially organized. It has a temple, Dadi Mata, dated to second to third centuries. Going further, by the close of this millennium, even the driest part in the west was settled, and it had also a political setup. However, all through the population was thin. An estimate for as late as the mid-seventeenth century informs that the present-day 280 mm rainfall tract had a population then that was less than one-third of that in existence in year 1941, thus showing that population density was low indeed over all this period. However, during this seventeenth-century period, the tract in 350–400 mm, watered further by the streams emanating from Aravalli Mountains, had a much higher density and well-established irrigated agriculture. For the subsequent period, an estimate of population for the early nineteenth century is available from Tod (1829–1832), who showed that in the tract with better rainfall, i.e. eastern part, the density was 30 persons/km^2, in the intermediate tract (~250–350 mm rainfall) ~14 persons and for the western dry part just 4.6 persons/km^2. Actually for this later mentioned dry area, 48% of the population was accounted for by the capital town of Jaisalmer alone, thereby giving a density for the rural area of just ~2 persons/km^2. Thus, a vast difference existed in the density of population in relation to aridity from east to west, just as it exists even today.

From year 1891, regular decadal census data are available. These show that the human population up to year 1921, like that during Late Mediaeval period, had hardly any net growth. In fact the situation was one of large fluctuations, with climatically favourable period leading to some growth only to be followed by severe

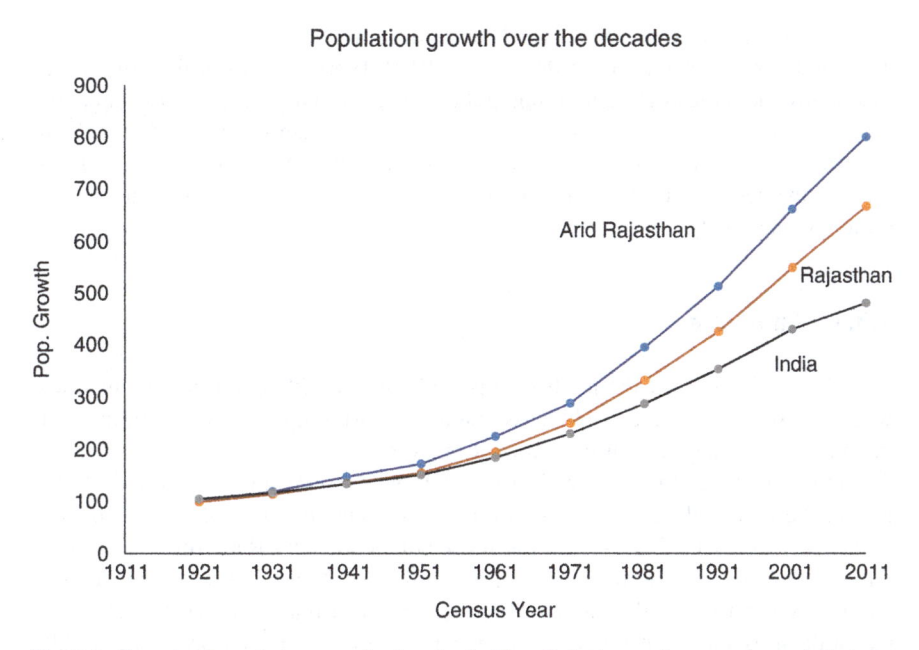

Fig. 7.2 Growth of population over the decades in Thar, Rajasthan and country as a whole (year 1921 = 100)

erosion during a successive famine period. A part of the nineteenth century, i.e. from years 1792 to 1854, experienced as many as 13 famines. For the remaining period of the nineteenth century, Narain et al. (2000) showed as many as seven famines and drought years, of which four lasted for two consecutive years. This situation of static (no net gain over a few decades) population shows how the harsh elements of nature, disease and a somewhat unstable/exploitative political system overwhelmed socioeconomic and demographic developments.

In comparison, the period from census year 1921 onwards is marked by one of continuous growth, the decadal growth rate being ~20% in the decades to year 1951 and ~30% from then on to census year 2001. It has been for the first time that the decade of 2001–2011 is marked by a decline to 20%. The growth rate all along has been much higher than that of Rajasthan and of India as a whole (Fig. 7.2).

Such an extraordinarily high growth rate during the past nearly a century is astounding indeed given that region is climatically challenged. Having a larger family had long been considered as a source of greater influence/strength and went with ability to command resources in the local context. Sociologists explain this as an outcome of early marriage of girls. Surveys of 1960s and 1970s showed that nearly 95% of girls in rural areas got married before they attained age of 15 years (Malhotra 1977; Bharara et al. 1984). The situation permitted a reproductive period of over three decades. Other reasons were the ethos as manifest in blessings of elders to a young woman to have numerous children (male in particular) and a social need to have a male child for performance of various religious ceremonies. To add to this was tenancy and land reforms, which brought in a degree of optimism in future. This situation prevailed during decades of 1920–1950s. But from then on, it was essentially the large decline in infant and maternal mortality rate. For example, the *infant mortality rate* per thousand live births in Rajasthan has fallen from 114 in 1961 to 83 by 1981 to just 60 in 2009 – a reduction by 50%. This rate currently in most of the desert districts is in fact 45–55 and thus is lower than the state average (Anonymous 2012).

7.3.1 Land Use

Records of land use for the mediaeval period are few, although it has been shown with a good degree of reasonableness that arable farming was in vogue in the Thar in an area adjoining present-day Gujarat in the fourth to fifth centuries. The arable farming from here spread to well-watered Luni River basin as the town of Pali located here was already a major commercial centre in early Mediaeval period. After the plunder of this town in the thirteenth century by Rathores, a section of Paliwals travelled all the way to Jaisalmer in the driest part and started niche agriculture by the fourteenth century. However, by far a major part of the Thar had pastoralism as the main pursuit. This situation continued even up to the middle seventeenth century since as inferred from Nainsi's data, the cropped area in the ~280 mm tract then constituted a very small fraction of geographic area then, whereas the Luni Basin (referred to above) had a flourishing agriculture comprising as many as 7,400 wells producing crops such as sugarcane, opium, cotton, aniseed, caraway, vegetables and cucurbits (Bhadani 1980). From then on till the 1920s, the region experienced little growth of cropping activity (Balak Ram et al. 2013). However, the period thereafter saw an expansion. For example, Dhir (1982) showed a considerable expansion in area during the period from 1930 to 1958 in the central part of the erstwhile Marwar State. Jodha and Vyas (1969) and Dhir (1982) attributed this growth as a response to some degree of tenancy reforms and moderation in the grip by feudal system.

With independence in the year 1947 and amalgamation soon after of princely states, the region underwent land reforms such as ownership of land and distribution of land beyond a ceiling limit to landless and poor. Also the system of land records and land use data in a 16-point format and 12 land use classes was introduced. This system was in conformity with the national system in existence then. Besides, a survey was made twice in a year on the crop grown on individual agricultural fields. Thus, a wealth of information, with a fair degree of reliability, is available for study of dynamics of land use and of cropping pattern. The data (Table 7.1) for the year 1956–1957 shows that the net sown area for the Thar region as a whole then stood at 36.6% with an interdistrict range of ~70% in eastern most districts, ~40% in the centrally located districts and only 2% in the driest Jaisalmer (Fig. 7.2). The high incidence of cropping in Ganganagar District, despite its location in drier part, is due to existence of irrigation facility. The fallow lands are those where a crop has

Table 7.1 Land use in Thar (in %) at three points of time over more than five decades

Land use class	1956–1957	1981–1982	2010–2011
Lands under forest	1.0	1.3	2.3
Lands not available for cultivation	2.8	3.4	4.7
Barren and unculturable lands[a]	11.8	5.1	4.6
Lands under permanent pastures	2.2	4.1	3.8
Culturable wastelands[b]	25.4	24.0	17.6
Lands under long fallows	11.7	7.4	4.8
Lands under current fallows	9.0	7.2	4.6
Net cropped area	36.3	47.8	57.4
[c]Area sown more than once in million hectare	0.23	0.96	3.73

Source: Revenue Board, Government of Rajasthan, Ajmer
[a]Unculturable lands: These are the wastelands which are not available for cultivation. These include barren rocky lands, steep sloping areas and areas covered by snow or glaciers
[b]Culturable wastelands: These are lands that are capable of being used for arable farming but are not being so used. The ownership of the lands rests with the government but their use for grazing etc is accessible to all
[c]Area sown more than once = Gross cropped area − net cropped area

been raised in the past but are under rest since then. These lands are called "current fallow" if the crop has been raised in the preceding year and "long fallow" if the period since last cropping is more than 1 year. These lands constituted 21% of the region in year 1956–1957 and are particularly extensive in western half. The culturable wastelands category in Thar constitutes 25% of the area or 5.3 Mha and had its main occurrence in dry districts of Jaisalmer and Bikaner.

But the period thereafter saw a considerable expansion of farming activity with the net sown area rising to 47.8% of Thar by year 1981–1982 and to 57.4% by year 2010–2011. This is due to three factors: (a) a reduction in the extent of fallow lands, (b) opening of new lands in the drier western half and (c) expansion of irrigation facilities. From the temporal distribution at district level data, it is seen that the eastern districts with higher rainfall had reached saturation as regards area suitable for cultivation in 1970s (Fig. 7.3).

The main trigger for expansion was population growth (Dhir 1988). This factor continued to exert its influence even during the 1980s and 1990s but even more important became the coming onto the scene of tractors for cultivation. These machines enormously enhanced the capacity of farmers to sow their holding in the small time span of adequate moisture in the plough layer of sandy soils. This period saw also a large expansion of arable farming to the drier western part (Dhir 2003). In the subsequent period, both these factors lost much of their significance, and further growth has been due to expansion of irrigation. In fact this is the main factor that explains also the vast expansion of area sown more than once. In year 1956–1957, this area was only 0.23 million ha (Mha), and it has risen to 3.73 Mha by year 2010–2011, i.e. a 15-fold increase (Table 7.1).

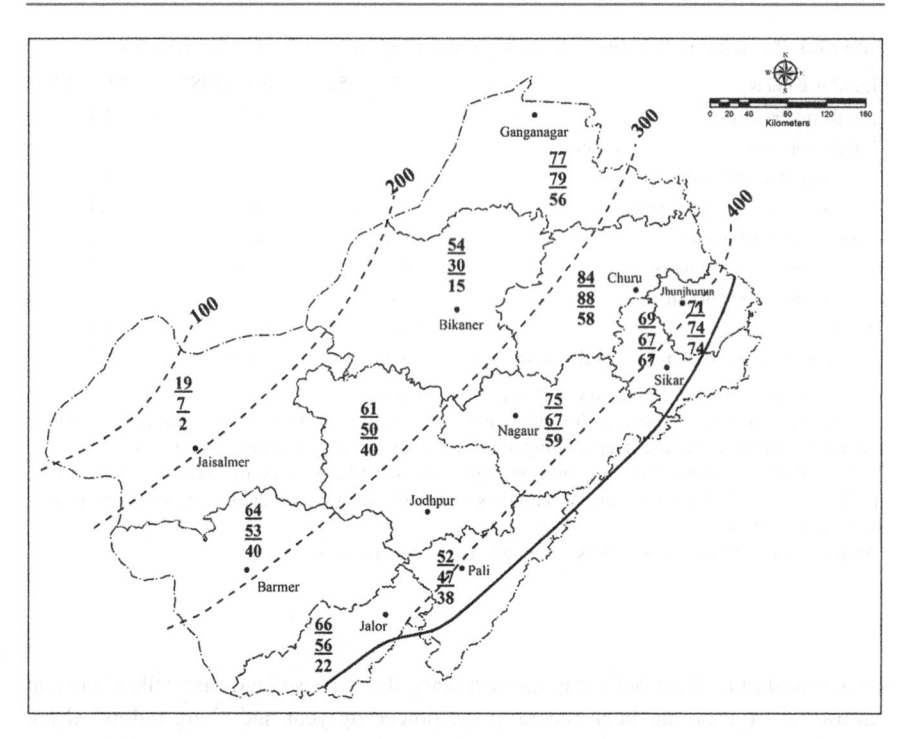

Fig. 7.3 District-wise net cropped area as percent of geographic area at three time periods, namely, years 1956–1957, 1981–1982 and 2010–2011. The numeral at the bottom corresponds to earliest period, i.e. 1956–1957, whereas middle and upper numeral is for successive periods, respectively

7.3.2 Irrigation

Water is a constraint underlying all problems in a desertic tract. This is particularly so for agricultural activity where irrigation can make all the difference in land productivity. Therefore, development of water resources has received a major attention all along. This has been the situation in the historical past, with the royalty, as part of its public welfare concern or charity, have undertaken construction of reservoirs across ephemeral streams or rocky catchments, tanks, step wells and *tankas* (a water harvesting-based small storage structure for individual family or few families) or even bringing water from a river outside the region at a huge investment. In the post-independence era, development of irrigation based both on surface or groundwater sources became a major program to enhance and stabilize food production and to meet domestic water needs. Thar is an outstanding example of this effort. Groundwater exploration activities, comprising exploration, increased access to drilling technology, credit and rural electrification, were started in 1960s and ever since these have only got strengthened. Even a more ambitious effort was bringing of water from the rivers up north all the way to Thar over a distance of several hundred kilometres through a canal with a capacity to carry 10,601 million cubic meters

(MCM) water annually and an objective to irrigate 1.5 Mha of parched lands of Thar. The project involved a huge monetary investment that included also the command area development and maintenance effort. As a result of these activities, the irrigated area has risen from 0.28 Mha under canal system and 0.22 Mha from wells in year 1956–1957 to 1.98 and 2.11 Mha from respective sources by year 2010–2011 – a nearly tenfold increase. A great effort indeed. In other words, of the total cropped area in Thar, 27% is irrigated by one or the other source now. Impact of irrigation in terms of agricultural production is still greater. In fact two-thirds of the value of crop production of the Thar is from irrigated lands.

7.4 Environmental Consequences of Land Use and Management

Arable farming and livestock rearing are the two major pursuits of desert dwellers. Arable farming both under irrigated and rainfed situations is essentially a fine blend of cropping and animal husbandry, i.e. a mixed farming enterprise. The crop residues and also some amount of green fodder sustain the livestock. In fact in the rainfed situation, livestock and their produce constitute almost the sole source of cash income. The system has several collateral benefits such as minimizing impact of drought, women empowerment and the much-needed increased employment of the farming household. Manure supply and fertilizing effect of animal penning are side benefits. There has been some change in herd composition; earlier it used to be largely cattle but with erosion of cow economy, it is now mainly goat and buffalo. Likewise, the pastoralists have shifted to a large extent from camel to sheep and goat. As explained above, the past six to seven decades have seen a large expansion of agricultural activity, a reduction of area under fallow lands as well as a tremendous loss of community grazing lands. This expansion of agricultural activity has occurred mainly in the western drier part, where the soils are sandy and wind regime is strong also. The use of a tractor as a means of cultivation has set in motion a process of accelerated wind erosion (Fig.7.4). This latter process causes removal of fertile soil particles as dust as well as near surface movement of sand (Dhir 1993). The sands redistribute locally and also travel some distance to encroach upon settlements or pile up against other obstacles, including sand dunes. In the process the lands lose some fertility and also levelness of the surface. The wind erosion is equally serious on community grazing lands, where overgrazing and consequential depletion of vegetation covers make these lands highly vulnerable also. The seriousness of problem of wind erosion has been mapped from time to time, and as per the latest estimate, 78% of the area of Thar is affected by this process, within which 16% is severely eroded. (Kar et al. 2009). The vegetational degradation problem is even more grievous. Though the past few decades have seen an enhanced dependence on crop residues and market purchased feeds, natural vegetation from designated grazing and other community lands have had a big role in meeting biomass needs of livestock and of fuel, wood and fencing material of desert dwellers.

Fig. 7.4 Accelerated wind erosion is an outcome of current land use and management and a major manifestation of desertification problem. (**a**) Saltating sand deposits within field means a somewhat reduced fertility and a serious loss of land level. (**b**) Drift sands pile over dune flanks and crest. (**c**). These encroach also onto dwelling units

However, due both to over grazing and agricultural land uses, the productivity and gross biomass availability have undergone severe erosion. The grazing lands have historically supported the traditional pastoralists. The livestock of farmers also used to substantially depend upon grazing on community lands in vicinity of the village and, during periods of scarcity, on distant lands in western thinly populated areas. The presence of high yielding perennial grasses and of shrubs and of multi-purpose trees had ensured a high productivity in good (rainfall-wise) years and some productivity even in poor years. The deep-rooted trees and shrubs here provided some quantity of utilizable material even during extended droughts. But progressive increase in biotic pressure on the one hand and shrinkage of the area due to expansion of cultivation on the other have resulted in persistent overgrazing and severe loss of biomass production ability. The perennial grasses have almost disappeared and are replaced by very low yielding annuals and ephemerals. Thus, not only is there loss of productivity but also a reduced duration of the grazing period. In fact these lands are producing today one-third to one-fifth of that from a well-managed grassland (Table 7.2).

Besides the grazing lands, the cropped lands invariably carry a good stand of highly useful shrub, like *Ziziphus nummularia*, and an equally important tree,

Table 7.2 Vegetation cover and yield under non-degraded and degraded condition

Grass cover type	Herbage cover (%)		Herbage yield (kg/ha)	
	Non-degraded	Degraded	Non-degraded	Degraded
Dichanthium annulatum -Desmostachya bipinnata	4–8	0.5–2	4,000	130–1100
Eleusine- Dichanthium- Desmostachya bipinnata	3–9	0.5–2	1,200–1,500	100–600
Sporobolus marginata- Dichanthium annulatum	4–7	1–3	1,400–2,600	300–500
Saccharum spontaneum-S. bengalense	4–7	1–3	1,400–2,600	300–500
Cenchrus ciliaris-C. setigerus	4–6	1–2	2,000–2,500	300–500
Eleusine compressa-Dactyloctenium sindicum	3–4	0.5–2	800–1,000	175–450
Lasiurus sindicus	5–14	2–4	4000	400–500
Lasiurus sindicus -Panicum turgidum	5–8	2–3	1,500–2,000	300–450
Aristida- Eragrostis- Cenchrus biflorus	2–3	1	500–800	100–200

Source: Suresh Kumar (1997)

namely, *Prosopis cineraria*, which are known to offer little competition for moisture for the growing annual crops and is an illustrious example of the traditional agroforestry system. The stand of shrubs used to be 300–500 per hectare in 1950–1970s in 300–400 mm rainfall tract. It provided 50–200 kg of highly nutritious fodder that can be stored for up to 2 years. Likewise, in equal measure came another significant contribution from trees, particularly *P. cineraria*. Tractorization of plowing operation has resulted in virtual ruination of the stand of the useful shrub. The stand of the tree is adversely affected also due to casualty of the young replacement population. A study based on aerial photos of years 1958 and 1985 of a stand of principal agroforestry tree, *Prosopis cineraria*, at 12 sites in Jodhpur district showed that whereas at 2 sites there was a population increase, the rest had a marked decline, with an overall decrease in stand of 14.4%. Similarly, the stand of other perennial vegetation such as *Calligonum polygonoides*, *Acacia jacquemontii*, *Leptadenia pyrotechnica* and *Haloxylon spp.* has suffered also. This depleted perennial vegetation cover under all land uses has meant also considerable loss of biodiversity. As per latest estimate (Pandey et al. 2012), the number of taxa in rare and threatened categories has reached to 43.

As mentioned earlier, irrigation is by far the biggest development, and of the total irrigated area, ~53% is from ground water sources. Withdrawal of water exceeds the recharge capacity and water tables are dropping. Current estimates show that water availability is 3,315 MCM, but demand is 5,056 MCM, i.e. 153% larger (Anonymous 2009) and many areas are facing a water table decline of 1–3 m per year. Clearly this is unsustainable and collapse of the system in inevitable, especially if climate change brings a warming trend and reduced or more erratic rainfall patterns.

7.5 Projections of Climate in a Global Warming Scenario

Increase in temperature is the most obvious outcome of current enhanced, and ever increasing, concentration of greenhouse gases. The Intergovernmental Panel on Climate Change (IPCC 2007) reports state that a rise of 0.8 °C in temperature of land and sea has already happened during the twentieth century. It is further projected that during the twenty-first century, there will be a further increase of 1.1–2.9 °C at the lowest emission scenario and 2.4–6.4 °C at the highest. Regarding Asia, the IPCC has predicted a temperature rise of 3 °C by end of 2050 and of 5 °C by 2080, and this warming is greater than that for the world as a whole. Looking to critical dependence of much of this region on monsoon, scientists have focused on its behaviour over time. Meehl and Washington (1993) employed a GCM model and projected an intensification of monsoon in a doubled CO_2 situation. A later study by Lu et al. (2007) found a similar outcome but predicted also an expansion of subtropical deserts, to which Thar Desert also belongs. The IPCC (2007) review also projects an increase in occurrence of extreme weather phenomena including heat wave and intense precipitation events in South Asia, East Asia and Southeast Asia.

Regarding India, efforts to predict the extent of climate change have been made using both projections from recent trends in climate parameters and from use of global climate models. The India Meteorology Department worked out state-level trends for the period 1951–2010 and showed that in Rajasthan mean annual temperature and rainfall have increased at a rate of 0.01 °C and 0.04 mm per year, respectively (Rathore et al. 2013). Concerning the Thar region, Pant and Hingane (1988) analyzed over a 100-year instrumental record of rainfall for the period to 1980s at several rain gauge stations in Thar and found non-significant trend of rainfall. Instead Dash et al. (2009) from study on duration and intensity of rainfall have suggested a weakening of the summer monsoon circulation over India.

Analysis for an extended period to 2010 by Rao and Roy (2012) again did not show any systematic trend in rainfall over time. However, there were intra-regional variation with Ganganagar (in an irrigated tract in the north) showing an increase of rainfall at a mean rate of 1.2 mm/year, but out of 65 locations studied in Thar, 37 showed a decline in monsoon rainfall. For the recent period (1971–2011), Poonia and Rao (2013) showed a mean annual rainfall increase of 0.56 mm per year for the Thar as a whole. But individual stations showed varied trend. Projection of these for the closing period of the twenty-first century suggests rainfall to increase from the present 252 to 308 mm at Bikaner, from 176 to 234 mm at Jaisalmer and from 487 to 613 mm at Pali, but for Jodhpur, the rainfall is likely to decrease from 325 to 275 mm. They showed also based on this recent period trend that the mean annual temperature may increase by 3.3 °C at Bikaner, 3.4 °C at Jaisalmer, 2.9 °C at Jodhpur and 2.5 °C at Pali. Goswami and Ramesh (2008) used weighted epochal trend ensemble approach in analysis to assess future scenario of Thar based on trend analysis of the period 1990–2003 and suggested that the area of the Thar Desert is poised to expand greatly by the end of the twenty-first century.

Global modelling of climate in progressively warming condition has been a major exercise. Rupa Kumar et al. (2006) using regional climate model, PRECIS, showed that the all-India summer monsoon rainfall will increase by about 20% during the 2071–2100 period. The annual mean surface air temperature during this time slot is set to rise by 3–5 °C in one scenario and between 2.5 and 4 °C in the other. However, the Thar Desert and adjoining parts of northwestern India are expected to show little change or, in fact, a decline in rainfall of 10–15%. Rajendran and Kitoh (2008) carried out a study using a super high-resolution GCM, which enabled a far greater spatial resolution and also brought out clearly the orographic effect. The study concentrated on the later half of the twenty-first century, more specifically on the 10 year slot of years 2075–2084 period. The results showed a widespread but spatially heterogeneous increase in rainfall – much of peninsular and northern plain region marked by an increase in rainfall of 1–3 mm/day for summer monsoon period, but others had much less increase. The overall projected increase is shown at ~10%, which is less than the value of Rupa Kumar et al. (2006). The Thar region is expected to experience a modest increase of 0.5 mm or less. The results also show a large increase in days with temperature over 45 °C during pre-monsoon summer period and of those with temperature over 35 °C during monsoon period, the effect being at its maximum in northwestern part of the country, including Thar. Likewise, temperature projections showed substantial inter-model variation. For Thar, while showing considerable spatial variation, most models show little or no improvement in rainfall, while temperature increase was close to all-India average.

Kar (2012) made a study specifically for northwest arid zone of the country. Of the 15 GCMs and their ensemble under A2 scenario, Kar (2012) found the two models to predict better the 2001–2010 period rainfall and hence used these to predict rainfall regime for the period to mid of the twenty-first century. The results from both these models, albeit with some difference in magnitude, showed a decline in rainfall for the period from 2016 to 2030 with a spill over continuing to the period ending in 2050. Importantly, he worked also on likely wind erosion scenario, which showed that for much of Thar, coming decades will see a substantial increase in wind erosivity and the same will happen again in late 2040s onwards. Summing up, with a degree of uncertainty, most predictions are for a conspicuous rise in temperature, increased erraticity in distribution of rainfall as also its intensity. Regarding rainfall amount, most of the results do not show a systematic trend but some show a decline.

7.6 Implications and Vulnerability of Thar

Because of an exploitative use of natural resources, the region is already under going environmental deterioration. Degradation of vast community grazing land is very severe, besides of course, the loss of useful perennial vegetation on croplands. Wind erosion problem in general is considerable also and most of the dunes, which occupy 45–50% of the area of the Thar are semi-stabilized. This includes some area also of shifting dunes and hummocks, particularly in the windy western half of the Thar. The Central Arid Zone Research Institute has already been working since the 1960s on a huge program of technology development that comprises grassland rehabilitation under extreme habitat situations, sand dune stabilization with mini-wind breaks, identification of fast growing trees and nursery techniques for tree multiplication especially for road-side plantation. Wind erosion control on croplands, soil and moisture conservation were part of this activity. Therefore, concerned with deteriorating environmental degradation, the government launched a major effort under the aegis of Desert Development Program in year 1977–1978. Road and canal side plantation, village energy plantation and sand dune stabilization were some of the salient components of this effort, and about 0.3 Mha have been treated so far with considered success. But efforts to rehabilitate the community grazing land did not have much success, not because of any deficiency of technology but because of lack on grazing control. In the area of water resources, some efforts were made to expand the traditional water harvesting systems both for drinking (*tankas*) and for agricultural lands, i.e. *khadin* system. Likewise, a program of drought/famine relief was started also. Over time, with increased supporting capacity of central and state governments, human deaths from starvation are a thing of the past. Fodder is imported from all over. Though this is able to keep the most productive of the cattle alive, distress sale of animals and some mortality have been unavoidable.

Right from the beginning of independence era, a socialistic pattern of society has been a consistent and major policy of the government. Over time with the capacity of providing succour to the numerous poor and handicapped having increased, there has been no let-up of social upliftment effort. Reach and content of the welfare programs have been enhanced to include safe drinking water supply, literacy, skill development, strengthening of public health services, subsidized housing, rural roads and communication. Public distribution of essential commodities has been a major activity, and over time the entitlement and quantity of subsidized food grains have increased. Further, there has been a major reduction in cost of grain and fuel supplies for the people below the poverty line. Simultaneously, rural employment generation has been taken up also from 1980's, and in fact from year 2006, it took the form of a watershed with launching of employment guarantee scheme that entitled every rural family a 100/150 days of employment at handsome wages. This had a profound effect not only in strengthening household economy but also in empowerment of women. Irrigation, as said above, is a major transformer of economic well-being of farmers, and to facilitate this, farmers are being supplied power for tube wells free or at nominal rates. Likewise, canal water supply is highly subsidized also. To cap it all and to minimize use of electric power, almost entirely generated by coal-based systems, solar power-based pumps are being supplied at 70–80% subsidy. New methods of irrigation are helpful also in economizing water use by crops. But, as mentioned earlier, the current rate of ground water withdrawal will cause exhaustion of most of the aquifers in another 10 years or so to undo the current gains.

The growth of a service sector and industry has been a contributing factor to a largely improved lifestyle and general well-being of desert dwellers. Liberal attitude and promotional effort have led to a fast development of solar energy utilization enterprises. From a modest start with just 5 MW capacity in year 2010–2011, there is an aggregated generation of 1,360 MW presently. Besides, projects amounting to another 15,000 MW capacity at a cost of 200 billion Indian rupees stand committed. But except for the solar energy, the support for socioeconomic developments, just mentioned, is externally sourced in Thar. Because of this conditionality, the local adaptive capacity based on indigenous biophysical and social resources is rather low. Little above the bare subsistence level living of rainfed farmers and pastoralists endows a poor buffering capacity of farmers to adjust to adverse environmental perturbations.

Rainfed farming and ground water-based irrigated farming are the pursuits of the largest part of population of the Thar. The predicted climate changes bode ill despite appropriate traditional management, comprising of mixed farming, well-adapted crops, mixed cropping and agro-forestry and the best technologies for both the farming systems. Besides the highly vulnerable indigenously sourced ground water supply for irrigation, even the precious canal-irrigated area, with its externally sourced water, is liable to suffer. Firstly, the increased atmospheric temperature will mean increased water requirements for crops. Rao and Poonia (2011) have shown that an increase of 3 °C will mean a 9–10% higher water need for monsoon season crops and 13–14% for winter season crops. Further, higher temperature during the most important winter season (productivity-wise) will reduce the length of growing season thereby adversely affecting the crop yields. Among the other biotic stresses, the problem of pests and diseases is going to be aggravated.

References

D.P. Agrawal, R.K. Sood, Ecological features of the Harappan civilization, in *Harappan Civilization, A Contemporary Perspective*, ed. by G. L. Possehl, (Aris and Phillips Ltd., Warminster, 1982), pp. 223–231

Anonymous, *Status of Ground Water Resources, Rajasthan Ground Water Department* (Govt. of Rajasthan, Jodhpur, 2009), p. 14

Anonymous, *Demographic and Health Scenario of Rajasthan from an Analytical Perspective, Directorate of Economics and Statistics* (Govt. of Rajasthan, Jaipur, 2012.) 236 p

R. Balak, J.S. Chauhan, Appraisal of landuse and agriculture in erstwhile Marwar State of Rajputana. Curr. Agric. **37**, 29–39 (2013)

B.L. Bhadani, Economic conditions in Pargana Merta, in *Some Aspects of Socio-economic History of Rajasthan*, ed. by G. L. Dewra, (Rajasthan Sewa Mandir, Jodhpur, 1980), pp. 113–129

S. Bhan, The sequence and spread of prehistoric cultures in upper Saraswati basin, in *Radiocarbon and India Archeology*, ed. by D. P. Aggarwal, A. Ghosh, (TIFR, Bombay, 1973), pp. 252–263

L.P. Bharara, W. Khan, V.N. Mathur, Some socio-economic characteristics of the upper Luni Basin in Rajasthan desert. Ann. Arid Zone **23**, 313–324 (1984)

P.D. Clift, A. Carter, L. Giosan, J. Durcan, G.A.T. Duller, M.G. Macklin, A. Alizai, A.R. Tabrez, M. Danish, S. Van Laningham, D.Q. Fuller, U-Pb zircon dating evidence for a Pleistocene Sarasvati river and capture of the Yamuna river. Geology **40**, 211–214 (2012)

S.K. Dash, M.A. Kulkarni, U.C. Mohanty, K. Prasad, Changes in the characteristics of rain events in India. J. Geophys. Res. **114**, D10109 (2009). doi:10110.11029/12008JD010572

R.P. Dhir, Human factor in ecological history in Thar, in *Desertification and Development-Dryland Ecology in Social Perspective*, ed. by B. Spooner, H. S. Mann, (Academic Press, London, 1982), pp. 311–332

R.P. Dhir, Flux in the Indian arid zone, in *Desert Ecology*, ed. by I. Prakash, (Scientific Publishers, Jodhpur, 1988), pp. 15–36

R.P. Dhir, Problem of desertification in arid zone of Rajasthan-a view. Ann. Arid Zone **32**, 79–88 (1993)

R.P. Dhir, Thar Desert in retrospect and prospect. Proc. Indian Natn Sci. Acad. **69A**, 167–184 (2003)

Y. Enzel, L.L. Ely, S. Misra, R. Ramesh, R. Amit, B. Lazar, S.N. Rajaguru, V.R. Baker, A. Sandler, High-resolution Holocene environmental changes in the Thar desert, North West India. Science **284**, 125–128 (1999)

D.Q. Fuller, M. Madella, Issues in Harappan archaeology: retrospect and prospect, in *Archaeology in Retrospect. Protohistory Part II*, ed. by S. Settar, R. Korisettar, (Manohar Publications, New Delhi, 2001), pp. 317–390

L. Giosan P.D. Clift, M.G. Macklin, D.Q. Fuller, D.Q. Fuller, S. Coinstantinescu, J.E. Durcan, T. Stevans, G.A.T. Duller, A.R. Tarbez, K. Gangal, R. Adhikari, A. Alizai, F. Filip, S. Vanlaningham, P.M. Syvitski. *Fluvial Landscape of the Harappan Civilization* (PNAS, 2012) (www.pnas.org/cgi/doi/10.1073/pnas.1112743109)

B.N. Goswami, K.V. Ramesh, The expanding Indian Desert: Assessment through weighted epochal trend ensemble. Curr. Sci. **94**(4), 476–480 (2008)

Inter-governmental Panel on Climate Change (IPCC), Fourth Assessment Report (AR4) Cambridge University Press. ISBN 978-0-521-88009-1 (pb: 978-0-521-70596-7) (2007)

N.S. Jodha, V.S. Vyas, *Conditions of Stability and Growth in Arid Agriculture* (Agro-economic Research Centre, Vallabh Vidyanagar, Gujarat, 1969), p. 21

A. Kar, GCM – derived future climate of arid Western India and implications for land degradation. Ann. Arid Zone **51**, 147–169 (2012)

A. Kar, P.C. Mohrana, P. Raina, M. Kumar, M.L. Soni, P. Santra, A.S. Ajai Arya, P.S. Dhinwa, Desertification and its control measures, in *Trends in Arid Zone Research in India*, ed. by A. Kar, B. K. Garg, M. P. Singh, S. Kathju, (Central Arid Zone Research Institute, Jodhpur, 2009), pp. 1–47

S. Kumar, Vegetation of the Indian Arid ecosystem, in *Desertification Control*, ed. by S. Singh, A. Kar, (Agro Botanical Publishers, India, 1997), pp. 71–79

J. Lu, G.A. Vechhi, T. Reichler, Expansion of the Hadley cell under global warming. (PDF) Geophys. Res. Lett. **34**(6), L06805 (2007). doi:10.1029/2006GL028443. Bibcode:2007GeoRL.3406805L

M. Madella, D.Q. Fuller, Palaeocology and Harappan Civilisation of South Asia: reconsideration. Quat. Sci. Rev. **25**, 1283–1301 (2006)

S.P. Malhotra, Socio-demographic factors of the arid zone, in *Desertification and Its Control*, ed. by P. L. Jaiswal, (ICAR, New Delhi, 1977), pp. 310–323

G.A. Meehl, W.M. Washington, South Asia summer monsoon variability in a model with a doubled atmospheric carbon dioxide concentration. Science **260**, 1101–1104 (1993)

V.N. Misra, Climate, a factor in the rise and fall of the Indus civilisation-evidence from Rajasthan and beyond, in *Frontiers of the Indus Civilization*, ed. by B. B. Lal, S. P. Gupta, (Books & Books, New Delhi, 1984), pp. 461–489

R. Mughal, Recent archaeological research in Cholistan Desert, in *Harappan Civilsation – A Contemporary Perspective*, ed. by G. L. Possehl, (Arris and Phillips Ltd., Warminister, 1982), pp. 85–96

P. Narain, K.D. Sharma, A.S. Rao, D.V. Singh, B.K. Mathur, U.R. Ahuja, *Strategy to Combat Drought and Famine in Indian Arid Zone* (Central Arid Zone Research Institute, Jodhpur, 2000), p. 65

R.P. Pandey, S.L. Meena, R.P. Pandey, M.K. Singhadiya, A review of depleting plants resources, their present status and conservation in Rajasthan. Biol. Forum Int. J. **4**, 231–220 (2012)

G.B. Pant, L.S. Hingane, Climatic changes in and around the Rajasthan desert during the 20th century. J. Climatol. **8**(4), 391–401 (1988)

S. Poonia, A.S. Rao, Climate change and its impact on Thar Desert ecosystem. J. Agric. Phys. **13**(1), 71–79 (2013)

K. Rajendran, A. Kitoh, Indian summer monsoon in future climate projection by super high-resolution global model. Curr. Sci. **95**(11), 1560–1569 (2008)

A.S. Rao, S. Poonia, Climate change impact in on crop water requirement in arid Rajasthan. J. Agrometeorology **13**, 17–24 (2011)

A.S. Rao, M.M. Roy, *Weather Variability and Crop Production in Arid Rajasthan*, vol 70 (CAZRI, Jodhpur, 2012)

L.S. Rathore, S.D. Attri, A.K. Jaswal, *State Level Climate Change Trends in India Meteor. Monog. ESSO?IMD/EMRC?02/2013* (India Meteorological Department, New Delhi, 2013), p. 147

V.N. Reu, *Marwar Rajya ka Itihas* (Rajasthan Granthgar, Jodhpur, 1935) (reprinted 1997). p358+Appendices

K. Rupa Kumar, A.K. Sahai, K. Krishna Kumar, S.K. Patwardhan, P.K. Mishra, J.V. Revadekar, K. Kamala, G.B. Pant, High-resolution climate change scenarios for India for the 21st century. Curr. Sci. **90**(3), 334–345 (2006)

G. Singh, R.J. Wasson, D.P. Agrawal, Vegetation and seasonal climate changes since the last full glacial in the Thar desert, NW India. Rev. Palaeobot. Palynol. **64**, 351–358 (1990)

C.J. Tod, *Annals and Antiquities of Rajasthan Vol. I&II* (Routledge and Kegan Paul Publishers, London, 1829–32) (reprinted 1957)

K.S. Valdiya, The River Saraswati was a Himalayan-born river. Curr. Sci. **104**(1), 42–54 (2013)

R.P. Wright, *The Ancient Indus Urbanism, Economy and Society* (Cambridge University Press, Cambridge, UK, 2010.) 396 p

S.B. Yashpal, R.K. Sood, D.P. Agarwal, Remote sensing of the "lost" Saraswati River. Proc. Indian Acad. Sci. Earth Planet. Sci. **89**, 317–331 (1980)

Part III

Northern Hemisphere Aridlands: Selected Examples

Five authors with specific knowledge and expertise provide input into this Part. Here the attention is switched to countries in the Northern Hemisphere with extensive areas of dryland.

In Chap. 8, *Victor Squires* turns the spotlight onto the problems and prospects in the vast North American continent. Much attention is paid here to water and the challenges faced already to satisfy competing needs for urban areas and for food production from irrigated farmlands. Desert ecology in such places as the Mohave and the Chihuahuan is under dire threat from invasive plants and risk of fire. Mitigation and adaptation strategies are being developed, but the outcomes are far from certain.

India with its extensive aridlands and huge population is under great threat from climate change. According to *Ajai* and *PS Dhinwa*, land degradation and desertization are major threats to livelihoods. Action plans to combat these scourges are already in place, but progress is being slowed by changing climates (more seasonal variability, longer and more severe droughts and water shortages). Irrigated agriculture – a major activity in the arid regions – is vulnerable to water shortages but a target for adaptive strategies including widespread use of water-saving agriculture and other measures.

The drylands of China, including the famous Gobi Desert and the Taklamakan Desert, plus many lesser-known ones present problems but some bright prospects as outlined by *Shiming Ma*. China has made great progress in the past 60 years or so to stabilize mobile sand dunes, reduce dust and sand storms, rehabilitate millions of hectares of degraded lands and lift millions of people out of poverty.

Around the shores of the Mediterranean Sea, there are huge tracts of drylands and some deserts. West Asian countries and the Middle Eastern countries are often grouped with North African countries in some regional efforts to mitigate and adapt to climate change. The WANA (West Asia and North Africa) region and the MENA (Middle East and North Africa) face some similar problems but also some that are unique to only certain near-neighbours. **Ahmed Mohamed** and ***Victor R. Squires*** outline the major issues that beset this group of countries. Reference is also made to countries on the north side of the Mediterranean and their efforts to deal with land degradation, biodiversity loss and threats to livelihood.

Dryland Resources of North America with Special Reference to Regions West of the 100th Parallel

Victor R. Squires

8.1 The North American Continent

North America, the third largest continent, is entirely within the Northern Hemisphere and almost all within the Western Hemisphere. It is bordered to the north by the Arctic Ocean, to the east by the Atlantic Ocean, to the west and south by the Pacific Ocean, and to the southeast by South America and the Caribbean Sea. There are many countries—the major ones are Canada, the USA, and Mexico (the subject of this chapter). North America is a vast region stretching from the Arctic Circle in the north to the Equator. It is home to more than over 550 million people. A significant part of the continent is arid or semiarid. Some of this is high elevation cold desert where precipitation is low and temperatures are often well below freezing point. The focus of this chapter though is with the arid lowlands west of the 100th parallel.

Human occupation of the arid and semiarid areas of North America has had a long history dating back to the end of the last Ice Age (Huntsinger and Starr 2006). Indigenous peoples learned to live with the constraints of aridity and low plant productivity. Each of the American Indian tribes (now called Native Americans) adapted ways which the meager resources could be used for human food and domestic animal feed. Most of the meat protein was derived from wildlife killed by hunting parties. Nuts, berries, and seeds collected from wild plants were stored for use in winter or other times of stress. Compared with present-day land use practices, the Native Americans lived in harmony with nature. Conquerors and settlers from Europe took control of the land either by means of land grants from a foreign power, such as the Spanish Land Grants in Mexico, or by the extension of sovereign power over large tracts of "public land." To encourage settlement of the western desert, the

---

V.R. Squires (✉)
Adjunct professor, University of Arizona, Tucson, AZ, USA
e-mail: dryland1812@internode.on.net

US Congress passed the Enlarged Homestead Act of 1911 and the Stock-Raising Homestead Act of 1916 that permitted entry of 320- and 640-acre[1] homesteads, respectively. Coming as they did from climates and vegetation dissimilar to that of the new lands on which they settled settlers had a slow and sometimes painful learning process. Misinformation of the basic productivity of the land and competition for use of relatively free "common" resources led to buildup of livestock inventories at levels far in excess of carrying capacity. The result was deterioration of land productivity through overgrazing of forage grasses, a decrease in high palatability plant species and accelerated erosion. In the late 1880s and early 1900s, sheep numbers in the USA reached peak levels and created intense competition for grazing land. Against this background of intense use of land, abuse, and lack of understanding, much damage was done. Although the aridlands were affected adversely by past malpractice, greater knowledge of ecological potential of individual plant species and the whole plant community and improved management practices give reason for optimism.

8.2   Aridlands of North America

North American aridlands represent one of the more extensive natural resource bases on the continent, yet they are one of the least understood and possibly the poorest managed. Extensive regions in the western half of northern America are relief deserts.[2] High mountain ranges (The Rockies, etc.) prevent moisture-bearing clouds from the Pacific reaching the interior. Another type of desert occurs along the west coast of Mexico's Lower California Peninsula that receives rain only every 5 years of so. The cold California current that comes from the Arctic prevents moist air masses from reaching the peninsula[3] and points west of there.

The ecological implications of the climatic and biological restrictions that are imposed on those seeking settlement and development of aridlands are seldom related to the lives of people who inhabit the aridlands. Equally as important as the physical environment, but usually not as well understood, are sociological and economic restrictions that are imposed in the aridlands. For instance, most aridland areas have a low population density, and as a result, the lack of support for public funding projects (development or environmental protection) and research institutions in these sparsely populated areas is not related to need so much as to the perception of priorities by people living in the high population areas where political powers are concentrated. People living in rural parts of aridlands often lack the ability to support its development on their own, and often the urban community does not view their own development as a high priority.

---

[1] 2.5 acres equals 1 ha.

[2] Relief deserts form in areas beyond high mountains or on high plateaus that clouds cannot reach. In rising, the clouds cool off and rain falls on the outer slope or the mountain range or plateau.

[3] Winds blowing across the oceans toward the land mass crosses the cold current which cools the air so that is unable to hold large amounts of water.

Fig. 8.1 Map showing the location of the "American West," including the 100th meridian

Support facilities and services are poor or lacking in many aridlands. Communication between individuals and groups of people is usually difficult. Public transport facilities are poor or entirely lacking in some areas; this influences access to market, education, health services, and other amenities. Social institution serving the aridland dwellers is usually weak. The combination of a low population with little political organization usually tends to have the aridland dwellers underrepresented in organizations and structures affecting decision making about development and/or land use. Sociological factors are further complicated by rather severe economic restrictions. The arid regions offer a high risk with low returns to investment. Until venture capital is available under conditions specifically designed for arid aridlands, little sound development can take place in aridlands.

In summary, the setting under which we must try to achieve the potential of aridlands is harsh. Precipitation is low, biological activity is erratic, and practically no social or economic institutions have developed to aid economic development. Because the distribution of rainfall is not uniform across the USA, water (or, more accurately, the lack thereof) defines the region known as the American West: 17 states located on and westward of the 100th meridian (see Fig. 8.1). Much land in the American West is publically owned (state and federal) or Indian (Native American)-owned land.

In the past, arid and semiarid regions were defined on the basis of climatic factors alone, e.g., Koppen's world map (published in 1923), and in fact such classifications are in use today such as those used by the UNCCD where drylands are defined as areas with an aridity index value of less than 0.65, that is, areas in which annual mean potential evapotranspiration is at least 1.5 times greater than annual mean precipitation.

However, since vegetation is the result of the responses of plants to many variables of the climatic factors, no single factor can characterize deserts or arid or semiarid regions. The key climatic data are normal average precipitation, potential evaporation, water deficiency, and aridity index. Sometimes it is that the highest aridity index is in cold deserts at higher elevations rather than the hot, dry lowland deserts.

8.3 Patterns of Occurrence of Aridlands of North America

The patterns of occurrence for North American aridlands are based on many historical factors extending through geological time. The North American region's geography and climate have undergone dramatic changes as a result of significant geological alterations. The present Cascade Sierra and Sierra Madre Oriental ranges are a part of the geologic mountain building upheaval that created a rain-shadow effect in the path of prevailing westerly wind currents, causing a gradual but significant change to drier climate east of the mountains. Axelrod (1940, 1981) speculated that during the Tertiary period the climate changed from temperate to warm-temperate to tropical and, finally at the end of the Tertiary, to a dry and relatively cool climate similar to the present. At higher elevation in the Rocky Mountain chain, glaciations occurred.

8.3.1 Evolution of Major Vegetation Types

The plant fossil record documents the progression of vegetation types during the climatic changes. Many mesic species failed to survive the drier conditions but those related to the present-day desert shrublands types were common in the Tertiary flora, and the more xeric species gradually became climax and now dominate vast desert areas that were formerly woodland in more mesic geological periods. There was evolution of semi-xeric shrub species during the Upper Cretaceous period when arid areas were relatively numerous and widespread. With the onset of widespread mesic, equitable climates at the beginning of the Tertiary, particularly in the Eocene, only the hardiest of these xeric and semi-xeric species could survive in specialized ecological islands that provided favorable habitats, e.g., western Oregon where outcrops of hard rock and shallow soil support localized populations of species that occur chiefly in drier regions. When during the Oligocene and /Miocene periods, more xeric islands appeared and increased in size; those species that had survived the mesic interlude in special situations were now able to spread and, in instances, to become widespread.

The evolution of "secondary shrubs" is, therefore, a good example of the general principle of evolutionary opportunities. Many monotypic or ditypic genera exist in the vegetation of the aridlands of western North America. A reasonable hypothesis of their taxonomic isolation makes them products of the extensive evolution of semi-xeric shrubs during past eras. The evolution of these large complex shrubby (in particular) genera is therefore an example of the principle of rapid evolution in small semi-isolated populations (Dozhansky 1970).

Marine influences from the great inland ocean are evident in landform and chemical composition of soils in western North America. Mountain faces show sedimentary stratification and marine deposits. Weathered rocks and soils yield salts from marine formations. These changes since the Pleistocene period have an especially influential on present-day conditions, especially vegetation cover. During the past 150 years, human impacts have seen more rapid changes than in any short period of history because of the degrading effects of exploitative grazing use. Intensive utilization of shrublands and grasslands by domestic livestock has shifted the balance from a shrub- and grass-dominated vegetation to shrub dominance in many sites and encouraged desertification processes. Heavy grazing by livestock and big game over the past 150 years has been responsible for reducing, and in some instances markedly opening, closed stands of shrubs and grasses on extensive areas in the western half of North America, especially in the USA and Mexico. Hutchinson et al. (2000) examined the case of southeastern Arizona where significant historical vegetation change has occurred (Bahre 1991). Using repeat aerial photography covering a period of approximately 50 years, the analysis considers the two primary hypotheses for vegetation change in the region: climate and land use. From both this analysis and a description of historical land uses in the area, evidence is presented that land use and not climate change is the primary driver of historical vegetation in the areas.

Some of the other uses related to human activities, of course, had sharper and quicker effects than grazing, but their impacts have not been so continuous nor have they affected such large areas. Important among them are clearing land for crops, mining, (and associated industry—mineral prospecting, seismic survey, fracking[4], etc.), infrastructure developments, harvesting of trees, and greater incidence of fire. Regardless of how the opening of closed communities resulted, it made a great many sites available for increase of shrubs (shrub encroachment). That many openings are still persisting is quickly recognized by rapid occupancy of them by annuals—most of them exotic species. Mention should be made of the invasion of European annual grasses into the herbaceous understory. The principal invader has been cheatgrass (*Bromus tectorum*). But medusahead (*Taeniatherum caput-medusae*), foxtail chess (*B. rubens*), and ripgut grass (*B. rigidus*) are common in certain types of habitats. Both cheatgrass and foxtail chess are widespread in the

---

[4] Also called hydraulic fracturing.

West, and medusahead continues to spread throughout the intermountain area.[5] Where conditions are favorable, these annuals have significant competitive ability and often dominate the understory if perennial vegetation is disturbed. In time, annuals may be replaced by perennials—including shrubs.

8.4 Major Vegetation Types

8.4.1 Prairie Grasslands

Prairies used to cover a huge area of land in Canada, the USA, and northern Mexico. The prairies are a type of grassland dominated by herbaceous plants principally grasses. Very few trees grow on the prairies and are usually widely scattered. The prairies form a triangular area from Alberta, Saskatchewan, and Manitoba down through the Great Plains to southern Texas and Mexico and approximately 1600 km from western Indiana westward to the Rocky Mountains. The rainfall in the prairies decreases from east to west. This creates different types of prairies, with the *tallgrass prairie*, known as the true prairie, in the wetter parts. Grasses such as big bluestem (*Andropogon gerardii*), and Indian grass (*Sorghastrum nutans*), and many species of flowers grow here. *Mixed-grass prairies* are found in the central Great Plains and *shortgrass prairie* toward the rain shadow of the Rocky Mountains. The rain shadow causes Pacific Ocean moisture to rise and cool, dropping as rain or snow on the western side of the mountains instead of on the prairies. The climate of the prairies is influenced by its mid-continental location and the sheltering effect of the Rocky Mountains. Being located far from the moderating effects of oceans causes a wide range of temperatures, with hot summers and cold winters. Strong winds blow across the endless plains during both summer and winter. Precipitation in the prairies can reach from about 300 mm in the shortgrass prairie to 600 mm in the tallgrass prairies.

The prairies were maintained in their natural state by climate, grazing, and fire. Rainfall varies from year to year in the prairies. There is usually a long dry period during the summer months. Every 30 years or so, there is a long drought period which lasts for several years. The most famous drought was in the 1930s, when the prairies were called the "Dust Bowl." Every 1–5 years fire would spread across any given area of land. These fires moved rapidly across the land and did not penetrate into the soil very far. They killed most saplings and removed the thatch of dead grasses (litter), allowing early flowering spring species to grow. Prairie plants have adapted to fires by growing underground storage structures and having their growth points slightly below ground surface. The soil under a prairie is a dense mat of tangled roots, rhizomes, bulbs, and rootstock. The plants die back every winter but are kept alive from year to year by the underground root system. Roots of prairie plants

---

[5] That area from the Southern Rocky Mountains on the east to the crest of the Sierra Nevada in the west and from southeastern Oregon, southern Idaho, and southwestern Wyoming southward through Utah and Nevada is often referred to as the Intermountain West.

can be longer than the plant is tall. The roots of big bluestem may be 2 m long, and switchgrass (*Panicum virgatum*) roots can be 3 m long. Two-thirds of the biomass of most prairie plant is below the ground. Some roots die each year and decompose, adding organic matter to the soil.

Before settlers moved west, the prairies were utilized by vast herds of grazing animals, such as buffalo, elk, deer, and rabbits. These animals increased the growth in prairies by adding nitrogen to the soil through urine and feces. Today very little of the original prairies survive, only 1–2%. Much of the land has been turned into agricultural uses, urban areas are moving in, and fires are being suppressed. The genetic and biological diversity of the plants are disappearing. The herds of thousands of buffalo were all but wiped out. There is a strong movement to educate people about prairies. Many states are rehabilitating what is left of their prairies and reintroducing the native wildlife and plants.

8.4.2 Salt Deserts

The salt desert shrub range in the cold desert and steppe climates of North America is a mosaic of communities in the plains, foothills, and valley edges of the Great Basin (see Fig. 8.2) and the northern part of the Colorado Plateau. Outliers are to be found in some arid areas of the Rocky Mountains and eastward on parts of the Great Plains. Low shrubs or half shrubs dominate. In some areas perennial grasses share dominance with low shrubs with a ratio (based on biomass yield) of 1:4. Usually 5–15% of the ground surface is shaded by living plant cover. The mosaic of plant communities is composed of homogeneous units ranging from 1 ha to many thousands of ha in size. The pattern is a function of variations in soil characteristics and moisture regimes.

This extensive ecological system includes open-canopied shrublands of typically saline desert basins, alluvial slopes, and plains across the intermountain western USA. Considered a matrix-forming system to the west of Colorado, this type also extends in limited distribution into the southern Great Plains, where it is a large patch system. Substrates are often saline and calcareous, medium- to fine-textured, alkaline soils but include some coarser-textured soils. The vegetation is characterized by a typically open to moderately dense shrubland composed of one or more *Atriplex* species or other salt-tolerant species.

As rangeland suited to livestock grazing, the salt desert shrub communities are unique among North American grazing lands. Due to the arid climate (100–200 mm of precipitation per year), herbage yields and carrying capacities are low. The salt desert shrub rangeland is a browse area (more than 90% of the forage for livestock is from browse). Shrubs or half shrubs provide most of the feed for livestock. About 0.6–1.2 ha are required to support a sheep for month and about 4–8 ha to support a cow for a month. Grazing is seasonal with water supply restricting use to winter when snow and metabolic moisture from living plants can meet daily water requirements of livestock (Squires 1987). It serves as a "holding area" for maintenance of breeding or gestating livestock: so nutritional requirements are relatively low.

Nonetheless the living twigs of dormant browse plants, some with persistent seeds (e.g., *Atriplex confertifolia*) make this rangeland more nearly adequate for animal maintenance than other cold-weather rangelands where the aerial parts are dead.

Throughout the past 150 years, the deterioration of North American salt desert shrub communities as a result of grazing leading to replacement of choice forage species has been duly documented (McKell 1989; Roundy et al. 1995).The first signs of deterioration are reductions in the size, vigor, and number of desirable forage plants. Concurrently soil stability decreases. Annual forbs and grasses become abundant in years favorable for their growth, and in time, less desirable species (shrubs and invasive plants – some of them toxic to livestock) increase and may attain dominance (Blaisdell and Holmgren 1984).

8.4.3 Sagebrush Steppe

This is a type of shrub steppe, a grassland characterized by the presence of shrubs and usually dominated by sagebrush, any of several species in the genus *Artemisia*. This ecosystem is found in the Intermountain West in the USA with a variant—sand sagebrush (*A. filifolia*) that occurs from Nevada east to South Dakota and from there south to Arizona, Chihuahua, and Texas. The most common sagebrush species in the sagebrush steppe in most areas is big sagebrush (*Artemisia tridentata*). Others include three-tip sagebrush (*Artemisia tripartita*) and low sagebrush (*Artemisia arbuscula*). Sagebrush is found alongside many species of grasses.

Sagebrush steppe is a diverse habitat that is a threatened ecosystem in many regions. It was once very widespread in the regions that form the Intermountain West, such as the Great Basin and Colorado Plateau (Figs. 8.2 and 8.3).

Fig. 8.2 Principal areas of sagebrush (*Artemisia*) shrub steppe in western USA

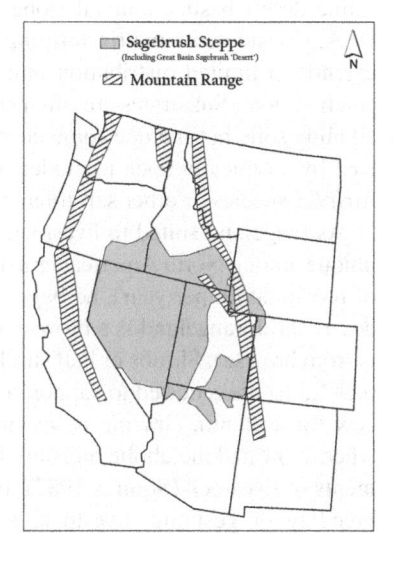

It has become fragmented and degraded by a number of forces. Steppe has been overgrown with introduced species and has changed to an ecosystem resembling pinyon pine (*Pinus cembroides*) and juniper (*Juniperus osteosperma*) woodland. This has changed the fire regime of the landscape, increasing fuel loads and increasing the chance of unnaturally severe wildfires. Cheatgrass (*Bromus tectorum*) is also an important introduced plant species that increases fire risk in this ecosystem. Other forces leading to these habitat changes include fire suppression and overgrazing by livestock. Besides severe fire, consequences of the breakdown of sagebrush steppe include increased erosion of the land and sedimentation in local waterways, decreased water quality, decreased quality of forage available for livestock, and degradation of habitat.

8.4.4 Desert Shrublands

Shrubs are the dominant growth form of deserts. They may be evergreen or deciduous, typically have small leaves, and frequently have spines or thorns and/or aromatic oils. Shallow but extensive root systems procure rainwater from well beyond the canopy of the shrub whenever it does rain. These are the true *xerophytes* adapted to tolerate extreme drought. They form an open canopy and, except after rains when annuals may cover the desert floor, the ground between shrubs is bare of vegetative growth. Water is not entirely lacking in the desert environment and several other growth forms represent strategies to reach water or to store water:

- *Phreatophytes* are plants with long taproots that may extend downward 7–10 m to tap groundwater supplies. Especially along intermittent streams or under dunes, underground water may be readily available. Mesquite is a good example in the SW (New Mexico, Arizona).
- *Succulents* store water accumulated during rains for use during the intervening dry spells. Different species store water in different parts of the plant; hence, we can recognize stem succulents, leaf succulents, root succulents, and fruit succulents. Many plant families have members that evolved succulence. Most prominent among stem succulents in the Americas are the Cactaceae. The agaves (Liliaceae) are examples of leaf succulents in the Americas. Most succulents do not tolerate freezing temperatures so they are essentially limited to the hot deserts.
- Another growth form adapted to desert conditions is the ephemeral. This is an especially short-lived annual forb that completes its life cycle in 2–3 weeks. The seeds are encased in a waterproof coating that prevents desiccation for years if necessary. These plants essentially avoid drought by occurring as seeds most of the time.
- *Perennial forbs* with underground bulbs store nutrients and water in underground tissues and also remain dormant most of the year. They can sprout rapidly after sufficient rains and replenish their underground stores.

Fig. 8.3 Map showing the geographical extent of the Great Basin—a region characterized by basin and range topography and an endorheic drainage system

8.4.5 The Great Basin

This basin is the largest area of contiguous endorheic[6] watersheds in North America. It is noted for both its arid climate and the basin and range topography. The region spans several physiographic divisions, biomes/ecoregions, and deserts The hydrographic Great Basin contains multiple deserts and ecoregions, each with its own distinctive set of flora and fauna. The ecological boundaries and divisions in the Great Basin are unclear. The US Environmental Protection Agency divides the Great Basin into three ecoregions roughly according to latitude: the Northern Basin and Range ecoregion, the Central Basin and Range ecoregion, and the Mojave Basin and Range ecoregion (Fig. 8.3).

---

[6] Endorheic is a closed drainage basin that retains water and allows no outflow to other external bodies of water, such as rivers or oceans, but converges instead into lakes or swamps, permanent or seasonal, that equilibrate through evaporation.

The climate and flora of the Great Basin is strongly dependent on elevation: as the elevation increases, the precipitation increases and temperature decreases. Because of this, forests occur at higher elevations. *Juniperus osteosperma/Pinus monophylla* (southern regions) and *Cercocarpus ledifolius* (northern regions) form open woodland on the slopes of most ranges. Stands of limber pine (*Pinus flexilis*) and Great Basin bristlecone pine (*Pinus longaeva*) can be found in some of the higher ranges. In riparian areas with dependable water, cottonwoods (*Populus fremontii*) and quaking aspen (*Populus tremuloides*) groves exist.

8.4.6 Coastal Desert

These deserts occur in moderately cool to warm areas such as Baja California. The cool winters of coastal deserts are followed by moderately long, warm summers. The average summer day temperature ranges from 13 to 24 °C; winter temperatures are 5 °C or below. The maximum day temperature is above 35 °C and the minimum is about −4 °C. The average rainfall measures 8–13 cm in many areas. The maximum annual precipitation over a long period of years has been 37 cm with a minimum of 5 cm.

The soil is fine-textured with a moderate salt content. It is fairly porous with good drainage. Some plants have extensive root systems close to the surface where they can take advantage of any rain showers. All of the plants with thick and fleshy leaves or stems can take in large quantities of water when it is available and store it for future use. In some plants, the surfaces are corrugated with longitudinal ridges and grooves. When water is available, the stem swells so that the grooves are shallow and the ridges far apart. As the water is used, the stem shrinks so that the grooves are deep and ridges close together. The plants living in this type of desert include the salt bush (*Atriplex* spp.), buckwheat bush (*Fagopyrum esculentum*), rice grass (*Oryzopsis hymenoides*), little leaf horsebrush (*Tetradymia glabrata*), black sage (*Salvia mellifera*), and *Chrysothamnus* spp.

Numerous vegetation types may be distinguished as a result of the large number of ambient factors involved, and this picture is further complicated by the interaction between people and grazing animals (Squires 2010). The depredation of overgrazing has altered the climax vegetation to such a degree in many areas that it will probably never recover, despite the conservation measures now being put into effect. A detailed discussion of vegetation types in this region is given by Stoddart et al. (1975) and Heady and Childs (2001).

8.5 Water: The Pulse of Life in the American West

Water is a tremendously important resource in the western USA and Canada.

Water has always been a vital, scarce, and variable resource in the American West, the source of both conflict and community long before this region became part of the USA. The American West contains the headwaters of the continent's

Fig. 8.4 The six major water divides in North America

major river systems (Fig. 8.4)—including the Columbia, Missouri/Mississippi, Rio Grande, and Colorado rivers—as well as the driest parts of the country: the Mojave, Sonoran, Great Basin, and Chihuahua deserts. A significant part of the region is owned by the federal government and managed as public land, including national forests, national parks, wildlife, and multiple-use public lands (Fig. 8.5).

Water development was essential to facilitate the expansion of the US population into the western regions in the late-nineteenth and early-twentieth centuries. In just a few generations, an ambitious campaign to harness the rivers of the American West transformed the region, attracting tens of millions of new residents and encouraging a major growth-oriented economy. The multiple-purpose water projects constructed through the early- to middle-twentieth century flattened the great variations in water availability from season to season and year to year, making possible extraordinary expansion of economic activity and quality of life for the new settlers. Moreover, the traditional life of indigenous peoples was adversely affected by this development. Today's challenges include reallocating water to meet new and changing demands—driven in large part by demographic shifts and legal mandates to

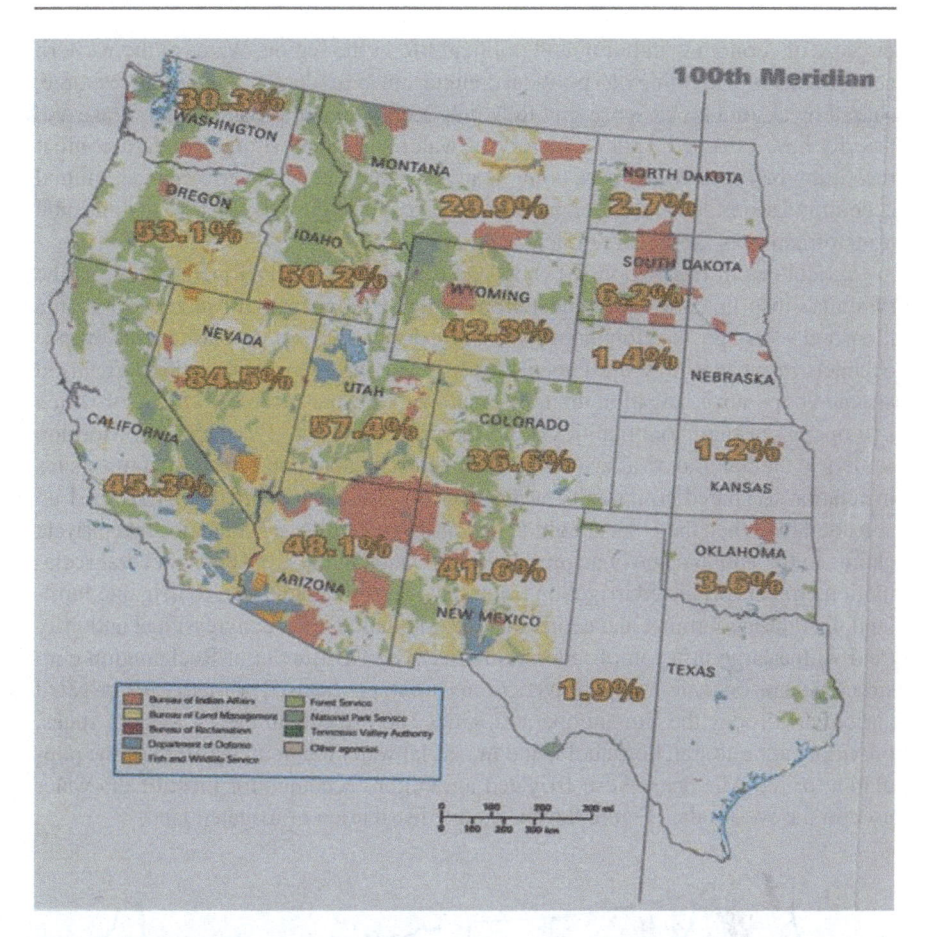

Fig. 8.5 Federal government ownership characterizes this vast region. The public estate (shown here in percentage of each state owned by the federal government) includes national forests, national parks, and multiple-purpose public lands

protect and recover endangered species—and addressing the realities of aging dams and other infrastructure. In short, the transformation continues.

Much of the industry in the West—including mining, mineral processing, manufacturing, and agricultural operations—would not have been possible without substantial public and private investment in water and power through multiple-purpose dams and related projects. Clean, reliable water supply remains a key factor in locating newer high-technology industrial sites. And the rapid expansion of both carbon-based and renewable energy development in the region requires access to water (more for some sectors than others). Much economic growth in the western states relied upon a legal system that recognizes private rights to put water to productive, economic use, while the water itself remains a public resource. In fact, a strong thread of public interest in water underlies a system that emphasizes private rights and decentralized decisions. For approximately 150 years, water issues have been at the core of economic, cultural, and political life in the region. Water in the western states presents something of a political conundrum: a fundamentally public resource, shared by all in a broad sense, but to which deeply valued private rights of use and priority have attached. From a history of water rights and straightforward conflict resolution has emerged a more complicated structure of water governance, mutual accommodation, and public engagement. Like the region itself, this institutional transformation is far from complete.

Agriculture in this region depended on diversion of water for irrigation, as the rainfall is insufficient to grow most crops without supplemental water. More than a thousand years ago, the Hohokam civilization developed extensive water conveyance channels in what is now the Phoenix valley, and early Hispanic settlers organized around cooperative irrigation ditch associations known as *acequias*, many of which still operate in rural communities of northern New Mexico. For their part, Mormon settlers in the Salt Lake Valley (now part of Utah) laid out homesteads and town sites in relation to shared community water sources. The scale of irrigation ramped up considerably when the USA sought to encourage agricultural settlers to stake private claims to the nation's newly acquired western lands in the late 1800s. Federal reclamation projects promised irrigation water to convert the arid desert to irrigable farmland, facilitating farming and economic development. Under congressional authority (and with a large infusion of federal funding), the US Bureau of Reclamation constructed dams on many western rivers to store water to be used primarily for irrigated agriculture. Today the Bureau operates about 180 projects in the 17 western states, providing agricultural, household, and industrial water to about one-third of the population of the American West. Irrigated agriculture accounts for most of the water used in the West today. Figure 8.6 shows the distribution of irrigated land.

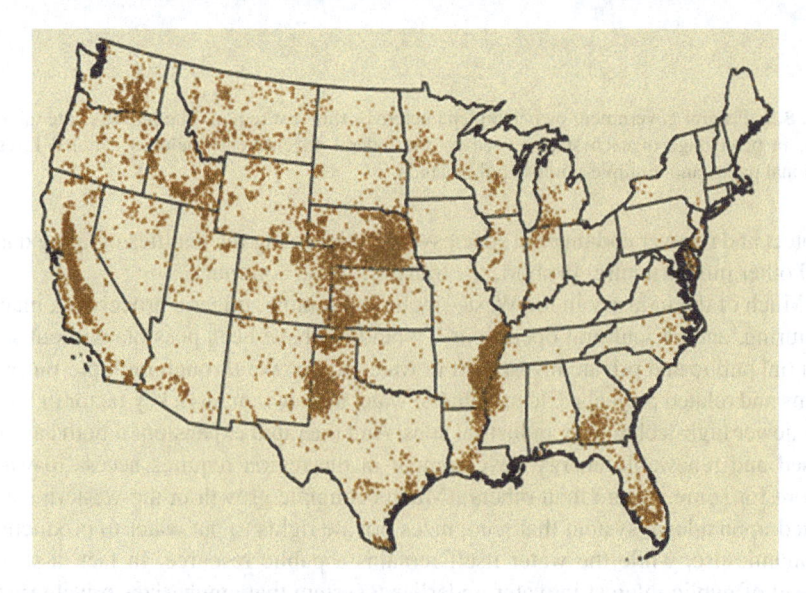

Fig. 8.6 The western lands of USA use a lot of water for irrigation but there is a significant number of irrigated farms in the eastern states

Basic human needs (drinking, cooking, bathing, washing, and sanitation) require about 49 l per person daily, but American households in many US cities use far more, typically averaging 750–1140 l daily. In the arid regions of the USA, a significant amount of domestic water use is for lawn and garden irrigation.

Clean, reliable water supply remains a key factor in locating newer high-technology industrial sites. And the rapid expansion of both carbon-based and renewable energy development in the region requires access to water (more for some sectors than others). In addition to the diverse out-of-stream uses of water, the past several decades have seen a revolutionary shift in scientific and public appreciation for the value of water flowing in its river of origin. "Environmental" or "instream" flows offer myriad values, ranging from commercially profitable recreation (boating and fisheries) to protection of tribal fishing rights and less tangible but important ecosystem services and aesthetic values to residents drawn to the region by its scenic beauty.

What is now the American West was inhabited by Indian tribes long before the formation of the USA. In ceding portions of their land, Indian nations reserved rights to hunt, fish, and continue other traditional uses of their historical homelands and reserved the water necessary to support those rights. They also reserved rights for irrigated agriculture and other uses necessary to a sustainable homeland. Nearly every western river is now harnessed by dams and diversions, to the point that natural flows are a matter of estimate or distant memory. Vast stretches of rivers have been inundated or depleted to achieve economic benefits, but ecological considerations often received limited attention in early decisions governing dam location and operation. Indeed, some dam operators today still follow management regimes that were worked out many decades ago to serve the objectives of navigation (primarily barge traffic), hydroelectric power generation, and water diversions for agricultural irrigation, industries, and municipalities.

8.5.1 Protecting River Ecosystem Values

Historically, water laws in the western states of the USA emphasized economic development by encouraging utilitarian approaches to putting the West's rivers to work—storing water for use during dry periods, moving water over the landscape to satisfy human demands, and maximizing a limited number of "beneficial" uses such as irrigation, municipal, industrial, and domestic supplies. As one US water official observed recently, "What worked is what we valued when we set up the system—water for new communities, irrigation, and flood control. Our challenges today arise from the consequences of what we didn't prioritize or plan for—water for ecosystem services and to fulfil promises to Indian tribes." At their most extreme, historical policy choices resulted in over-appropriated and dewatered streams, depleted aquifers, and degraded river ecosystems. Figure 8.7 is a map of western USA showing the percentage of total water that comes from groundwater aquifers. In some cases, water withdrawals and land development practices compromised water quality in both surface and ground water. Demographic shifts have changed the very

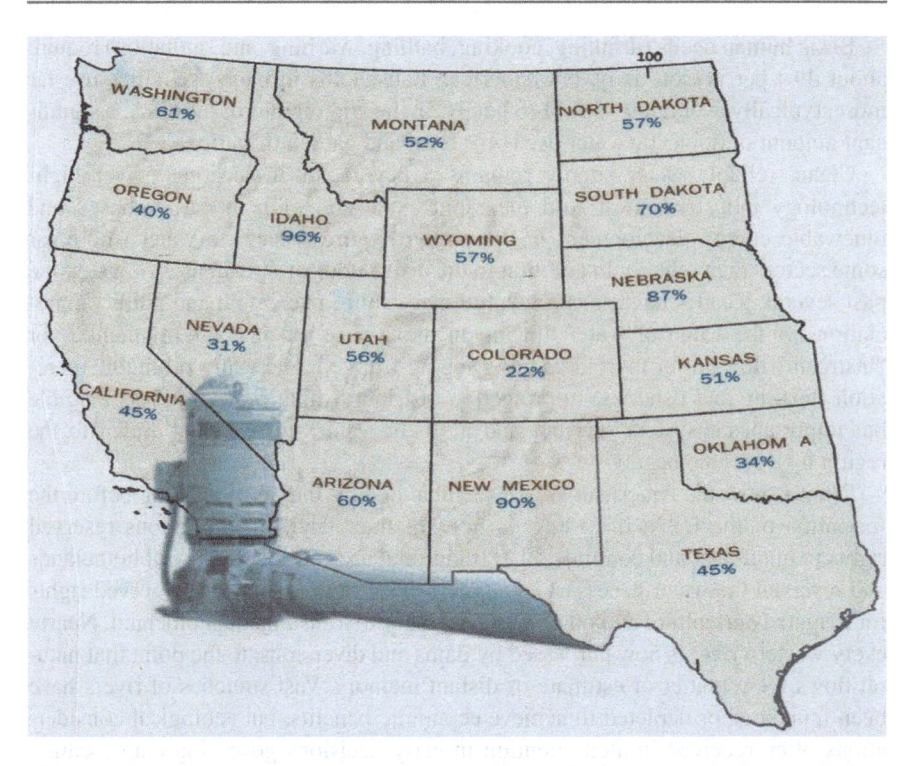

Fig. 8.7 Percentage of population of each state in the contiguous western USA dependent on groundwater for domestic needs. Groundwater contributes a large proportion of total water extracted in each state in the USA west of the 100th meridian. New Mexico is especially dependent on groundwater

structure of demand for water in the West, but institutions continue to reflect historical distribution patterns.

Over time, societal values have shifted and recognized the value of water flowing in rivers, where it supports natural ecosystem functions (including water quality) and supports a diversity of living creatures. In some cases, new information revealed important linkages between development practices and the security of water for future generations, raising additional concerns and desires for protection. Projected impacts from climate change have sparked a broader and more focused conversation about necessary measures to ensure healthy and resilient watersheds and rivers in the coming decades. One piece of US federal legislation alone—the Endangered Species Act—arguably prompted more creative, innovative, ecosystem-based solutions than any other factor in this region. River restoration includes multi-stakeholder partnerships that work out mutually beneficial plans for operating water projects differently—in some cases, decommissioning them. Like air, rivers are a quintessential shared resource, linking people and landscapes through mutual reliance on precious water.

8.6 Adaptive Strategies for Water Scarcity and Variability

Several climate scientists have observed that, "Water will be the delivery mechanism for climate change in the West." The American West is likely heading into a period of less reliable water, with greater variability between wet and dry periods. This will challenge managers of fully (or over-)allocated rivers, who will need every possible tool to encourage efficient use of existing supplies and to facilitate transfers of water from existing uses to where it is most critically needed to respond to shortages. Important tools to support this adaptation—better modelling at finer scales, improved reservoir operations, and more detailed forecasts of risks to human safety from larger hydrological fluctuations—will require sustained investment in long-term water data: stream-gaging, diversion records, and groundwater monitoring (Squires et al. 2014).

Climate change and other hydrological variability require adaptive institutions and technologies capable of responding to new and changing information. Emerging policies should reflect the new concerns about the consequences of development practices and climate change. As surface water supplies have become fully appropriated, many water users have turned to groundwater, sometimes pumping at unsustainable rates or impairing the flows of adjacent rivers and impairing the rights of nearby water rights holders. Western states are beginning to address the relationship between surface and groundwater through new legal rules and practices, but there is little consistency and many loopholes remain.

Challenges of managing water in the American West have given rise to innovative and adaptive management strategies, many of which continue to evolve today.

Some examples of emerging adaptations include:

- Incentives and mandates to reduce energy and water consumption
- Clear guidelines to review potential water transfers
- Cooperative efforts to protect and restore watersheds, wetlands, and floodplains to maximize their ability to provide valuable ecosystem services
- Emerging efforts to share data and prioritize data needs among federal and state and local agencies and Indian tribes and across international boundaries
- Aggressive development of technological adaptations, including desalination of ocean and brackish groundwater, and wastewater treatment and reuse.
- Attention to the impacts of pumping groundwater and improved methods of conjunctive management of surface and groundwater

The key strategies developed in the arid inland western states are described in six categories that relate to "water solution themes laid out for discussion at the 6th World Water Forum".

Box 8.1 relates these strategies to the 6th World Water Forum themes and conditions for success.

> **Box 8.1 Western USA: Water Adaptive Strategies and Conditions for Success (CS)**
>
> | | |
> |---|---|
> | *CS – Managing water as a scarce and variable resource* | |
> | 1.4 – Prevent and respond to water-related risks and crisis | |
> | Protecting river ecosystem values | |
> | 2.4 – Promote green growth and the value of ecosystem services | |
> | 3.1 – Improve the quality of water and ecosystems | |
> | Honoring indigenous water rights | |
> | 1.1 – Guarantee access to water for all and the right to water | |
> | Engaging diverse stakeholders in developing new solutions | |
> | *CS 1 – Good governance* | |
> | Managing water across state and international boundaries | |
> | 1.5 – Cooperation and peace through water | |
> | 2.1 – Balance multiple uses through integrated water resources management | |
> | *CS 2 – Financing water for all innovative tools for water infrastructure financing* | |

8.6.1 Water-Energy Nexus

Energy development poses new challenges to water quality and quantity. Pumping and transporting water require a great deal of energy; producing energy sometimes consumes a great deal of water. This relationship has gained new attention in recent years. With the exception of the Columbia River Basin, where integrated planning for hydropower, flood control, and fish and wildlife needs is long established, public understanding of the water and energy nexus is still in its infancy. At the very least, managers are recognizing that water conservation saves energy (and vice versa) and are beginning to calculate energy costs of new water projects. Large portions of the American West are experiencing rapid energy development, both renewable (solar, wind, and geothermal) and carbon-based (including shale-oil development through hydraulic fracturing). New technologies may impair the reliability and quality of local water supplies. Federal and state regulatory agencies are working to study and develop new regulations to address these potential impacts. In many cases new technologies outpace regulations.

8.7 Land-Based Resource Utilization

Land-based activities include mining, ranching, forestry, tourism, and recreation and conservation reserves and national parks. The focus in this chapter will be on ranching. Reader's interested in the other aspects are referred to the specialist literature on those topics. See, for example, Yeldell and Squires (2016) for a discussion on mining in western USA.

8.7.1 The Range Livestock Industry

The land dedicated to extensive grazing is, in general, located in the 17 states of the USA, west of the 100th median and in Prairie provinces in Canada (Saskatchewan and Alberta and parts of British Columbia) and in northern Mexico (Huntsinger and Starr 2010; Jemison and Raish 2000). These lands are characterized by low rainfall (20% receives less than 250 mm and an additional 25%, i.e., 45% in all, less than 380 mm). In some of these areas, the greater part of precipitation falls during winter, the moisture, consequently, only becoming available as the snow melts. Nonetheless, these vast lands have supported range-based ranching. The rangeland livestock industry in Mexico, the USA, and Canada had its origin nearly a century after Columbus first brought animals to the Western Hemisphere. Livestock imported from various European countries and elsewhere were important to then new colonies as sources of food, fiber, and draught (draft) power.

After independence in 1776, the US government took steps to encourage agriculture at a time when over 80% of the population lived on farms. A similar pattern of events took place in Canada where the first school of agriculture was established in1789. Early in the nineteenth century, agricultural societies were formed and agricultural statistics were collected. In the USA, as long ago as 1862, legislation was enacted to establish agricultural colleges and a national Department of Agriculture with a strong extension mandate. Similar activities were not long in following in Canada. From early on, agricultural associations and shows (county fairs, etc.) were popular, and this stimulated widespread awareness of the importance of superior stock and improvement of breeds. Throughout the past 200 years, the livestock industry in both countries has been well organized from a solid producer base. The industrialization from the mid-1800s, especially in the USA, helped to create strong markets for livestock products which, in turn, served as a strong incentive to producers.

Climate change is emerging as a major factor in several regions. Population size and growth, relative to land availability, also have major implications for the evolution of ranching systems as change (global change—not just climate change) overtakes them. Change is not new. For example, in the middle 1880s, much of the northern Chihuahuan Desert of southwestern USA and Mexico was dominated by semiarid grasslands. Historical studies have shown that desert-scrub species have been steadily expanding their geographic range since 1858. In the past 130 or more years, there has been a documented shift from mosaic of grasslands with desert scrub to a desert landscape with remnant pockets of grass. Since that time, land degradation in this region, identified as the replacing of grasses with desert shrub and bare soil, has been widespread and well documented. Using high-temporal resolution satellite data, it was possible to detect the plant community composition shift from grasses to scrub indicative of land degradation (Peters et al. 2012).

8.7.2 Major Ranching Systems

In North America, beef is produced from cattle raised on forage that other animals generally do not use. Range and pasture lands are located in all 50 states of the USA. Privately owned range and pasture lands make up over 27% of the total acreage of the contiguous 48 states, and these lands constitute the largest private land use-category, exceeding both forest land (21%) and crop land (18%). Most grazing lands are considered either range or pasture, but grazing lands also include grazed forest lands, grazed croplands, haylands, and native/naturalized pasture. These other land use types make up an additional 26 million hectares of privately owned grazing lands or about 17% of the total US grazing lands. These other types of grazing lands provide a significant forage resource for US livestock production.

In most areas where beef cattle are raised, beef production offers the most efficient use of resources—labor, land, and capital. In terms of land use, ranching is peripheral to the sedentary agricultural resource-use systems such as rainfed cropping. Extensive beef cattle operations are characterized by large-scale ranching, mostly on native and naturalized pastures. Ranch size is largely a function of the density of human populations, location relative to urban centers, range carrying capacity, and socioeconomic conditions. Cattle ranches may aim to produce a specialized type of turnoff. They can be differentiated as (a) ranches whose function is to raise calves (cow-calf operation) with young stock being sold off for finishing elsewhere (b) whose operation involves both breeding and, where possible, fattening for slaughter (c) fattening ranches, where animals are bought in for finishing to slaughter weight. Systems of raising beef vary greatly both within countries and within them. In some cases, whole regions are dominated by one production system. Cattle systems evolve as a result of a host of factors, some of which are stimulants while others are constraints. The systems are not static but changes over time in response to technological, political, and economic forces. Some of the key factors which simulate or constrain cattle production systems are land, production complementarity, tradition, social conditions, location, level of effective demand, price of beef, and government policies. Ranching systems in the drier regions are based on cow-calf operations. Offspring may be sold at or soon after weaning or may be raised to almost 300 kg liveweight for later fattening in regions with better quality pastures or to specialized feedlots. Occasionally, and depending upon the available forage/fodder resources, the whole cycle of breeding, raising, and fattening is found in a single property. Grazing practices in these extensive systems are variable and opportunistic rather than planned. Continuous grazing is most generally practiced when grazing native pastures, but some form of alternate or rotational grazing is frequently found on sown pastures (Huntsinger and Starr 2010; Brown and McDonald 1995).

8.8 Brief Overview of Ranching Regions in Mexico

An examination of the grazing industries and ranching systems in the various examples given below will show that the basics are similar throughout the drylands of USA and Mexico. Mexico has extensive areas of drylands where 77% of the people live and produce 87% of GDP but receive only 31% of the precipitation. Mexico is also an oil-exporting country with the fourth most important reserves of shale gas. Land use changes, overexploitation of aquifers, chaotic urbanization, and export agribusiness in drylands have increased the impacts of global environmental and climate change with severe soil erosion and biodiversity loss. The overexploitation of aquifers in coastal regions due to export agriculture in arid climate and a growing urban population have produced intrusion of seawater into the aquifers in several coastal areas. This intrusion is increasing the ongoing salinization process in agricultural land, triggered further by higher evapotranspiration and evaporation due to climate variability. Less precipitation and longer and more intensive droughts are affecting also the natural cover of biota but require also more water for irrigation. Crop yields get reduced due to brackish water and soil get increasingly salinized, while the return water into aquifers raises further the amount of salt content in groundwater.

Lands devoted to grazing by domestic animals in northern Mexico occupy nearly 100 million ha. Almost most are classified as desertic, arid, or semiarid zones; these lands provide some of the most nutritious and palatable species of forage plants. In these regions, a prosperous extensive livestock industry makes an important contribution to the national economy. Over eight million head of cattle, two million sheep, one million horses and mules, and many million head of wild animals depend on the vegetation of these arid zones. Ecological characteristics vary within these 100 million ha; although large areas are occupied by pure, open grasslands, in most places shrub species are dominant. In arid zones we should not expect to have a solid cover of grasses. Shrub or cactus species in northern Mexico are very important.

In the natural pasture lands of northern Mexico, six vegetative types illustrate the importance of shrub species in the livestock economy of this area.

(i) *Larrea-Flourensia*

Characteristic of the Chihuahuan/desert, this plant community is found in over 10 million ha in northern Mexico, in areas where annual rainfall varies from 180 to 250 mm. It is found mainly in the States of Chihuahua, Durango, and Coahuila. Soils where this community prospers are generally light colored and calcareous, ranging from sandy loam to clay loam; topography is flat or slightly rolling. Vegetation is dominated by *Larrea tridentata* and *Flourensia cernua*. Other woody species present are *Acacia, Parthenium, Fouquieria,* and *Prosopis* spp.

(ii) *Atriplex-Prosopis-Sporobolus*

Most of this vegetative type is found in the northern part of the State of Chihuahua, in the Vizcaino region of Baja California, and in the Liguna region around Torreon. Annual rainfall is less than 200 mm (in some areas 60 mm). Dominant shrub species are *Atriplex canescens* and *Prosopis juliflora* and the soils are sandy, forming dunes in most areas.

(iii) *Agave-Yucca-Bouteloua*

This plant community covers numerous scattered rocky hills and sierras throughout northern San Luis Potosi. The hills are generally surrounded by the microphyllous desert, where rainfall seldom reaches 300 mm annually. Soils are rocky and calcareous; topography is very steep and some places are inaccessible for animals.

Vegetation is dominated by rosette-type fibrous plants, such as species of *Agave*, *Yucca*, and *Dasylirion*.

(iv) *Opuntia-Acacia-Bouteloua*

This vegetation type covers a considerable area of the States of Durango, Zacatecas, and San Luis Potosi where annual rainfall may average up to 500 mm. It is found exclusively on volcanic, somewhat rocky, reddish soils, both in plains and low hills.

(v) *Acacia-Leucophyllum*

This community is almost exclusively found in the northeast part of the State of Coahuila where annual rainfall is from 350 to 400 mm. Soils are calcareous and shallow in foothills and ridges and a little deeper in depressions. The physiognomy of this vegetation type is one of "chaparral," where medium-height woody species are dominant.

(vi) *Cercidium-Olneya-Bouteloua-Cathestecum*

This woody plant-dominated community, a component of the Sonoran Desert, is found in the north-central part of the state of Sonora, where annual rainfall averages 280 mm in plains with deep, sandy loam soils.

The grazing areas on Mexico are basically divisible into two: first, the higher land in the north and center of the country, which forms an extension to rangeland found in southwest USA and where the better grasses include *Bouteloua* and *Hilaria* spp., and second, the tropical Caribbean coastlands of the Gulf of Mexico, where introduced species, notably *Digitaria* and *Panicum*, predominate. Here the utilization of rangeland is often improved by the use of supplementary pastures which provide superior grazing for certain classes of livestock; a highly productive example occurs in the north of Mexico where fertilized forage oats or Italian ryegrass (*Lolium multiflorum*) is used to fatten calves which have been weaned off the rangeland.

8.9 Impacts of Climate Variability and Longer-Term Climate Change on Human and Natural Systems in the Deserts of Southwest USA

The focus here is on the interaction between land and water use (both historical and contemporary) and climate change. After brief description of the North American deserts, attention turns to an assessment of the likely impact of temperature rise and disruption to precipitation in these desert ecosystems.

8.9.1 North American Deserts

These deserts fall into several categories: hot and dry, cold and dry, and coastal. Semiarid areas, on the margins of deserts, may also be included in some classifications. In western USA, aridlands are located in the subtropical hot deserts of the southwest, comprised of the Mojave, Sonoran, and Chihuahuan deserts, and the temperate cold deserts of the Intermountain West. Coastal deserts are special case (see below). Annual precipitation is low (<400 mm), but the seasonality of precipitation differs substantially among hot deserts. The Mojave Desert is dominated by winter precipitation; thus biological activity is greatest during the cool season. The Chihuahuan Desert is dominated by summer precipitation with biological activity during hotter conditions. The hottest of the three deserts, the Sonoran, is intermediate, receiving both winter and summer precipitation. Each of these deserts is characterized by low productivity and slow plant growth, both of which are primarily water-limited (Archer and Predick 2008).

Coastal deserts occur in moderately cool to warm areas such as Baja California. The cool winters of coastal deserts are followed by moderately long, warm summers. The average summer day temperature ranges from 13 to 24 °C; winter temperatures are 5 °C or below. The maximum temperature is about 35 °C and the minimum is about −4 °C. The average rainfall measures 8–13 cm in many areas. The maximum annual precipitation over a long period of years has been 37 cm with a minimum of 5 cm.

8.9.2 Likely Impacts of Temperature Rise and Disruption to Precipitation Cycles

Rangelands have undergone—and continue to undergo—rapid change in response to changing land use and climate. There is broad consensus among climate models that the arid regions of the southwestern USA will become drier in the twenty-first century and that the transition to a more arid climate is already underway. The western USA has been identified as a region expected to undergo significant changes in climate over the twenty-first century, including increases in temperature (e.g., Intergovernmental Panel on Climate Change (2007), aridity (Seager et al. 2007), and moisture variability (e.g., Diffenbaugh et al. 2008). Arid and semiarid ecosystems across the western USA face widespread ecological alteration as a consequence of global climate change (Archer and Predick 2008). Projections suggest an acceleration in the rate of climate change across the western USA throughout the twenty-first century, yielding pressing concerns over the reliability of future water supplies (e.g., Mote et al. 2005), incidence of wildfire (e.g., Westerling et al. 2006), and the stability of natural resources and fragile ecosystems (e.g., Adams et al. 2009). A chief concern in the arid and semiarid rangelands of the western USA is the rapid conversion of shrublands and desert into annual grassland through the spread of invasive annual grass species (Mack 1981; Brooks and Esque 2002; Bradley and Mustard 2005) and the negative impacts this conversion will have on wildfire regimes, surface hydrology, and loss of critical habitat for threatened and endangered species (Brooks et al. 2004; Wilcox and Thurow 2006). While the encroachment of invasive annual grasses throughout rangelands of the western USA is of concern even in the absence of climate change, direct and indirect impacts of climate change are hypothesized to hasten the rate of land-cover conversion, resulting in an increasingly challenging problem for land and resource managers (Scheffer 2009).

The deserts of the western USA are characterized by a high diversity of floral and faunal species adapted to arid niches (Rickerts et al. 1999). Over the past three decades ending 2010, there has been a significant increase in the abundance and extent of invasive annual grass species in the southwest, including cheatgrass (*Bromus tectorum*) across the Great Basin Desert Bradley and Mustard, 2005), red brome (*Bromus rubens*) across the Mojave Desert (Brooks and Berry 1999), and buffel grass (*Pennisetum ciliare*) across the Sonoran Desert (Betancourt 2007). This change has coincided with an increase in the number of large fires and area burned across the arid and semiarid western USA that have been linked to the increased presence of invasive annual grasses (Brooks et al. 2004), wildfire management (Swetnam et al.1999), and climate change (Westerling et al. 2006). While the forested uplands of the southwest have historically experienced frequent wildfires (Swetnam and Betancourt 1998), fire is historically rare in the desert shrublands and grasslands because of insufficient fuels to carry surface fire (Brooks and Matchett 2006; Mensing et al. 2006). Invasive annual grasses not only increase fuel continuity and loading (Brooks et al. 2004) but also are lighter, flashier fuels with lower fuel moistures that cure earlier in the growing season and increase wildfire potential compared to the dense, waxy shrubs and cacti native to the deserts.

Evidence and implications of climate change on the spread of invasive annual grasses are poorly understood and have only recently begun to be addressed comprehensively at landscape scales. These studies are constrained by limited geographic data on the spread of invasive species, resulting in a diminished ability to quantify and model the factors contributing to invasive species expansion. Evidence suggests, however, that observed changes in climate have assisted in invasions across the deserts of the western USA both directly through changes in temperature and precipitation and indirectly through their influence on wildfire. First, warming across the southwest over the twentieth century has extended the length of the frost-free season (Weiss and Overpeck 2005), thereby increasing climatically suitable ranges for cold-intolerant species such as *B. rubens* and *P. ciliare*. Second, while precipitation variability is inherent in the southwest, a significant increase in cool-season (November–March) precipitation variability over the past half century in the Sonoran Desert may be responsible for assisting the colonization of invasive annual grasses and increasing fine fuel loading. Finally, increased temperature and reduced humidity during spring and summer associated with recent protracted drought conditions since 2000 (Weiss et al. 2004) have doubled the frequency of extreme fire danger in the Sonoran Desert, resulting in an earlier start and lengthening of the fire season. Although some of these changes (e.g., increasing temperature) are attributable to anthropogenic climate change, it is important to note that other observed changes (e.g., precipitation variability and timing) are likely products of natural modes of climate variability (Ramirez 2015).

The current prognosis for global climate is for an increased frequency of extreme weather events (heat waves, droughts, and floods). Furthermore, warmer nights and projected declines in snow pack, coupled with earlier spring snow melt, will reduce water supply, lengthen the dry season, create conditions for drought and insect outbreaks, and increase the frequency and intensity of wildfires. Temperatures currently considered unusually high will occur more frequently. These model-based projections align with observations made in the region over recent years.

Because plants and animals in the deserts of SW of the USA are living at the extreme edge of their physiological capacity, these desert areas are vulnerable to long-term climate change. The entire ecosystem can easily become uncoupled, for example, if seasonality of rainfall is accompanied by rising temperatures. Archer and Predick (2008) list a number of likely impacts of rising temperatures and disruption to seasonality of precipitation that are predicted to occur under various climate change scenarios:

- *The geographic boundaries of the southwestern deserts will likely change*
- *Vegetation composition, diversity, and growth will all likely be altered*
- *Net primary production (NPP) is likely to decline*
- *Intensification of the hydrologic cycle due to atmospheric warming is expected to reduce rainfall frequency, but increase the intensity and/or size of individual precipitation events*
- *Water availability likely will be reduced in arid ecosystems*
- *Native fish diversity will likely decline*
- *Soil erosion losses are expected to increase*
- *Nonnative species are likely to increase in abundance*

But as Archer and Predick (2008) say "The response of arid lands to climate change will be strongly influenced by interactions with non-climatic factors, such as land use and non-native species abundance. Climate change will influence both the types of future land use and the impacts of a specific land use. Climate change will occur concurrently with other environmental factors. Some of these factors might reinforce and accentuate climate effects (e.g., livestock grazing, nitrogen deposition); others might constrain, offset, or override climate effects (e.g., atmospheric CO_2 enrichment, fire, non-native species). Because many factors are affecting ecosystems simultaneously, it is difficult, and in many cases impossible, to separate the magnitude of each factor separately. The synergistic effects of these interacting factors often differ from the sum of the separate effects. Thus, climate effects should be viewed as the backdrop against which other factors act, and simple generalizations regarding climate effects should be viewed with caution."

8.9.3 Land Use Change: A Key Driver

Grazing has traditionally been the most pervasive and extensive land use in aridlands. Although livestock grazing remains an important land use in aridlands, there has been a significant shift to ex-urban development and recreation, reflecting dramatic increases in human population density since 1950. Areas of Nevada, Arizona, and New Mexico are currently among the fastest growing in the USA. In addition to grazing, aridlands are now being exposed to new suites of environmental pressures such as air pollution, atmospheric nitrogen deposition, energy development, motorized off-road vehicles, feral pests, and invasion of nonnative plants.

In desert regions in the SW of the USA, urban expansion has led to a major change in the aesthetics of the landscape. The structural, visual, and functional continuity of the desert landscape are threatened not only by increasing pressure these by rapid urbanization and industrialization and its attendant infrastructure (airports, highways, power transmission lines, etc.). Rapid growing populations demand more and more land and water for this expansion and thereby enforce more drastic and far-reaching changes on the open landscape than any agricultural or pastoral economy had previously done. Powerful bulldozers and other earth-moving equipment have razed the living sponge of woodlands, shrubs, and grasslands. They are replaced by buildings, installations, asphalt highways, and stone and concrete bedded waterways and pipelines. All these are devoid of the absorptive powers of the original vegetation canopy: they are visible by ugly wounds of denuded slopes and embankments and abandoned quarries that increase the hazards of flooding, soil erosion, and environmental pollution over wide areas and thus spoil the landscape.

8.9.4 Plant Communities: Projected Trajectories Under Climate Change

Peters et al. (2012) speculate on some trajectories of plant communities in the Chihuahuan desert.

> Future trajectories under climate change with no change in annual precipitation (average of 25 cm year^{-1}) and well-managed livestock grazing, Chihuahuan Desert landscapes are expected to continue to be dominated by desertified shrublands with isolated patches of perennial grasslands⋯ Under a directional decrease in precipitation, the areal extent of mesquite (*Prosopis* sp.) and creosotebush (*Larrea* sp.) shrublands are expected to increase ⋯ Most of the increase in mesquite and creosotebush is expected to occur at the expense of upland grasslands and tarbush (*Flourensia cernua*) shrublands, similar to historic trajectories.

Desert landscapes consist of a heterogeneous suite of ecosystem types that differ in vegetation, soils, historic legacies, and connectedness with other ecosystems. Predicting future trajectories of change at the landscape scale requires a consideration of both wetter and drier conditions. There is increasing evidence that climate change is resulting in more ecosystems crossing thresholds of change (Peters et al. 2012). Combining century-long patterns of ecosystem change with long-term data in response to variation in rainfall is one approach to elucidating pattern–process relationships across heterogeneous landscapes. Monitoring is an important part of developing early warning as part of an adaptation strategy to cope with climate change and other drivers.

8.9.5 Monitoring Change: Climatic and Non-climatic Drivers

Although the deserts of North America have been the sites of many important ecological studies, there have been relatively few long-term monitoring studies that provide the opportunity to observe changes in ecosystem structure and function in response to climate change per se. Coordinated measurements at sites spread across the deserts would enhance our ability to detect change in ecosystem structure and function in relation to climate change. Current observation systems are inadequate to separate the effects of changes in climate from the effects of other drivers (e.g., land use). There are few observing systems for monitoring wind and water erosion and for examining interactions among climatic and non-climatic drivers. To identify climate effects would require a broad network, with many measured indicators, coupled with a network of controlled experimental manipulations. A coordinated national network that monitors ecosystem disturbance and recovery would greatly contribute to attributing disturbances to a particular cause and identifying the consequences of those disturbances. However, no such network currently exists. Time-series satellite observations, coupled with ground-based measurements, can identify disturbance, changes in productivity, and changes in land use over time and over large, remote areas. However, ascribing changes in these metrics to climatic vs. non-climatic factors is challenging.

8.10 Summary and Conclusions

The likely impact of global change (including climate change) on the lands and its peoples will be profound, but as a "slow-onset" phenomenon, there is time to adapt. Initially, we will see coping mechanisms brought into play but ultimately adaptation must occur. From an ecological viewpoint, we can be certain that within existing ecological communities, some species will persist, some species will depart, and new species will arrive, or evolve. Most ecological models are based upon modern observations and so might fail to accurately predict ecological responses to future climates occurring in conjunction with elevated atmospheric CO_2, nitrogen deposition, and nonnative species introductions as well as the impact of urbanization, new technologies, and burgeoning population in water-depleted areas. In aridlands, sustaining ecosystem processes in the face of climatic variability requires a sound foundation of monitoring and research, as well as a good working relationship between people and organizations with diverse goals and interests. Many challenges lie ahead in the pluralistic societies that are becoming the norm.

References

B.W. Adams, G. Ehlert, C. Stone, M. Alexander, D. Lawrence, D. Willoughby, C. Hincz, A. Burkinshaw, J. Carlson, K. France, *Rangeland Health Assessment for Grassland, Forest and Tame Pasture* (Alberta Sustainable Resources Department. Public Lands Division, Edmonton, 2009.) Pub. No. T/044

S.R. Archer, K.I. Predick, Climate change and ecosystems of the southwestern United States. Rangelands **30**(3), 23–28 (2008)

DI. Axelrod, *A Miocene Flora from the Western Border of the Mohave Desert.* PhD dissertation, Universityof California, Berkeley) 1940)

D.I. Axelrod, Holocene climatic changes in relation to vegetation disjunction and speciation. Am. Nat. **117**, 847–870 (1981)

C.J. Bahre, *A Legacy of Change: Historic Human Impact on Vegetation in the Arizona Borderlands* (University of Arizona Press, Tucson, 1991)

J. L. Betancourt, From fireproof desert to flammable grassland: buffelgrass invasion in the Sonoran desert. Eos. Trans. AGU **88**(52) (2007). Fall Meeting Supplement, Abstract B34A-02

J.P. Blaisdell, R.C. Holmgren, *Managing Intermountain rangelands S –alt-Desert Shrub Ranges*, USDA Forest Service General Technical Report INT-163 (Intermountain Forest and Range Experiment Station, Ogden, 1984.) 52 p

B.A. Bradley, J.F. Mustard, Identifying land cover variability distinct from land cover change: cheatgrass in the Great Basin. Remote Sens. Environ. **94**, 204–213 (2005)

M.L. Brooks, K.H. Berry, Ecology and management of alien annual plants in the CalEPPC News. Calif. Exot. Pest Plant Counc. **7**(3/4), 4–6 (1999)

M.L. Brooks, Effects of protective fencing on birds, lizards, and black-tailed hares in the western Mojave Desert. Environ. Manag. **23**, 387–400 (1999b)

M. L. Brooks, Bromus madritensis subsp. rubens (L.) Husnot [B. rubens L.] , Foxtail Chess (Red Brome), in *California Deserts. Noxious Wildland Weeds of California*, ed. by C. Bossard, M. Hoshovsky, J. Randall (University of California Press, 2000)

M.L. Brooks, T.C. Esque, Alien annual plants and wildfire in desert tortoise habitat: status, ecological effects, and management. Chelonian Conserv. Biol. **4**, 330–340 (2002)

M.L. Brooks, J.R. Matchett, Spatial and temporal patterns of wildfires in the Mojave Desert, 1980–2004. J. Arid Environ. **67**, 148–164 (2006)

M.L. Brooks et al., Effects of invasive alien plants on fire regimes. Bioscience **54**, 677–688 (2004)

J.H. Brown, W. McDonald, Livestock grazing and conservation on southwestern rangelands. Conserv. Biol. **9**(6), 1644–1647 (1995)

N.S. Diffenbaugh, R.J. Trapp, H. Brooks, Does global warming influence tornado activity? EOS Trans. Am. Geophys. Union **89**, 553–554 (2008)

T. Dobzhansky, *Genetics of the Evolutionary Process* (Columbia University Press, New York, 1970)

T.C. Esque, R.H. Webb, C.S.A. Wallace, C. van Riper III, C. McCreedy, L. Smythe, Desert fires fueled by native annual forbs: effects of fire on communities of plants and birds in the Lower Sonoran Desert of Arizona. Southwest. Nat. **58**(2), 223–233. http://www.bioone.org/doi/full/10.1894/0038-4909-58.2.223

H.F. Heady, D. Childs, *Range Ecology and Management* (Westview Press, Boulder, 2001)

L. Huntsinger, P. Starr, Grazing in arid North America; a biogeographical approach. Secheresse **17**(1–2), 219–233 (2006)

C.F. Hutchinson, J.D. Unruh, C.J. Bahre, Land use vs. climate as causes of vegetation change: a study in SE Arizona. Glob. Environ. Chang. **10**, 47–55 (2000)

L. Huntsinger, P. Starr, Rangeland Grazing, in *North American Commercial Ranching*, Range and Animal Sciences and Resources Management. Encyclopedia of Life Support Systems, ed. By V. Squires, (EOLSS/UNESCO, Eolss Publishers, Oxford, 2010), pp. 60–91

IPCC, *Fourth Assessment Report: Climate Change (AR4)*, IPCC, Cambridge (2007).

R. Jemison, C. Raish (eds.), *Livestock Management in the American Southwest: Ecology, Society, and Economics: Developments in Animal and Veterinary Sciences 1e* (Elsevier Science, Amsterdam, 2000), pp. 523–554

R.N. Mack, Invasion of Bromus tectorum L. into western North America: an ecological chronicle. Agro-Ecosystems **7**, 145–165 (1981)

C. M. McKell, E. Garcia-Moya, North American shrublands. in *The Biology and Utilization of Shrubs*, ed. by C. M. McKell (Academic Press, Inc., 1989)

S. Mensing, S. Livingston, P. Barker, Long-term fire history in Great Basin sagebrush reconstructed from macroscopic charcoal in spring sediments, Newark Valley, Nevada. West. North Am. Nat. **66**, 64–77 (2006)

P. W. Mote, A. F. Hamlet, D. P. Lettenmaier, Variability and trends in mountain snowpack in Western North America. (cses.washington.edu/db/pdf/moteetalvarandtrends436.pdf)

D.C. Peters, J. Yao, O.E. Sala, J.P. Anderson, Directional climate change and potential reversal of desertification in arid and semiarid ecosystems. Glob. Chang. Biol. **18**, 151–163 (2012)

F. Ramirez, Flammable deserts: understanding the impacts of fire on southwestern desert ecosystems of USA. Dissertation, Iowa State University (2015)

T.H. Ricketts, E. Dinerstein, D.M. Olson, C.J. Loucks, W. Eichbaum, D. DellaSala, K. Kavanagh, P. Hedao, P.T. Hurley, K.M. Carney, R. Abell, S. Walters, *Terrestrial Ecoregions of North America: A Conservation Assessment* (Island Press, Washington, DC, 1999), p. 485

B.A. Roundy, E.D. McArthur, J.S. Haley, D. K. Mann, comps, Proceedings: Wildland Shrub and Arid Land Restoration Symposium; 1993 October 19–21; Las Vegas, NV. Gen. Tech. Rep. INT-GTR-315 (Ogden: U.S. Department of Agriculture, Forest Service, Intermountain Research Station, 1995). 384 p

R. Seager, M. Ting, I. Held, Y. Kushnir, J. Lu, G. Vecchi, H. Huang, N. Harnik, A. Leetmaa, N. Lau, C. Li, J. Velez, N. Naik, Model projections of an imminent transition to a more arid climate in southwestern North America. Science **316**, 1181–1184 (2007)

M. Scheffer, *Critical Transitions in Nature and Society* (Princeton University Press, Princeton, 2009)

V.R. Squires, Water and its functions, regulation and comparative use by ruminant livestock. in, *The Ruminant Animal: Its Digestive Physiology and Nutrient Metabolism*, ed. by D.C. Church, Prentice-Hall, Harlow, UK (1987)

V.R. Squires, People in rangelands: their role and influence on rangeland utilization and sustainable management, in *Range and Animal Sciences and Resources Management. Encyclopedia of Life Support Systems*, ed. By V. Squires, (EOLSS/UNESCO, Eolss Publishers, Oxford, 2010), pp. 36–59

V.R. Squires, H.M. Milner, K.A. Daniell, *River Basin Management in the Twenty-First Century: Understanding People and Placed* (CRC Press, Boca Raton, 2014)

L.A. Stoddart, A.D. Smith, T.W. Box, *Range Management*, 3rd edn. (McGraw-Hill Book Co., New York, 1975), p. 532

T.W. Swetnam, J.L. Betancourt, Mesoscale disturbance and ecological response to decadal climatic variability in the American Southwest. J. Clim. **11**, 3128–3147 (1998)

T.W. Swetnam, C.D. Allen, J.L. Betancourt, Applied historical ecology: using the past to manage for the future. Ecol. Appl. **9**, 1189–1206 (1999)

J. Valdes, T. Maddock, in *Water and Sustainability in Arid Regions*, ed. by G. Schneier-Madanes, M. F. Courel. III Conjunctive Water Management in the US Southwest, (Springer Science + Business Media B.V., Netherlands, 2010), doi:10.1007/978-90-481-2776-4_14

J.L. Weiss, D.S. Gutzler, J.E.A. Coonrod, C.N. Dahm, Long-term vegetation monitoring with NDVI in a diverse semi-arid setting, central New Mexico, USA. J. Arid Environ. **58**, 249–272 (2004)

J.L. Weiss, J.T. Overpeck, Is the Sonoran Desert losing its cool? Glob. Chang. Biol. **11**, 2065–2077 (2005)

A.L. Westerling, H.G. Hidalgo, D.R. Cayan, T.W. Swetnam, Warming and earlier spring increases western U.S. forest wildfire activity. Science **313**, 940–943 (2006)

B.P. Wilcox, T.L. Thurow, Emerging issues in rangeland ecohydrology: vegetation change and the water cycle. Rangel. Ecol. Manag. **59**(2), 220–224 (2006)

A. Yeldell, V. Squires, Restoration, reclamation, remediation and rehabilitation of mine sites: which path do we take through the regulatory maze? in *Ecological Restoration: Global Challenges, Social Aspects and Environmental Benefits*, ed. By V. Squires, (NOVA Press, New York, 2016), pp. 37–56

Further Readings

M.L. Brooks, Alien annual grasses and fire in the Mojave Desert. Madrono **46**, 13–19 (1999a)

M.L. Brooks, Habitat invasibility and dominance by alien annual plants in the western Mojave Desert. Biol. Invasions **1**, 325–337 (1999b)

M.L. Brooks, Competition between alien annual grasses and native annual plants in the Mojave Desert. Am. Midl. Nat. **144**, 92–108 (2000)

M.L. Brooks, Effects of high fire frequency in creosote bush scrub vegetation of the Mojave Desert. Int. J. Wildland Fire **21**, 61–68 (2012)

Desertification and Land Degradation in Indian Subcontinent: Issues, Present Status and Future Challenges

9

Ajai and P.S. Dhinwa

9.1 Introduction

Desertification occurs in all the continents except Antarctica, affecting the livelihood of millions of people all across the world. It has been recognized as one of the major environmental problems, having global concern, affecting about 250 million people directly. In addition, over one billion people in more than hundred countries are at risk (Adger et al. 2000). Deserts have been known to mankind since the time immemorial; however the term 'desertification' is not very old and was first used by Lavauden in 1927 to describe severely overgrazed lands in Tunisia (Dregne 2000) and was later brought back to use in 1949 by a French forester (Aubreville 1949) to show excessive soil erosion due to deforestation in the West Africa and also to explain the way in which the Sahara Desert was expanding to engulf savannah grasslands on desert margins. The Food and Agriculture Organization (FAO) was the first international agency to use the word 'desertification' in 1962 (Amin 2004). The term 'desertification' was raised as a major environmental issue at the United Nations Conference on Human Environment, Stockholm 1972, at which the United Nations Environment Program (UNEP) was established. Desertification got the attention of the international audience in the 1970s when this issue received importance on the plight of the drought – stricken Sahel zone of Africa, south of the Sahara. One outcome of this attention was the United Nations Conference on Desertification (UNCOD), held at Nairobi in 1977 (UNCOD 1977). The deliberations in this conference clearly indicated that desertification was not just happening in Africa alone but active all over the world. Following the conference, UNEP was

Ajai (✉)
E S- CSIR, Space Applications Centre, ISRO, Ahmedabad, India
e-mail: drajai_in@yahoo.com

P.S. Dhinwa
TALEEM Research Foundation, Ahmedabad, India

entrusted with the responsibility of coordinating the implementation of the United Nations Plan of Action to Combat desertification which was adopted by the conference and endorsed by the United Nation General Assembly.

Often, the terms 'desert' and 'desertification' are used interchangeably. However, we need to understand that 'desert' is used to describe a particular state of the land, whereas 'desertification' is a process which may turn a 'productive land' to a 'desert', if remained unchecked. We shall discuss the difference between the terms 'desert' and 'desertification'. In fact, there has been ongoing debate on not only defining the term 'desert' but also 'desertification' (Behnke and Mortimer 2016). So far, there is no single definition of 'desert' which is accepted by most of the experts and scientists. People have different perceptions and definitions of the 'desert'. According to Cambridge Dictionary, desert means 'A region, often covered with sand or rocks, where there is little rain and not many plants'. Ecologists believe deserts are ecosystems, whereas soil scientists say that deserts are the areas with low potential for production. Climatologists define deserts as areas with high temperature and low precipitation (less than 250 mm).

The same is the case with the term 'desertification'. It has been a pretty difficult task to find out a precise and well-accepted definition of the term 'desertification'. Considerable controversy exists over the definition of 'desertification' for which Helmut Geist (2005) has identified more than 100 formal definitions. In most general term, desertification is understood to be a geo-environmental hazard triggered by either or a combination of natural or man-made factors in dryland regions. It is a complex phenomenon resulting from the interaction of physical, meteorological, biological, socioeconomic and cultural factors.

One of the most widely accepted definitions of desertification is from the 'Princeton University Dictionary'; it states 'the process of fertile land transforming into desert typically as a result of deforestation, drought or improper/inappropriate agriculture. The definition, given by the UNCOD conference, held in Nairobi in 1977, is 'the diminution or destruction of biological potential of the land, (which) can lead to desert like conditions' (UNCOD 1977). Definitions of desertification and the evolution of the desertification concept have been reviewed by many authors (Thomas and Middleton 1994; Puigdefabregas 2009; Eswaran et al. 2001; Reynolds et al. 2007, 2011; Brabant 2008, 2010; Safriel 2007; Ajai et al. 2007a, 2009; DSD 2009). However, the most comprehensive and widely accepted definition of 'desertification' is from the United Nations Convention to Combat Desertification (UNCCD) which defines it as 'land degradation in arid, semi-arid and dry subhumid areas resulting from factors including climatic variations and human activities (UNCCD 1994).

'Land degradation' also has a number of definitions which emphasize different aspects of reduction in quality of the land through degradation processes. Land degradation is a process that diminishes or destroys the agricultural-crop or livestock and forest production capacity of the land. It is induced by human activities or can be a natural phenomenon aggravated by the affects of human activities (Brabant 2008, 2010). Land degradation occurs slowly and cumulatively and has long-lasting impacts on rural people who become increasing vulnerable (Muchena 2008).

Another definition used both by the UNCCD and OECD states: 'land degradation is the reduction or loss of agricultural, biological or economic productivity of rain-fed cropland, irrigated cropland, or range, pasture, forest, and woodlands resulting from land use or from a combination of processes arising from human activities and habitation patterns' (UNCCD 1994). Land degradation is a process which may turn a productive land in to "degraded land" or "waste land". There are a large number of land degradation types (or processes) which are found in various parts of the world. Details on the types of degradation processes are discussed by (Brabant 2008, 2010; Ajai et al. 2007a, b, 2009).

9.2 Global Status of Desertification

As per United Nations Environmental Program (UNEP 1977a, b), about 39.7% of the world's land area (or 5172 million ha (Mha)) falls under dry lands and is susceptible to desertification. This excludes 978 Mha area which falls under 'hyperarid'. Out of the total dryland area, arid land is about 26%, semi-arid 38% and dry subhumid 21%.World desert area is about 5 million km^2 and an additional 40 million km^2 is arid or semi-arid. About 33% area of earth's surface exhibits desert like conditions to some degree or other. Additional desert like conditions are added at the rate of 60,000 km^2 per year (The World Environment Directory, 1998). The areas peripheral to the present desert are directly under the threat of desertification. In Asia, about 35% of the productive land seriously faces desertification hazards. About 70% of the total land of the world is under degradation (Dregne and Chou 1994).

As per the 'Assessment of Human Induced Soil Degradation (GLASOD)' (GLASOD 1990), Asia has the highest loss of productivity due to desertification in the world. Africa has, by far, the greatest amount of hyper-arid land, mainly the Sahara Desert, followed by Asia, Australia and New Zealand. Australia has the greatest amount of drylands, but none of it is hyper-arid. Approximately 3% of the world's drylands are irrigated, 9% are rainfed cropland, and 88% are rangelands. The dominance of rangelands, with their high percentage of degraded land, is why the global desertification level is a high 70%.

9.3 Indian Subcontinent: Issues Related to Desertification

The Indian subcontinent or South Asia comprises of seven countries, namely, Bangladesh, Bhutan, India, Maldives, Nepal, Pakistan and Sri Lanka, and it occupies the major land masses of south Asia. The Indian subcontinent covers a geographic area of 4.4 million km^2 having a population of about 1.749 billion (2013). One of the major environmental concerns in the entire Indian subcontinent is land degradation and desertification as a very large area of this subcontinent is affected by land degradation/desertification. In addition to the above seven countries, Afghanistan, just adjoining Pakistan in the west, is also included in this chapter because it has similar physiography, geo-environmental setup as well as the problems and issues related to desertification and land degradation. Major processes of land degradation in this geographically and physiographically homogeneous landmass of above eight countries are: vegetal degradation, water and wind erosion, water logging, salinization, mining and mass wasting. The main causes of land degradation and desertification in these countries are deforestation including shifting cultivation, overgrazing, unsustainable agricultural practices, cultivation in marginal lands and steep slope areas, industrialization/urbanization and mining. Most of the above-mentioned drivers of land degradation, triggered and accelerated by anthropogenic activities, are mainly due population pressure, poverty, illiteracy and lack of sufficient resources. Apart from the above-mentioned anthropogenic causes, there are several natural causes for land degradation and desertification. They include, the physiographic setup, topography, the extremities in weather conditions and natural disasters including recurrent droughts and floods as well as climate change. Details about the desertification and land degradation processes active in the above eight countries and the major causes for land degradation are discussed below.

9.3.1 India

India occupies only 2.4% of world's geographical area yet supports about 16.2% of the world's human population. It has only 0.5% of the world's grazing area but supports 18% of the world's cattle population. India is endowed with a variety of soils, climate, biodiversity and ecological regions. About 228 Mha (69%) of its geographical area (about 328 Mha) fall within the drylands (arid, semi-arid, dry subhumid). In India alone, arid zone covers about 320,000 km^2 (12% of the geographical area). Besides, there is an additional 70,300 km^2 area of cold desert. Rajasthan occupies the greater part of the Indian arid landscape and about 60% area of India, lying west of the Aravallis, followed by Gujarat (19%), Punjab and Haryana (9%) and Karnataka and Andhra Pradesh (10%). According to NBSSLUP, about 50.8 Mha (15.8%) of the country's geographical area is arid. About 123.4 Mha (37.6%) of the country's geographical area consists of the semi-arid region (NBSSLUP 2001). About 54.1 Mha (16.5%) of the country's geographical area falls within the dry subhumid region. The desertification status mapping carried out for India at 1:500,000 scale by Space Applications Centre (ISRO), Ahmedabad, India using multidate IRS-AWiFS data reveals that 105.48 Mha (32.07% of the total geographical area of the country) is undergoing processes of land degradation (Ajai et al. 2009). Area undergoing desertification is 81.4 Mha.

9.3.2 Pakistan

About 90% of the land area of Pakistan falls in the arid and semi-arid (dryland) category and is vulnerable to desertification. Human population pressure has been found to be the leading cause of desertification. The annual increase in population at 2.1% has led to over exploitation of land and water resources (Shakeel et al. 2010). The per capita cultivated land is only 0.16 ha. Only 3.77% land area of Pakistan is under forest cover. Pakistan has about its 26.77% of land area as grazing land. The drylands in Pakistan are characterized by high velocity wind, huge shifting and rolling sand dunes, high diurnal variations of temperature, scarce rainfall, intense solar radiations and high evapotranspiration.

The process of desertification in Pakistan falls into three categories: (i) expansion of the great deserts of Thar and Cholistan, (ii) increase in desertification due to deforestation and (iii) land degradation due to salinity, water logging, overgrazing, urbanization and unsustainable management practices (Shakeel et al. 2010). Land area undergoing the process of water erosion is 13.0 Mha whereas the wind erosion in 6.1 Mha area. The third major process of desertification in Pakistan is water logging. About 9.37 Mha land is under water logging. Overgrazing and deforestation are the other major causes of desertification in Pakistan (Shakeel et al. 2010).

9.3.3 Bangladesh

Major types of land degradation that occur in Bangladesh are water erosion, soil fertility depletion, salinization, water logging, pan formation and deforestation. Among them, water erosion and fertility depletion are the main factors (Hasan and Ashraful Alam 2006). The soils of hilly area are the most susceptible to water erosion in which sheet, rill and gully erosion occurs. About 75% of the hilly areas have very high susceptibility to erosion, 20% have high susceptibility, and 5% have moderate susceptibility to erosion (BARC 1999). Faulty 'Jhum' cultivation in hilly area causes gully erosion and losses of soil ranges from 10 to 120 t/ha/year (Farid et al. 1992). Decline of soil fertility occurs through and combination of lowering of soil organic matter and loss of nutrients. The average organic matter content of top soils (high land and medium high land situation) has decreased from about 2 to 1% over the last 20 years due to intensive cultivation (Miah et al. 1993).

9.3.4 Bhutan

Bhutan, a country from the Himalayan region, has a total geographic area of 3.8394 Mha. About 70.46% of its geographic area is under forest cover and 10.81% area is under shrub. Agricultural land is only 2.93%, of which 31% cropland is located in more than 51% slope. More than 60% arable land is rainfed. 23% of the country's population lives below the poverty line. A large portion of the country's land is affected by land degradation and desertification. As of now, there is no systematic spatial inventory available for Bhutan on land degradation and desertification and the types and severity of degradation. The major processes of desertification active in Bhutan are vegetal degradation, mass movement, water and wind erosion and mining. The major causes of land degradation are forest fire, deforestation including shifting cultivation, overgrazing and unsustainable agricultural practices including steep slope agriculture, industrialization and mining (Bhutan-NAP 2014).

9.3.5 Nepal

Nepal, located in the north of India in the Himalayas, has a total geographic area of 147,181 km^2. About 46.12% of the land area of Nepal is under forest cover (includes scrub), as of the year 2000 (FAO 2010). Agricultural land is about 28.1%, and grassland is about 12%. The major processes of land degradation in Nepal are vegetal degradation, water erosion, wind erosion, mass movement, soil degradation due to chemical process and water logging. About 45.5% of the land area is affected by water erosion (Acharya and Kafle 2009). About 4% of the geographic area is effected by wind erosion. Major causes of land degradation and desertification in Nepal are fragile geological structures, deforestation including shifting cultivation, overgrazing and unsustainable agricultural practices including steep slope agriculture, industrialization and mining (Acharya and Kafle 2009).

9.3.6 Maldives

Maldives is situated in the Indian Ocean, south-southwest of India. It consists of approximately 1190 coral islands grouped in a double chain of 26 atolls with a total land area of 298 km^2 and population of 294,000. About 10% of the land area is under agriculture. There are three environmental constraints that underpin the Maldives' vulnerability to land degradation (Maldives 2009). The first constraint is that 1192 islands are highly dispersed across the chain of 20 atolls. Second, these islands are very small and vary in size from 0.5 km^2 to around 5.0 km^2. Third, all islands are very low lying, with more than 80% of the land has an elevation of less than 1 m above mean sea level. Even the highest elevated islands are not more than 3 m above mean sea level. The underlying causes of land degradation in the Maldives can be better understood in the context of these three constraints. The causes and extent of land degradation vary across islands depending on the environmental and socioeconomic problems unique to the island. The key factors responsible for land degradation in this country are the following: (i) population distribution in islands is not proportionate with land capacity, (ii) increased pressure on groundwater, (iii) poor agricultural practices and (iv) frequent natural disasters.

9.3.7 Sri Lanka

The total land area of Sri Lanka is 65,610 km^2 and the population is over 19 million. Increasing pressure of population on land has led to a rapid decline in the per capita land availability from 2.5 ha in 1880 to 0.38 ha in 2000. Pressure on land has led to various kinds of land degradation such as vegetal degradation (deforestation), soil erosion, salinization, urbanization/industrialization, etc. In addition to this, mass movement/landslides are the land degradation processes occurring in the highly undulating areas of the country. Forest cover in the country has been steadily declining over the years, from 80% of the land area in 1881 to 24% in 1992 (MENR 2002). The dense forest cover in Sri Lanka has decreased by 20%, mostly in the dry zones, during 1956–1992. The average rate of deforestation during 1956 and 1992 has been over 36,000 ha per year. Reasons for deforestation are planned deforestation for irrigation, settlement purposes, illicit felling, firewood collection and shifting cultivation (also called *Chena*) with inadequate fallow periods, and neglected soil conservation measures is responsible for degrading about 1 Mha of land in the country. About 44% of agricultural land has undergone land degradation of some or the other form (MENR 2002). It is estimated that about 30% of the drylands in Sri Lanka is degraded and unsustainable for agriculture, while 30% of tea lands in the wet zone are considered to be marginal and uneconomical for cultivation. The major causes for land degradation in Sri Lanka are deforestation including shifting cultivation, overgrazing and unsustainable agricultural practices including steep slope agriculture, industrialization and mining.

9.3.8 Afghanistan

Afghanistan has about 45% of its land area (30 Mha) as rangeland and 2% under forest cover. According to the GLASOD program (Global Assessment of Soil Degradation), about 16% of Afghanistan's land area is severely affected due to anthropogenic activities. About three fourths of Afghanistan's geographic area is vulnerable to desertification which is among the highest in the world. The geological, physiographic, topographic and climatic features prevailing in the country make it highly susceptible to land degradation. On the other hand, human activities such as farming on steep slopes and marginal lands, deforestation and de-vegetation of lands and unsustainable use of scrub and grasslands make it more vulnerable to desertification. Land degradation is, in some places, so severe that recovery is impossible without human intervention. One of the most threatening impacts arising from loss of soil and vegetation is desertification and increased floods.

Afghanistan's many rangelands are in poor condition due to overgrazing problem. In the mountains, overgrazing is the main factor leading to soil erosion and forest degradation, hampering forest regeneration.

9.4 Causes of Desertification

Desertification is caused by a combination of social, political, economic and natural factors which vary from region to region. Policies that can lead to unsustainable use of resources, inappropriate land use practices and lack of infrastructure are major contributors of land degradation and desertification.

Human activities are the main factors triggering and accelerating the desertification processes. Some of the human activities that can cause desertification are:

- Cultivation of soils that are fragile or exposed to erosion by wind or water
- Reduction in the fallow period of soils and lack of organic or mineral fertilizers
- Overgrazing
- Overexploitation of forests, in particular for fuel wood and fodder
- Uncontrolled use of fire for regenerating pasture/forest and shifting cultivation
- Agricultural practices that destroy the soil structure, especially the use of unsuitable agricultural machinery
- Wrong agricultural practices, including cultivation on marginal and high-slope lands
- Diversion of rivers to create irrigation schemes
- Irrigation of soils prone to salinization, alkalinization or water logging, especially in irrigation command areas
- Developmental activities such as industrialization, urbanization, mining, etc.

All the above activities derive from two root causes: (i) poverty, illiteracy and underdevelopment and (ii) 'modern' development that disregards the impact of the technologies used on land sustainability.

Climatic conditions like frequent droughts, natural denudational processes like erosion/geological hazards and human interference are the major contributors to the process of desertification. The causes of desertification are as follows:

- Drop in frequency and amount of rainfall (frequent droughts)
- Very high variation in temperatures
- Dwindling vegetal cover
- Climatic change
- Lowering of groundwater table
- Frost shattering and frost heaving in cold arid regions
- Mass movement in mountainous areas

Degradation of vegetation, most often by human activities, accelerates land degradation and may lead to desertification. When a soil loses vegetation cover, it becomes more susceptible to wind and water erosion. In a process called deflation, wind entrains the finer soil particles and organic material from the soil. A layer of coarse sand with little or no structure is left behind. The loss of organic material decreases soil aggregation and stability, so the soil is left even more susceptible to aeolian activity. In such cases, the plants and vegetation are more likely to be 'sandblasted' and buried. The water-holding capacity and the nutrient content of the soil are reduced when organic material is lost, which is an additional strain on vegetation survival. A soil is most vulnerable to deflation and wind erosion in arid, hot and windy climates, at the end of the dry season or when vegetation is removed for cultivation or as a result of overgrazing (Dregne 1985; Kishk 1986; Mainguet 1994; Sabadell et al. 1982).

9.5 Consequences of Desertification

The consequence of desertification and land degradation depend on the following factors that may vary by region, country and year:

- The severity and extent of land degradation
- The severity of climatic conditions (especially annual rainfall)
- The number and diversity of affected populations
- The level of development in the country or the region

Desertification reduces the biological productivity of the land and may lead to reduction in the ecosystem services. It may finally lead to the degradation of the ecosystem and may lead to economic loss to the society. Thus it may ultimately create economic and resource stress to the people living in the area who face desertification and land degradation. Forced to attend to the most urgent things first, populations resort to survival strategies that unfortunately make desertification worse and prevent any development.

The most immediate and generally widespread of these survival strategies is to intensify over exploitation of the available natural resources but at the cost of enormous effort. The second strategy may be to sell off everything, including agricultural equipment to cope with the basic monetary needs such as for buying food, schooling, medical help, fulfilling social obligations, etc. The third strategy is increasing rural migration: this may simply involve men and young people leaving for a job in other parts of the country, especially to the urban areas, or the migration may take on the proportions of a population exodus in search of better living conditions (ecological refugees). These survival strategies are often accompanied by breakdowns in the integrity of communities and sometimes of families. To summarize, the consequences of desertification and land degradation are: it affects the socioeconomic conditions of the people, the natural resources, ecosystem services and the environment.

9.6 The Processes of Desertification

Desertification of forests, rangeland, grassland and cropland begins with deterioration of vegetal cover due to deforestation, overgrazing and improper agricultural practices. Once the plant cover is disturbed, soil degradation starts operating in the form of wind and water erosion, soil compaction, surface soil crusting, loss of humus and soil fertility, salinization and water logging (Dregne 1985). It is clear from the above that the major processes of desertification/land degradation can be vegetal degradation, soil erosion due to wind and water (also called aeolian and fluvial processes, respectively), salinization and/or alkalization. In addition to these, there are other desertification processes such as water logging, soil compaction and those arising from man-made impacts. There are desertification/land degradation processes which are specific to cold mountainous arid areas. These are frost heaving, frost shattering, mass movement and landslides.

Vegetal degradation is observed mainly as deforestation/forest-clearing/shifting cultivation and degradation in grazing/grassland as well as in scrubland. At places, agriculture is observed within forest lands; this is also considered under vegetal degradation. Vegetal degradation processes are found in both cold and hot regions.

Wind erosion pertains to the aeolian activities. It denotes the spread of sand by virtue of lift and drift effect of wind, even up to lofty altitudes of the Himalayas. Various categories of sand cover and their severity are classified based on the depth and spread of sand sheet/dunes and barchans.

Water erosion has similar consequences as wind erosion. 'Virtually every medium-to-fine textured cultivated soil having a slope greater than 1 degree is subject to water erosion' (Sabadell et al. 1982,). According to Mainguet (1994), rain splash is the most important process of water erosion. When soils are low in organic matter and high in silt and are without vegetation cover, they are particularly vulnerable to surface crust formation due to rains splash. Silt clogs the pores and creates a thin crusted soil surface, which is impenetrable to water. Crusted surfaces result in low infiltration rates, increased runoff and interference with seedling development. Certain improper grazing and cultivation practices also increase crusting. With increased runoff, rills form, causing further water erosion (Mainguet 1994; Sabadell et al. 1982).

Salinization/alkalinization is another process, which degrades the fertile lands and leads to desertification. Salt may originate from dryland saline seeps, irrigation waters, seawater, paleo-lagoons, inland lakes or sediments or groundwater, which rise up in artesian water. In few cases, wind also deposits salts. Salinization increases the osmotic pressure of the soil water, so a plant's ability to absorb water decreases or inhibits. Salinization may degrade the entire soil profile, resulting in an unstable, compacted structure, decreased permeability and porosity, decreased biological activity and changes in vegetation characteristics. Consequently, vegetation is reduced or eliminated depending upon the degree of salinization (Mainguet 1994; Sabadell et al. 1982).

Water logging degrades soils in several ways, including through salinization. Water logging 'arises when the infiltration rate of the soil exceeds the permeability (hydraulic conductivity) of some horizon within the soil' (Sabadell et al. 1982) or when all the pores of the soil are filled with water. Over irrigation in agricultural area is the most common cause of water logging. The combination of excess water and salt brought to the surface inhibits root growth and all aerobic life. An accumulation of organic matter occurs because the microorganisms responsible for mineralization and humidification of plant debris are reduced or eliminated. Soil becomes more acidic, and clay and iron are leached more easily (Mainguet 1994; Sabadell et al. 1982).

There are several other land degradation processes which are induced directly or indirectly by human intervention only and are not natural. These are anthropogenic desertification processes. They include mining/quarrying, brick kiln, industrial effluents, city waste, urban agglomeration, etc. These processes may occur across various land use/land covers.

In addition to the above desertification and land degradation processes, there are certain processes which are found only in mountainous cold arid regions. These are described below:

Mass wasting is predominantly observed in mountainous areas, especially in cold dryland regions, and it is defined as the process of desertification which leads to the down slope movement of rock, regolith and debris through the action of gravity for example, landslides, scree slopes, etc.

Frost heaving is a process of intense frost and freezing of water operating in glacial and periglacial environment and evolves peculiar forms of rock, regolith and soil.

Frost shattering is defined as the freeze and thaw action operating mostly in periglacial environment. When water, passing through the crevices and pores within the rock freezes, it expands by almost ten times. This puts enormous pressure on the surrounding rocks as at −22 °C; ice can exert a pressure of 3000 kg on an area of half a square inch. The process is most active where the periglacial environment exists, usually in areas adjoining glacial margins, with long cold winters.

9.7 Desertification Mapping and Monitoring

The important elements of desertification mapping, monitoring and assessment using remote sensing data are (i) indicator system, (ii) classification system for desertification status mapping (DSM), (iii) development of methodology for DSM mapping and (iv) methodology for desertification monitoring using multi-temporal satellite images. These elements are discussed in the following sections.

9.7.1 Indicators of Desertification

One of the important requirements for desertification status mapping, monitoring and assessment is to identify/define the indicators and the frequency interval at which these need to be recorded/monitored. An 'indicator' is most often defined, according to the OECD (2003), as 'a parameter, or a value derived from parameters, which points to, provides information about and describes the state of a phenomenon, with a significance extending beyond that directly associated with any given parametric value', where a parameter is a property that is measured or observed. Sometimes the choice of indicators is limited by the constraints in terms of the available technology and infrastructure for collection/measurement and processing of the data needed to generate a particular indicator. The indicator system for desertification monitoring and assessment are classified into four categories viz.:

1. *Pressure Indicators*: Climatic and socioeconomic
2. *State Indicators*: Hydrological, physical, biological, etc.
3. *Impact Indicators*: Socioeconomic, migration, unemployment, etc.
4. *Implementation Indicators*: The changes in the land cover conditions owing to mitigation measures aimed at combating desertification

Table 9.1 Indicator system for desertification monitoring and assessment

1.1 Pressure indicators	
Climatic	Rainfall, temperature, wind, humidity potential evapotranspiration, solar radiation, cloud cover
Socioeconomic	Population density, education status, poverty, livestock density, forest felling, fuel and fodder consumption/supply, collection of medicinal plants, shifting cultivation, diminishing water resources, land management practices
1.2 State indicators	
Physical and hydrological	Erosion status of the land, salinity/alkalinity, shifting in sand sheet/sand dunes, water logging, soil moisture, soil types and properties stone, coverage/barren rocky area, number and spread of water bodies groundwater status, turbidity of water bodies, soil compaction etc.
Biological	Types of vegetation, species composition of vegetation, condition and coverage of vegetation, biomass and productivity of vegetation, crop area and yield
1.3 Impact indicators	
Socioeconomic	Income, migration, mortality rate, health conditions, unemployment, illiteracy, food security and malnutrition, prices of food grain, energy consumption, infrastructure, gender specific issues, living standard
Eco-environmental	Air and water quality, occurrence of dust storm and sandstorm, land pollution, degradation of natural resources, ecosystem services
1.4 Implementation indicators	
Action indicators	Economic input for combating desertification, investment level, state of the development and implementation of action plan to combat desertification, state of the legislation and execution related to combating desertification, people participation, NGO involvement
Effect indicators	Proportion of desertified land rehabilitated, socioeconomic standard of the people, improvement in ecosystem services, natural resources and environmental conditions

Source: Ajai et al. (2007a)

The details of the above-mentioned four categories of indicators are given in Table 9.1

India has also developed its indicator system for desertification monitoring and assessment (DMA). The indicator system, developed for India, is in harmony with the indicators as adopted in Asian TPN-1 for desertification monitoring and assessment in Asian region, and thus the above-mentioned indicator system (Table 9.1) will be applicable to all the other countries of the Indian subcontinent. TPN-1(Thematic Programme Network) is one of the six TPNs identified by UNCCD (United Nations Conventions to Combat Desertification) under its Regional Action Programmes (RAPs).The overall objective of TPN-1 has been to enhance the desertification monitoring and assessment capacities in the region through the establishment of a network and the harmonization of approaches for its conduct in the region. China was identified as the host country to coordinate Asian TPN-1 activities among the member countries in establishing the Asian Regional Desertification Monitoring and Assessment Network (TPN-1). In India, Space

Applications Centre (ISRO), Ahmedabad was identified as the national focal institution to coordinate TPN-1 activities within the country and establish the national network for desertification monitoring and assessment.

Many of the above indicators are amenable to remote sensing and therefore can be deciphered and monitored using satellite images.

9.7.2 Classification System

The basic requirement in preparation of desertification/land degradation status map, using satellite data, is a 'comprehensive classification system' which is able to take care of all the types/processes of land degradation existing or active in the study area for which the mapping is to be done. One of the most comprehensive classification systems for desertification and land degradation status mapping, using satellite data, has been evolved and standardized by Ajai et al. (2007a, 2009). The classification system is based on the above-mentioned status indicators (Table 9.2) which are amenable to remote sensing. The classification system is a three-level hierarchical classification system. Level one comprises of land use/land cover, level two deals with the process (type) of land degradation, and the level three represents the severity of land degradation. The details of the classification system are given in Table 9.2.

The above-mentioned classification system has been used to prepare national-level DSM maps for India on 1:500,000 scale and also for detailed level mapping for districts/watersheds on 1:50,000 scale. Asian TPN-1 has adopted a classification system which is very much similar to the above classification system, for land degradation and desertification status mapping of the Asian region. As the processes of land degradation found in the other countries of the Indian subcontinent (mentioned in Sect. 9.3) are all covered in the classes as given in Table 9.2, the above classification system can be very well used for DSM mapping for each of the countries of the Indian subcontinent. Details of the above-mentioned classes of land degradation/desertification processes, their legends for mapping based on satellite images and the benchmarks (threshold values) for the three severity levels are provided in Ajai et al. (2007a).

9.7.3 Desertification Status Mapping

The methodology for preparation of Desertification Status Map (DSM) using satellite data, as developed and standardized by Ajai et al. (2007a, 2009), is given in Fig. 9.1. Satellite images of the study area pertaining to three season, namely, winter (November–April), summer (May–June) and monsoon (July–October), are used to prepare land degradation and desertification maps.

Satellite images are geometrically corrected and geo-referenced. The geo-referenced satellite images are interpreted/analyzed along with the ancillary data and ground truth information to delineate land use/land cover classes (level 1) existing in the study area. The use of three season data helps in better discrimination of

Table 9.2 Classification system evolved and adopted for desertification status mapping

Level 1: land use/land cover		
The following categories have been identified:		
Agriculture – unirrigated	(D)	
Agriculture – irrigated	(I)	
Forest	(F)[a]	
Grassland/grazing land	(G)	
Land with scrub	(S)[b]	
Barren/rocky area	(B/R)[c]	B(Sc) indicating scree areas in cold deserts
Dune/sandy area	(E)	
Waterbody/drainage	(W)	
Glacial/periglacial (in cold region)	C/L	
Others (urban, man-made, etc.)	(T)	
Level 2: processes of degradation		
Types of processes resulting in degradation:		
Vegetal degradation	(v)	
Water erosion	(w)[d]	
Wind erosion	(e)	
Waterlogging	(l)	
Salinization/alkalinization	(s/a)[e]	
Mass movement (in cold areas)	(g)	
Frost heaving (in cold areas)	(h)	
Frost shattering (in cold areas)	(f)	
Man-made (mining/quarrying, brick kiln, industrial effluents, city waste, urban agg, etc.)	(m)	
Level 3: severity of degradation		
This level represents the degree and severity of the degradation		
Slight	1	
Moderate	2	
Severe	3	

Source: Ajai et al. (2007a)
[a]Rocky areas within forest can be annotated as only FV3-R in the map
[b]Vegetal degradation in land with scrub around periphery of notified forests can be delineated as SV_
[c]Barren and Rocky areas to be delineated separately as B or R and shown in others category of the legend
[d]Gully/ravines should be shown as Xw3, where X is the land use/cover class of surrounding area
[e]Salinization or alkalinization should be shown as 's' or 'a' separately. Where both occur, they should be shown together, i.e. $s_x\ a_y$, where x and y are respective degree of severities

the classes which are otherwise difficult to distinguish, e.g. wasteland and fallow fields. In the next step, land degradation/desertification processes are identified on the satellite images based on their signatures, indicators, interpretation key and ground truth. Assessment of severity of degradations (level 3) is carried out based on the signature on image as well as ground truth including soil sample analysis if

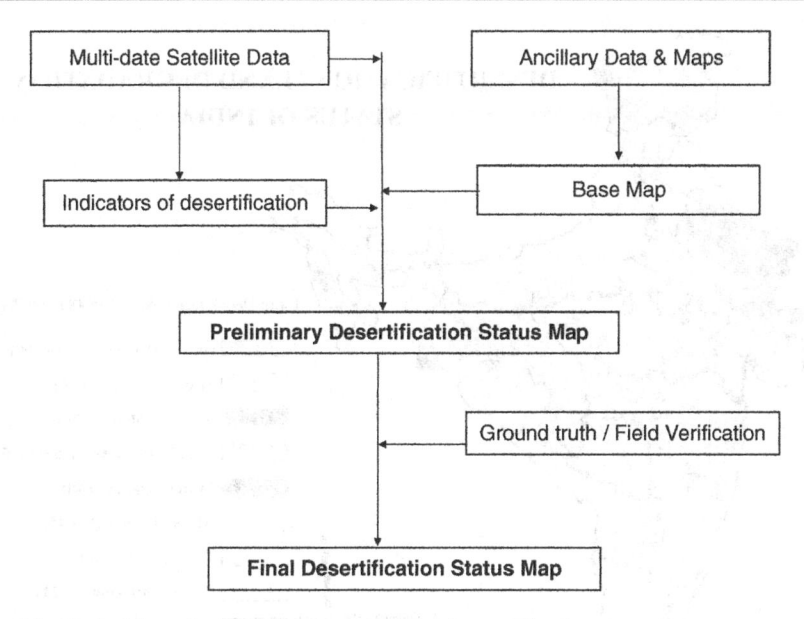

Fig. 9.1 Methodology for desertification mapping using satellite data

required. In some cases, like low-moderate salinity/alkalinity, soil sample analysis is needed as the signatures may not be very clear on satellite images.

Standardization of image interpretation key for various land degradation processes and their severity is one of the important steps in preparation of desertification status map through visual analysis of satellite images. It may require a reconnaissance survey of the study area, to broadly assess the type, extent, cause and possible consequence of the desertification (Ajai et al. 2007a). In case of large-scale mapping (1:25,000/50,000), it is possible to identify three severity levels. Whereas for national (1:500,000) or regional level (1: 1/5 million scale) DSM mapping, only two severity levels would be possible to discriminate on satellite images.

Once the preliminary desertification map is prepared, its ground truth verification is carried out through field surveys. Ground truth data is then used to finalize the DSM maps. Ground truth data is also used to assess the accuracy of the maps prepared. The next step is to create digital data base in GIS environment, and the area statistics for each of the classes of land degradation/desertification are generated.

Using the above-mentioned classification system and methodology, national-level desertification status map has been prepared for India at 1:500,000 scale using IRS (Indian Remote Sensing Satellite) AWiFS data of 56 m spatial resolutions (Ajai et al. 2007b, 2009). Desertification/Land Degradation Status Map of India is given in Fig. 9.2.

The mapping results show that the total land area undergoing the process of desertification and land degradation in India is 105.48 Mha, which is 32.07% of the total geographical area (TGA) of the country (Ajai et al. 2007b, 2009). The land

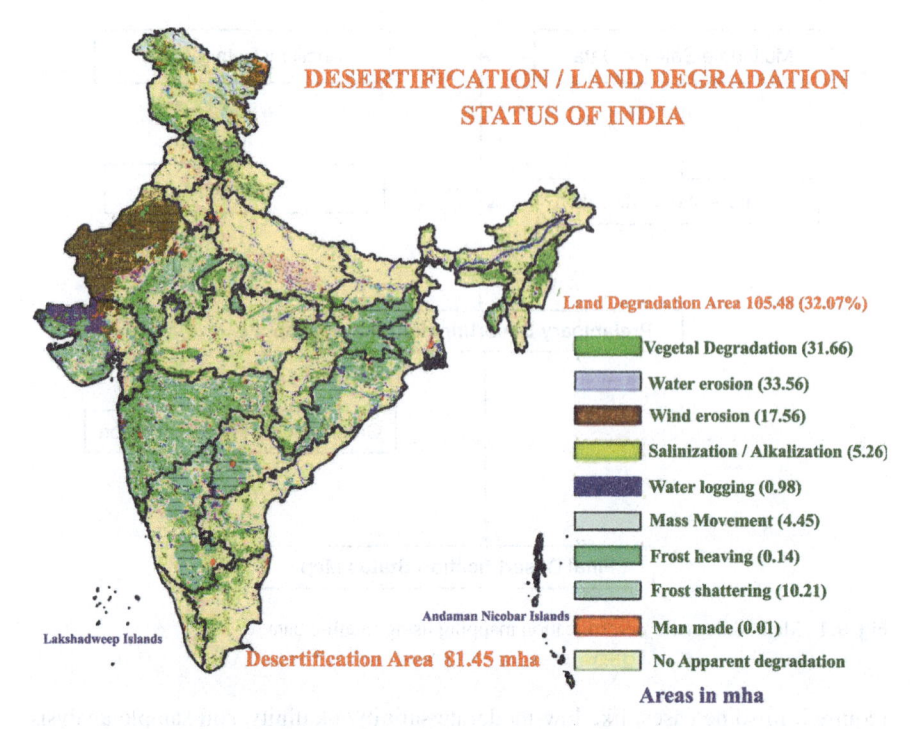

Fig. 9.2 Desertification/land degradation status map of India

area undergoing the process of desertification is 84.6 Mha which is about 25% of the total geographic area. Among the states, highest percent of desertification/degradation is observed in the Gujarat state (68.43%) followed by Rajasthan (67%) and J&K (60.73%). Among the NE states that fall outside the dryland boundary, the maximum land degradation is seen in Mizoram (78%) (Ajai et al. 2009).

Process wise, per cent of the total geographic area undergoing water erosion is 10.21%, vegetal degradation is 9.63%, and wind erosion is 5.34%. These are the major land degradation processes in India.

Examples of large-scale maps at watershed/district level on 1:50,000 scale are given in Fig. 9.3. Figure 9.3a is the DSM map for Ballia district (hot semi-arid region), Uttar Pradesh state of India. Figure 9.3b is the DSM map for cold arid region (Leh-south watershed) in Jammu and Kashmir State of India.

9.7.4 Desertification Monitoring

Studies have been carried out for monitoring desertification status using multi-temporal satellite images in a few hot spot areas. One such example for Nathusari block of Sirsa district in Haryana state (Northern India) is given in Fig. 9.4. IRS satellite images of 2000, 2005 and 2009 show significant increase in salinity (shown in white and blue colour) and waterlogging (dark blue colour) during a decade.

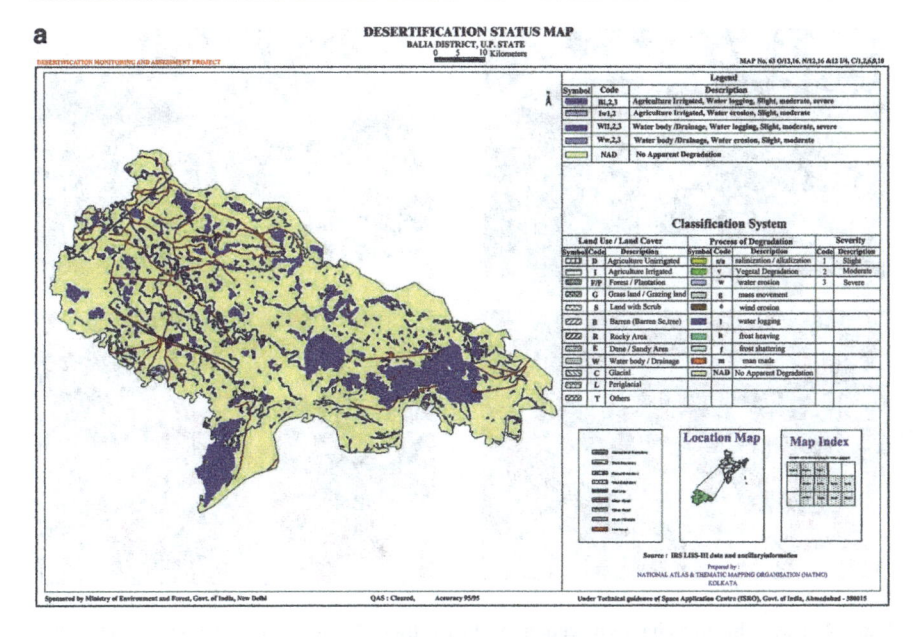

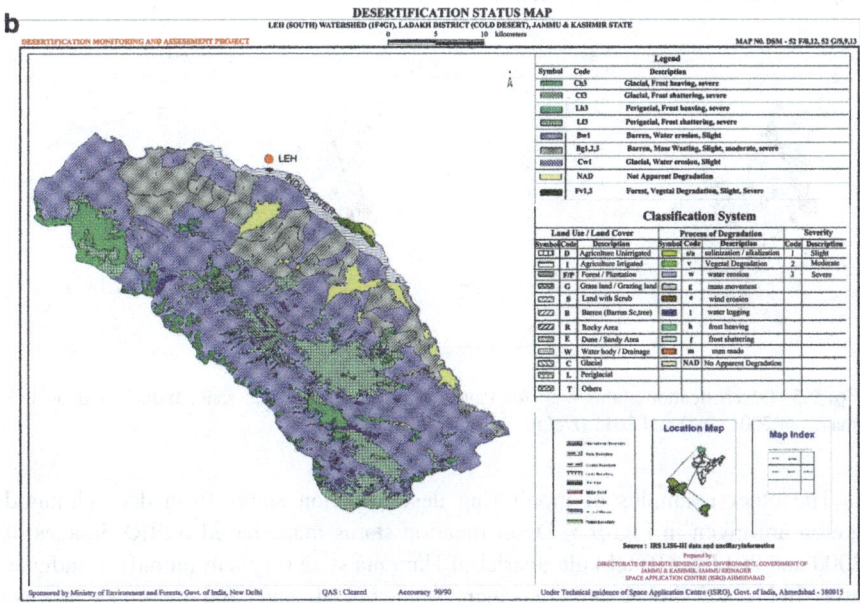

Fig. 9.3 (a) Desertification Status Map for Ballia district, Uttar Pradesh, Northern India. The major process of land degradation is water logging. (b) The Desertification Status Map of Leh-south watershed in Jammu and Kashmir State in Northern India. Major processes of land degradation are frost shattering, frost heaving, mass wasting and water erosion

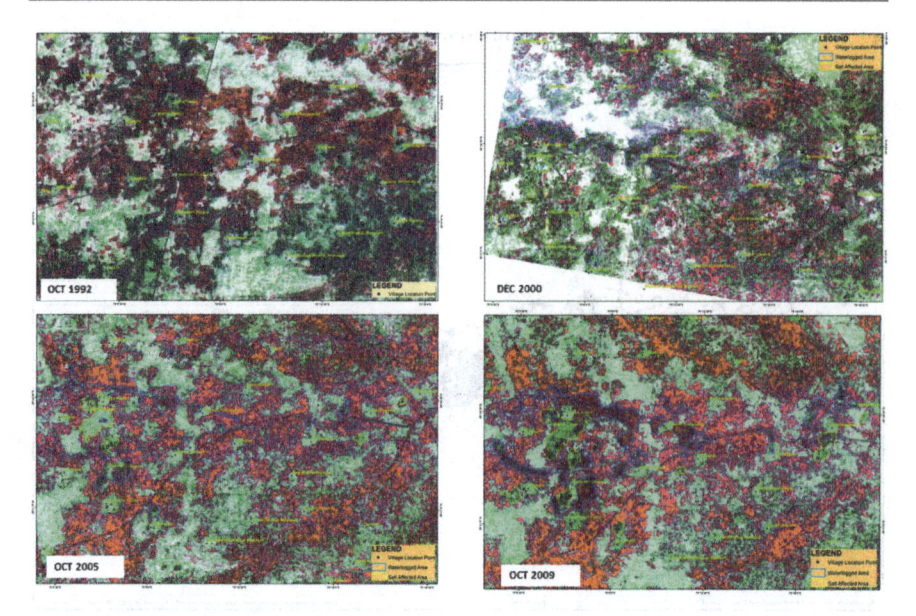

Fig. 9.4 Desertification change detection (Nathusari Block Sirsa district of Haryana state, India)

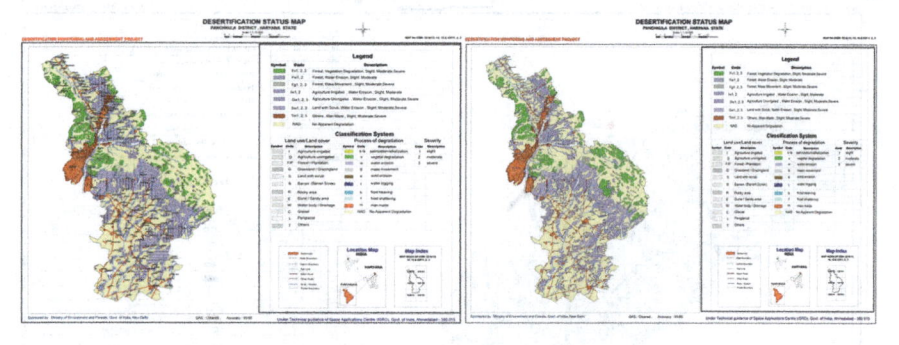

Fig. 9.5 Desertification status map for Panchkula district, Haryana state, India based on IRS images of 2001 (*left*) and 2011 (*right*)

The other examples of monitoring desertification status from dry subhumid region are given in Fig. 9.5. Desertification status maps based on IRS images of 2000 and 2011 for Panchkula district of Haryana state (dry-sub-humid) of India as given in Fig. 9.5 show that there is increase in vegetal, water and man-made process of degradation during the last one decade.

9.8 Future Challenges

One of the major environmental concerns in the entire Indian subcontinent (comprising of Bangladesh, Bhutan, India, Maldives, Nepal, Pakistan, Sri Lanka and Afghanistan) is land degradation and desertification. Very large proportion of the geographic area of this subcontinent falls under drylands which are highly vulnerable to land degradation. A large percentage of the geographic area of these eight countries are affected by desertification/land degradation. Major processes of land degradation in this subcontinent are vegetal degradation, water and wind erosion, water logging, salinization, mining and mass wasting. The main causes of land degradation and desertification in these countries are deforestation including shifting cultivation, overgrazing, unsustainable agricultural practices, cultivation in marginal lands and steep slope areas, industrialization/urbanization and mining. Most of the above drivers of land degradation, triggered and accelerated by anthropogenic activities, are mainly due population pressure, poverty, illiteracy and lack of sufficient resources. The human population in this subcontinent has been steadily increasing during the past hundred years and thereby increasing the pressure on the land-based natural resources, ecosystem services and the environment. Per capita land availability (arable land) has been continuously decreasing during the past many decades. Forest cover has also been decreasing in all the above-mentioned eight countries. Considering the emerging problems, feeding the continuously increasing population of this subcontinent and to meet their basic requirements and aspirations for better quality of life, from the resources of the finite land area would be difficult in the coming decades, especially as the agriculture and allied activities will also face the challenges of climate change.

In order to meet the above-mentioned challenges, especially the demand for food and water for the increasing population, we need to adopt the sustainable land management practices which include the development of, both, water and land resources. This will require appropriate strategies towards soil and water conservation. The strategies will vary from region to region, depending upon the existing ground realities including the status of the land and the socioeconomic profiles of the people living in the area. Problems related to land degradations need to be addressed towards: (i) reclamations of the already degraded lands, (ii) arresting the ongoing process of land degradation and (iii) appropriate measures for the lands which have potential to get degraded in future. The action plans for the above-mentioned tasks need to be locale specific and should be based on the modern technologies including the geospatial technology. Involvement of local people is essential in implementation of the sustainable land management practices towards combating the desertification/land degradation as well as for soil and water conservation. One of the prerequisites for preparation of action plans, for combating desertification, is a spatial inventory (mapping) of the land undergoing the process of desertification/land degradation along with the information on the type and severity of the desertification processes. Space data have been successfully used for making such spatial inventory of the status of desertification and land degradation at local, national, regional and global scales. The use of data from Earth Observation Satellites along with field information and the traditional knowledge and geospatial technology is a viable option in preparation of locale specific action plans for sustainable land and water management towards combating desertification and land degradation.

Acknowledgement Authors are thankful to CSIR, India for support under ES scheme and to Mr Tapan Misra, Director, Space Applications Centre for his guidance and support

References

A.K. Acharya, N. Kafle, Land degradation issues in Nepal and its management through agroforestry. J. Agric. Environ. **10**, 115–123 (2009)

W. N. Adger, T. A. Benjaminsen, K. Brown, H. Savarstad (eds.), *Advancing a Political Ecology of Global Environmental Discourse. Centre of Social and Economic Research on the Global Environment* (University of EastAnglia, London, 2000)

A.S. Ajai, P.S. Arya, S.K.P. Dhinwa, K. Ganeshraj, Desertification/land degradation status mapping of India. Curr. Sci. **97**(10), 1478–1483 (2009)

Ajai et al., Desertification monitoring and assessment using remote sensing and GIS: a pilot project under TPN-1, UNCCD. Scientific Report No: SAC/RESIPA/MESG/DMA/2007/01, Space Applications Centre, ISRO, Ahmedabad, (2007a)

Ajai, A.S. et al., Desertification/land degradation Atlas of India, Space Applications Centre, Ahmedabad. ISBN No. 978–81–909978-2-9, (2007b)

A.A. Amin, The extent of desertification in Saudi Arabia. Environ. Geol. **46**(1), 22–31 (2004)

A. Aubreville, Climats, Forets et Desrertification de l'Frique Tropicale Societe d' Editions Geographiquesm Maritimes et Coloniales, Paris. P 255, (1949)

BARC, Land degradation situation of Bangladesh. Soil Sc. Division. Bangladesh Agricultural Research Council, Farmgate, Dhaka, (1999)

Bhutan-NAP, National Action Program to combat land degradation, Ministry of Agriculture and Forest, Royal Government of Bhutan, (2014)

R. Behnke, M. Michael (eds.), *The End of Desertification* (Springer-Verlag, Berlin, 2016), p. 205

P. Brabant, Activities humaines et degradation desterres, collection Atlas Cederom, Indiateurs et methode, IRD Paris, Published under the International Year of Planet Earth (IYPE) Planet Terre label (cartographie.ird.fr/degra_PB.html), (2008)

P. Brabant, A land degradation assessment and mapping methodology, Standard guideline proposal, Comite' Scientifique Francais De la desertification les dossiers thematiques. *CSFD* Issue 8, Agropolis International, (2010)

H.E. Dregne, Aridity and land degradation. Environment **27**, 8 (1985), p. 16–20 and 28–33, (St. Louis)

H.E. Dregne, Desertification: problems and challenges. Ann. Arid Zone **39**, 363–371 (2000)

H.E. Dregne, N.T. Chou, Global desertification dimensions and costs, in *Degradation and Restoration of Arid Lands*, ed. By H.E. Dregne, (Texas Technical University, Lubbock, 1994)

DSD, Integrated methods for assessment and monitoring land degradation processes and drivers (land quality), White paper of the DSD working group 1, version 2. Http://dsd-consortium.jrc. ec.europa.eu/documents/wg1 white- paper draft-2 20090818.pdf, (2009)

H. Eswaran, R. Lal, P.F. Reich, Land degradation: an overview, in *Response to Land Degradation*, ed. By E.M. Bridges, I.D. Hannam, L.R. Oldeman, F.W.T. Pening De Vries, S.J. Scherr, S. Sompatpanit, (Science Publishers, Enfield, 2001), pp. 20–35

FAO, Landuse policy planning. FAO Report, Nepal, (2010)

A.T.M. Farid, A. Iqbal, Z. Karim, Soil erosion in the Chittagong hill tract and its impact on nutrient status of soil. B.J. Soil Sci. **23**(122), 92–101 (1992)

H. Geist, The causes and progression of desertification, Ashgate Publishers, (2005) pp 258

GLASOD, Global Assessment of Human induced Soil degradation, 1990 International Soil Reference and Information Centre (Wageningen Netherlands) and UNEP (Nairobi, Kenya) 2004, http://lime.isric.nl/index.cfm contentid_158, (1990)

M.K. Hasan, K.M. Ashraf Alam, Land degradation situation in Bangladesh and role of agroforestry. J. Agri. Rural. Dev. **4**(1&2), 19–25 (2006)

M.A. Kishk, Land degradation in Nile Valley. Ambio **15**, 226–230 (1986)

M. Mainguet (ed.), *Desertification: Natural Background and Human Mismanagement* (Springer-Verlag, New York, 1994.) 314 p.

Maldives, *Maldives National Capacity Self-Assessment Report and Action Plan for Global Climate Change, Biodiversity and Land Degradation Conventions, Ministry of Housing, Transport and Environment Government of Republic of Maldives* (2009)

MENR, Second National Status Report on Land Degradation Implementation of UNCCD in Sri Lanka, Ministry of Environment and Natural Resources, Sri Lanka, (2002)

M.M.U. Miah, A.K.M. Habibullah, M.F. Ali , Depletion of organic matter in upland soils of Bangladesh. In *"Soil resilience and sustainable land use"*. Proceedings of international symposium, heldon 28 September to 2 October 1992, Budapest, Hungary, (1993) pp. 70–79

F.N. Muchena, *"Indicators for Sustainable Land Management in Kenya's Context"*. GEF Land Degradation Focal Area Indicators (ETC-East Africa, Nairobi, 2008)

NBSS&LUP. 2001. Agro-ecological sub-regions of India for planning and development, NBSS&LUP Publications, ICAR, Nagpur

OECD, *OECD Environmental Indicators-Development, Measurement and Use*. Reference Paper 2003, http://oecd.org/dataoecd/7/47/24993546.pdf, (2003)

J. Puigdefabregas, G. del Barrio, J. Hill, *Advances in studies on desertification*, Contributions to the International Conference in memory of Prof. Johm B. Thornes. -Murcia: Universidad de Murcia, editum. Ecosystemic Approaches To Land Degradation, (2009), pp. 77–87

J.F. Reynolds, A. Grainger, S. Smith, D.M. Bastin, G.L. Garcia-Barrios, R.J. Fernández, M.A. Janssen, N. Jürgens, R.J. Scholes, A. Veldkamp, M.M. Ver-straete, G. Von Maltitz, P. Zdruli, Scientific concepts for an integrated analysis of desertification. Land Degrad. Dev. **22**, 166–183 (2011)

J.F. Reynolds, D.M. Stafford, E.F. Smith, B.L. Lambin, I.I. Turner, M. Mortimore, S.P.J. Batterbury, T.E. Downing, H. Dowlatabadi, R.J. Fernandez, J.E. Herrick, E. Huber-Sannwald, H. Jiang, R. Leemans, T. Lynam, F.T. Maestre, M. Ayarza, B. Walker, Global desertification: building a science for dryland development. Science **316**, 847–851 (2007)

J.E. Sabadell et al., *Desertification in the United States* (U. S. Government Printing Office, Washington, DC, 1982.) 277 p.

U.N. Safriel, The assessment of global trends in land degradation, in *Climate and land degradation*, ed by M.V.N. Sivakumar, N. Ndiang'ui, (Springer, Berlin, 2007), pp. 1–38

A.A. Shakeel, W. Long-chang, S.M.F. XueLan-lan, W. Guo-xin, Z. Cong-ming, Desertification in Pakistan: causes, impacts and management. J. Food Agric. Environ. **8**(2), 1203–1208 (2010)

D.S.G. Thomas, N.J. Middleton, *Desertification: Exploding the Myth* (Wiley, Chichester, 1994)

UNCCD, *United Nations Convention to Combat Desertification in those Coun-tries Experiencing Serious Drought and/or Desertification, Particularly in Africa*. UNEP. Final text of the Convention, (1994), www.unccd.int/convention/text/pdf/conv-eng.pdf

UNCOD, *Round-Up, Plan of Action and Resolutions*. United Nations Conference on Desertification, Nairobi, (1977), p 43

UNEP, *World Atlas of Desertification*, 2nd edn. (United Nations Environment Programme, Nairobi and Edward Arnold, London, 1977a)

UNEP, *UNEP and Fight Against Desertification*. Nairobi, Kenya, (1977b)

Drylands of China: Problems and Prospects

10

Shiming Ma

10.1 Introduction

The dryland in China is defined as the areas where the annual precipitation is less than 200 mm and the humidity index is smaller than 0.20. Therefore, the arid areas in China are located in the central part of Eurasian continent, including all of Xinjiang, Hexi Corridor in Gansu, and the west of the Helan Mountain of Inner Mongolia. Its geographical location is between longitude 73–125° E and latitude 35–50° N. This region occupies approximately 2.02 million km², accounting for more than 21% of China's total terrestrial area (Ci et al. 1994).

The Kunlun-Altun-Qilian Mountains are the southern boundary of the area which combined with the Qinghai-Tibet Plateau blocks the water vapor from the Indian Ocean. Altai Mountain is the northern boundary, blocking the water vapor from the Arctic Ocean (Fig. 10.1).

Tian Shan Mountains lie in the middle of the region and form the landscapes of alpine, inland basins, Gobi Desert, and widespread sandy desert. The average annual temperature is 6.1 °C (1971–2000), and the average annual precipitation is only 120 mm (1971–2000), both with uneven spatial and temporal distributions. The extreme minimum temperature −51.5 °C occurred in the Altai Mountain, and the extreme maximum temperature 48.9 °C occurred in the Turpan Basin.

This inland region is controlled by continental climate, with little rainfall throughout the year and strong evaporative potential. The average annual precipitation is only 130 mm, with uneven spatial and temporal distributions. The highest rainfall occurs at the eastern end of the Tian Shan Mountains in the Ili River Valley,

---

S. Ma (✉)
Institute of Environment and Sustainable Development in Agriculture (IEDA),
Chinese Academy of Agricultural Sciences (CAAS),
No. 12 Zhongguancun Nandajie, Haidian District, Beijing 100081,
People's Republic of China
e-mail: mashiming@caas.cn

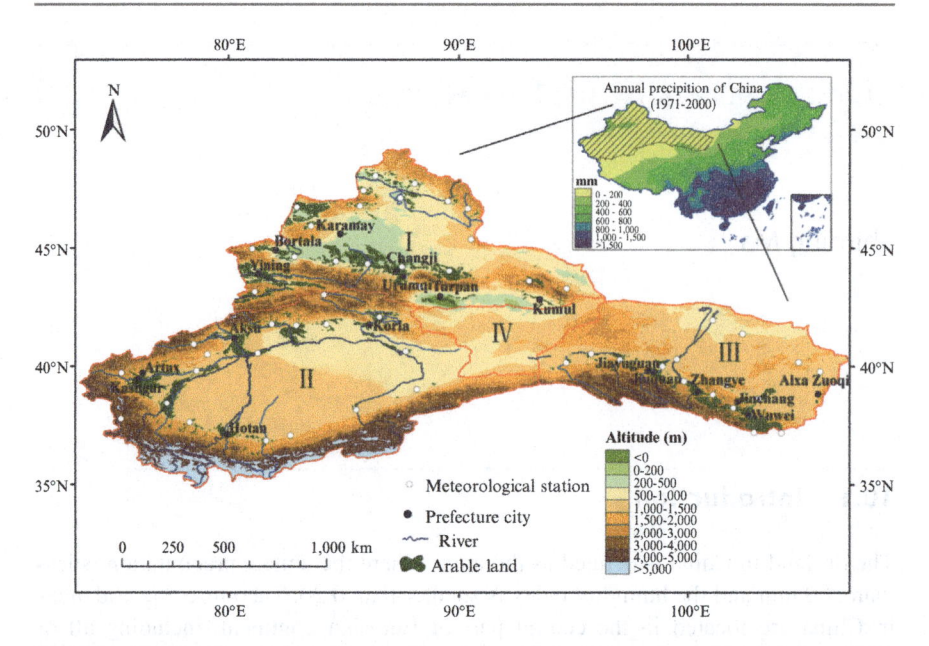

Fig. 10.1 Distribution of arid and semiarid regions of China (Shen et al. 2013) (The embedded figure shows the distribution of annual precipitation, precipitation in the arid area is generally less than 200 mm/year)

which could amount up to 800 mm/year or above; annual precipitation is generally less than 50 mm in the artificial oasis areas, where most agricultural activities exist. However, the potential evaporation in this arid region can be as high as 3,200 mm/year, eight to ten times the annual rainfall. The hyper-arid region is thus one of the most arid regions in the world (Shen and Chen 2010).

10.2 Physical Nature of Arid Lands

10.2.1 Climate

The annual average temperature in the arid areas of China varies from region to region. It is higher in the southern parts of the region and in the basins. For example, it is with 8–10 °C in Junggar Basin, with 10–13 °C in Tarim Basin, and with 7–9 °C in Alashan Plateau (Table 10.1).

Arid land of China lies in the middle latitude of the Northern Hemisphere, where it is affected by westerly wind. Due to complex influences of various factors including location (in the hinterland of Eurasia), rapidly uplift of the Qinghai-Tibet Plateau, effects of middle-scale landform, and feedback of earth surface on climate, precipitation in the region is rare and unevenly distributed (Chen 2010).

Generally, precipitation in the western part of the arid land is more than that in the eastern part, for example, precipitation in Ili River Valley is about five to six

Table 10.1 Annual average temperature of different decades in the different regions of the arid areas in China (in °C)

	Junggar Basin	Tarim Basin	Tian Shan Mountain	Hexi Corridor	Alashan Plateau	Arid areas of North China
1960s	6.0	11.0	0.6	7.2	7.1	6.4
1970s	6.0	11.1	0.8	7.3	7.4	6.5
1980s	6.5	11.4	0.7	7.4	7.8	6.8
1990s	6.9	11.6	1.2	8.0	8.4	7.2
1961–2000	6.4	11.3	0.8	7.5	7.7	6.7
2001–2007	7.4	12.0	2.0	8.6	8.7	7.8

Table 10.2 Annual average precipitation of different decades in the different regions of the arid areas in China (in mm)

	Junggar Basin	Tarim Basin	Tian Shan Mountain	Hexi Corridor	Alashan Plateau	Arid areas of North China
1960s	176.6	39.8	316.5	102.4	93.9	145.8
1970s	173.0	43.7	315.2	112.7	102.8	149.5
1980s	202.4	51.2	319.8	109.3	83.1	153.2
1990s	213.8	55.1	356.2	111.3	101.2	167.5
1961–2000	191.5	47.5	326.9	108.9	95.2	154.0
2001–2007	226.8	57.5	366.2	121.5	99.7	174.3

times of that in Alashan Plateau. A higher rainfall belt consisted of many isolated centers with mean annual precipitation of 500–1,000 mm lies between arid land of central Asia and arid land of China (Table 10.2).

This higher rainfall belt is the result of enforced uplift of the westerly airstream on the windward slopes of the Tian Shan Mountains and the Pamirs. In basin and plain, annual precipitation is usually less than 50 mm, and even less than several millimeters, for example, at Tuokesun station in the Turpan Basin, the recorded mean annual precipitation is only 7.6 mm. However, there are plentiful solar and heat energy resources in the basin and plain. In the area including basins and plains in Xinjiang Uygur Autonomous Region, the Hexi Corridor in Gansu, and the Alashan Plateau in Inner Mongolia Autonomous Region, the global solar radiation ranges from 170 to 210 W/m^2, taking second rank only to the Qinghai-Tibet Plateau. The photosynthetically active radiation (PAR) is about 80–100 W/m^2, and the mean annual total duration of sunshine reaches to 2,550–3,500 h, taking the first place in China.

Arid land of China has also plentiful wind energy resources. Spatially, the amount of wind energy resources in the northern part, the plain, and the Gobi and the fringe of basins in arid land of China are generally larger. The narrow mountain passages and mountain passes where the air current passes through are the places with the most plentiful and larger exploitable wind energy resources. It should be pointed out, however, that various disastrous weather and climatic events, such as drought, cold wave, gale, sandstorm/salt storm, blizzard, freeze injury, hail, frost, rainstorm and mud-rock flow, dry-heat wind, etc. occur frequently, which result in a lot of losses of properties in arid land of China. In recent 30 years, arid land of China experiences significant increase in temperature and a slightly increase in precipitation under the background of global warming. Since 2000, increasing trends in temperature and precipitation have kept or even have strengthened to a certain extent. However, the intrinsic nature of arid climate in drylands of China is still sustained.

10.2.2 Geomorphology

Arid land of China is located in the hinterland of Eurasia and the northern fringe of the Qinghai-Tibet Plateau (or the high Asia territory). Unique geographic location creates conditions for formation and evolution of the particular landforms, vegetation covers, and pedo-types. The geomorphologic and climatic distribution patterns featured by mountain alternating with basin enrich greatly natural landscape in arid land of China and provide ideal bases for living and developing of human being in arid land and hence prompt the formation, development, and evolution of the artificial oases by storing and diverting water from inland rivers.

Geological and geomorphologic patterns in arid land of China are mainly comprised of grand high mountains and large lower inland basins, featured by mountain alternating with basin. The Altai Mountains, the Junggar Basin, the Tian Shan Mountains, the Tarim Basin, and the Kunlun Mountains-Arjin Mountains distribute from the north to the south in turn (Chen 2010).

Various grand mountains are usually called as "wet islands" and are water sources in arid land. Snow-icing, periglacial, and fluvial action are common external forces sculpturing various alpine landforms. Basin is the water consumption region and is very dry. Formation of main landforms in basins is related to aridness. Sand dune, Gobi, and bad land can be seen everywhere, which form the typical desert landscape in arid land of China. However, the loneliness landscape is occasionally interrupted by oasis dotted in them. The arid geomorphologic pattern is seldom seen in other arid land of the world. The distinctive landforms in arid land of China, such as Yadan, Mud volcano, Karez, Desert Highway, etc., are praiseworthy.

The major rivers include the Tarim River, Ili River, and Heihe River. The Tarim River is the largest inland river in China and originates from the confluence of the Aksu River, Hotan River, and Yarkand River. The Ili River is located inside Tian Shan Mountain and is affected by westerlies, so more rain falls in this basin than in the other areas in Northwest China. The annual runoff is 17 billion m^3 in the Ili River, which is the largest in all of the river basins in the arid areas of China.

Except the Erjis River, not any surface and ground runoff in arid land of China is connected with remote oceans and other watersheds in semiarid and humid area. The flows of mass, energy, and information in arid land of China have obvious locality. Every river basin in the region has its tail lake and its constant circling system in lower air layer. Water circulation is characterized by local circle in each of river valley. The formation area and the consumption area of the water resources are separated by vast desert or Gobi. Water flow as a tie links the mountain area and the basin. As a result, a unique and completed local inland water circling system, or an organic and interacted mountain in forest and grassland-oasis-desert system, is formed by taking water circulation as the main and the bio-system in a tight link.

The amount, quality, and spatiotemporal distribution of water resources are obviously uneven in different geographic environment of the arid land. Water resources are comprised mainly of snowmelt water. Glacier and snow, groundwater, and lake and surface runoff as different types of water present jointly in a shorter-arid watershed. Water resources are sensitive to the influence of global warming. However, presently, water resources in arid land of China sustain a roughly constant condition owing to the unique water circling process.

Accumulated snow and glaciers play an important role in the discharge regulation of these inland rivers and reduce the variations in the inter-annual discharge at the mouths of mountain valleys. Owing to the alternately distributed terrain of high mountains and basins, all the rivers form into centripetal stream systems. Some of the rivers with small discharges soon disappear in the desert region after flowing out of the mouths of the mountain valleys, while a few larger rivers flow to depressions to form inland lakes. Such distribution characteristics of water resources decide the close interconnection of surface runoff and groundwater resources and also lay a basis for the economic development and hydrological regimes in the middle and lower reaches of the river basins. Under the influence of such water resource distribution and dry climatic conditions, the middle and lower reaches are macroscopically eco-environment fragile belts and thus seriously influenced by the hydrological regimes (Chen 2010). Since the 1940s, great changes have taken place in both the hydrological regimes and ecological environment in these areas due to population growth, socioeconomic development, and large-scale exploitation of water and land resources. Such changes seriously affect the sustainable development of human habitats.

10.2.3 Terrestrial Ecology or Vegetation

In recent decades, land use, global climate change, etc. have exerted enormous effects on habitat of various plant species in arid land of China, and as a result, plant diversity is seriously threatened and some ecological and geographic systems verge on collapse. To get endangered ecosystem out of dangerous condition, it is necessary to protect effectively and utilize rationally these ecosystems so as to realize sustainable development of society and economy in arid land of China (Chen 2010).

Arid land of China is located in the eastern section of the Africa-Asia desert. Varied natural geographic types and distinctive and extreme habitats are presented in the region in the special regional environment. Under this habitat condition, latitude, longitude, and altitude zonality of vegetation stamped with the brand of "desert" are clear.

During the historical period, the natural geographic environment in the region has changed repeatedly, which provided conditions for the contact, mix, and particularization of the components of florae. As a result of these, the flora of the region presents striking xerophytic and distinctively temperate characteristics. The flora shows a great disparity and distributes unevenly. The flora with an ancient origin has fewer endemic genera, but has more endemic species, in which seed plants reach to 109 species. These seed plants occupy an important position in the vegetation component and constitute xerophytic and extra-xerophytic shrub communities.

The desert vegetation constitutes the base belt of the vertical belts of mountain vegetation. The desert base belt is structurally simple and differs obviously in different aspects. In addition, after long-term evolution, the region carries within itself the plentiful and distinguished resources of plant species. These halophytic, xerophytic, and short-lived vegetations, wild fruit trees, etc. have irreplaceable advantages in aspects of resisting adverse circumstance, raising efficiencies of photosynthesis and water use, and breeding of fruit trees, and hence provide plentiful gene resources for human being.

Vegetation growth in most areas of arid and semiarid regions in Western China experienced an overall increase over the past two decades, which supported the conclusion that part of Northern China has undergone reversal of desertification since the 1980s (Zhang et al. 2003). According to the national monitoring data on desertification in Western China, the annual desertification rate decreased from 1.2% in the 1950s to 0.2% in 2003 (Piao et al. 2005).

The increases of temperature and pan evaporation may bring some adverse ecological effects. Recent research results show that the vegetation coverage and NDVI in the arid region of Northwest China exhibit an increased trend before 2000. However, the trends reversed to decrease since 2000 (Fig. 10.2) (Zhao et al. 2011; Li and Chen 2014).

Average rainy season NDVI in arid and semiarid regions both increased significantly during the period 1982–1999, while that in the Taklimakan (Takla Makan) Desert did not show statistically significant trend over time ($R2 = 0.05$; $P = 0.366$), suggesting the stability of the NDVI time series used in this study to identify interannual variations in green vegetation in arid and semiarid regions (Fig. 10.2). Over the past 18 years, average rainy season NDVI in arid regions increased by 0.0006 per year, with an annual increase rate of 0.5% ($R2 = 0.31$; $P = 0.017$). The relative annual increase rate of 0.8% in rainy season NDVI ($R2 = 0.58$; $P < 0.001$) in semiarid regions exceeded that of arid regions (Piao et al. 2005).

10.2.4 Soil

In the formation process of natural soil, two material circulations, the grand geological circulation as the base and the small biological circulation as the condition. These must coexist and interact before the soil with a certain profile and fertility level forms. The soil formation process in arid land has obvious vertical zonality. During geological historical period, especially since early Pleistocene, a glacial epoch alternated with an interglacial stage. On alpine glacial deposits at different times, organisms and generated soil evolved simultaneously, and accumulated an organic horizon with different properties. On the old and new moraine

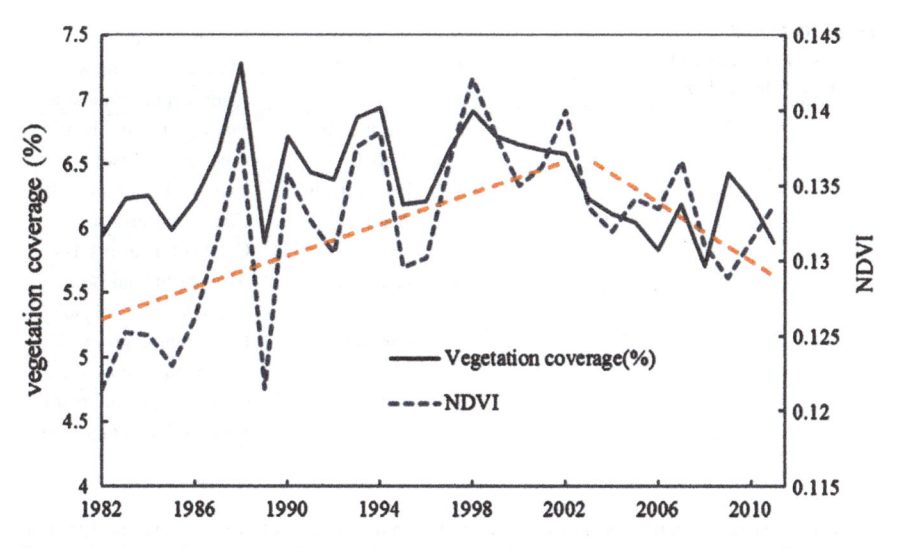

Fig. 10.2 Dynamic changes of vegetation area and NDVI in the Northwest China (Chen et al. 2015)

parent materials, primitive soil and climax soil developed separately. During interglacial stage, large area of melted water from alpine snow and glacier conveyed not only lots of solid runoff materials but also solvable salts and animal remains to the plain. In soils on shoals of ancient lakes, old river terraces, and fringes of alluvial fans, single gypsum layer or multiple gypsum horizons formed, and xerosol, halogenic soil, and anthropogenic soil developed separately. It is worth noting that in the deep soil profile, many soil genetic horizons developed and were preserved. These are the formative layer for determining ancient environment under global change (Chen 2010; Zhang et al. 2014).

The soil classification in the arid areas of China is shown in Table 10.3. Based on the pedogenesis, the soils in the arid areas of China are classified to 8 soil orders and 13 soil great groups (Chen 2010).

10.3 Anthropogenic Nature of Arid Lands

10.3.1 Land Use

Land use and cover change (LUCC) is increasingly recognized as a sensitive indicator of earth systems (Zhou et al. 2008). In the arid zone of Western China, rapid and frequent LUCC has been observed around the oasis where the majority of the regional population is concentrated. The observed changes largely reflect the impact of human activities on the fragile natural environment. Changes became most noticeable from the early 1990s when local governments launched and supported an agricultural development plan (Hu and Li 2010). It is a commonly held belief that

Table 10.3 Soil classification in the arid areas of China (Chen 2010)

No.	Soil order	Soil great group
1	Primosols	Gelic-orthic primosols
		Aridi-sandic primosols
2	Cambisols	Matti-gelic cambisols
3	Alfisols	Calci-ustic alfisols
4	Isohumisols	Calci-ustic isohumisols
5	Aridisols	Calci-cryic aridisols
		Calci-orthic aridisols
6	Halosols	Aridi-orthic halosols
		Takyri-alkalic halosols
7	Gleysols	Histi-stagnic gleysols
		Histi-orthnic gleysols
8	Anthrosols	Gleyi-stagnic anthrosols
		Siltigi-othnic anthrosols

Table 10.4 Status of land desertification in several main regions in northwest China (unit: 1,000 ha)

Region	Desertified land area	Total land area (%)	Oasis area (%)	Severe desertified land (%)	Potential desertification area	Total land area (%)
Xinjiang	304.71	1.82	43.14	24.99	154.96	0.92
Hexi Corridor	252.22	9.15	19.6	31.82	168.94	6.13
Alxa plateau	118.0	6.1	–	61.28	60.22	3.1

the rapid expansion of farmland (irrigated land, primarily cotton fields) and its subsequent abandonment or neglect would lead to increasing soil salinity and ultimately to desertification (Zhang et al. 2003; Ma et al. 2011).

In the arid zone of Western China, a widely accepted belief is that rapid farmland expansion and the subsequent abandonment of the farms, or their mismanagement, would lead to soil salinization and desertification (Sun and Zhou 2016).

Land desertification is developing rapidly due to the erodible surface of sandy deposits, frequent wind erosion, and over-exploitation of water and land resources. As shown in Table 10.4, the desertified land accounts for about 19–43% of the total. The severely desertified land has now reached 25–61.3% of the total desertified area. Besides the desertified land in Northwest China, there are about 3.8 Mha of potential desertification area (Wang and Cheng 2000).

There are three factors mainly responsible for the rapid development of modern desertification in arid Northwest China, namely, wind erosion due to the lack of protection, destruction of vegetation due to misuse of water resources and soil salinization, and sand dune encroachment of the oasis. Of particular importance are the latter two factors. For example, 83% of the total desertified land area in the Tarim Basin has been attributed to the latter two factors. From variations of the desertified land area during the period of 1949–1993 in the middle and lower reaches of Tarim and Heihe River basins, it can be seen that the development rates of land desertification are different for different periods.

The most rapid development period of desertification for the midstream regions of the inland river basins occurred in the 1970–1980s, with the quickest expansion rate of desertification occurring in the most recent half century (0.76% in the middle reach of Tarim River basin and 2.6% in the middle reach of Heihe River basin). Since 1990 the rate of desertification in the Tarim and Heihe River basins was reduced by 0.27% and 0.5%, respectively, and in local places, desertification has been arrested or even reversed. However, desertification in the lower reaches of inland river basins still tends to increase, and, in the lower reach of the Tarim River basin, the area of moving sand dunes has increased from 44.34% in 1958 to 64.47% in 1993 with the severe and very severe desertified sand area increasing by 3.12% and 3.56%, respectively. Since the 1960s the desertification development rate in the lower reach of the Heihe River basin has increased from 5.0% to 6.8% (Wang and Cheng 2000).

In addition, the oasis in the arid Northwest China is facing severe problems of soil salinization. In the Xinjiang Uygur Autonomous Region, there are 1.2 Mha of salinized cultivated land, occupying 31.1% of its total area. In the Hexi Corridor region, there are now 0.926 Mha of salinized cultivated land, occupying 14.7% of the total area (Wang and Cheng 2000).

10.3.2 Irrigation

Water is one of the essential resources in arid and semiarid regions especially for agriculture. The arid region in Northwestern China is one of the mostly water-stressed regions in the world (Shen and Chen 2010). This region is also an important food and cotton-producing region in China, but the imbalance between water supply and demand is also very prominent in the region.

Agricultural water consumption accounted for more than 95% of the total water use in the region, becoming the main factor affecting water resource allocation into the socioeconomic sectors and ecosystems in the region. The shortage of water resources has become a major limiting factor for the socioeconomic development in the areas. Even in places with relatively abundant water resources, it is not sufficient to meet the water demand of crop growth all the year round. With increasing acreage of crop growing, unreasonable crop planting structure, high irrigation quota, and low water use efficiency, the shortage of water resources is becoming increasingly serious.

Crop and irrigation water requirements for five main crops, including wheat, corn, cotton, oilseed, and sugar beet, from 1989 to 2010 were calculated, and the spatiotemporal variations were analyzed (Shen et al. 2013). The results suggested that the demand of irrigation water in the arid areas of China showed increasing trend during the past two decades, which mainly resulted from fast increase in cotton cultivation areas, because irrigation water requirement for cotton was much larger than other crops. The changes in cotton-growing area significantly affected the spatial pattern of water demand (Hu et al. 2013).

A total of 44.2 billion m^3 water was withdrawn for irrigation in year 2010. Larger amounts of water were consumed for crops in Northern Xinjiang and Tarim River basin than in Qilian-Hexi region.

Irrigation water requirement reaches its maximum in July and August. It is revealed that the critical period for water supply is during April and May through comparing the monthly irrigation water requirement with water availability, i.e., river discharge. Even though the annual water resources are much larger than the requirement, for some basins, there is severe physical water shortage during the critical water use period in April and May. The water resource supply is expected to be facing more difficulties in the future.

The irrigation water requirement in 2010 has reached 44.2 km^3, with a net increase of 12.9 km^3 compared with that in 1989. Among the five major crops, cotton was the most water-consuming crop in the region. The irrigation water demand for cotton reached 20.7 km^3 in 2010, accounting for 46.8% of the total water consumption. In terms of water irrigation intensity, sugar beet consumed the most with 7,100 m^3 ha^{-1} year^{-1}, followed by cotton, corn, wheat, and oilseed crops with 6,700 m^3 ha−1 year−1, 4,920 m^3 ha−1 year−1, 4,560 m^3 ha−1 year−1, and 3,520 m^3 ha−1 year−1, respectively (Shen et al. 2013).

10.3.3 Crops and Cropping Pattern

Abundant radiation and heat resources in this region make it an ideal place for thermophilous and photophilous crop growth, so it becomes an important high-quality and also high-yield cotton production zone (Xu et al. 2011). Other crops like wheat and corn are also widely cultivated in this region. All arable cropland is in artificial oases established at the edge of the deserts through which inland rivers flow. Oases rely on surface water and groundwater for irrigation (Shen et al. 2013). At the end of 2010, the total area of farmland was 4.9 million ha, effective irrigation reached 4.682 million ha, and the total crop growing area amounted to 5.55 million ha. Cotton-growing area was 1.5 million ha, and its acreage and production accounted for 31% and 42.7% of the country's total, respectively. According to the Bureau of Xinjiang Water Resources (2011), the annual withdrawal for irrigation reached to 48.83 billion m^3, accounting for 91.3% of the total water consumption in Xinjiang in 2010.

The irrigated area in northwestern arid regions increased from 2.7 million ha in 1989 to about 4.4 million ha (Fig. 10.3) in 2010, and the fastest increase goes to cotton-growing land. In the irrigated farmland, wheat- and corn-growing acreage was about 2 million ha, with little change in the total over the past 20 years. Beet and oilseed acreage was approximately 0.4–0.5 million ha, which slightly declined in recent years. The cotton-growing area showed the fastest increase, from only 0.4 million ha in 1989, mainly distributed in Tarim River basin, to 1.5 million ha by 2008. In 2010, the cotton-growing area slightly declined to 1.3 million ha, still accounting for about one third of the total cropping areas.

Fig. 10.3 The variations of irrigated crop-growing area in the arid areas of China (Shen et al. 2013)

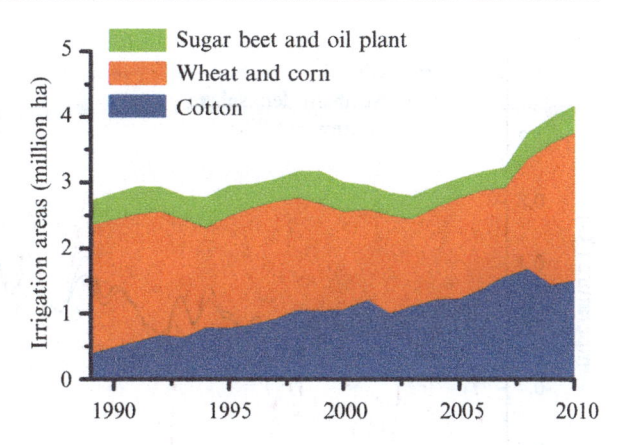

10.4 Climate Variability and Impact on Livelihood

Average temperature has risen above the global average. Over the past half century, the rate of temperature rise in the northwest arid area is up to 0.34 °C/10a, which is significantly higher than the global average by 0.12 °C/10a (IPCC 2013). Analysis showed that a significant temperature rise occurred in the late 1980s. During 1960–1986, the average temperature increased to a lesser extent. In 1987, the average annual temperature in the northwest arid area showed an abrupt change, and the elevated rate of 0.517 °C/10 years showed an increasing trend to accelerate (Fig.10.4).

Temperature experienced a "sharply" increase in 1997; however, this sharp increasing trend has turned to an apparent hiatus since the twenty-first century. The dramatic rise in winter temperatures in the northwest arid region is an important reason for the rise in the average annual temperature, and substantial increases in extreme winter minimum temperature play an important role in the rising average winter temperature (Chen et al. 2015).

In the arid region of China, temperature and precipitation were "sharply" increasing in the past 50 years. The precipitation trend changed in 1987, and since then, in a state of high volatility, during the twenty-first century, the increasing rate of precipitation diminished.

Glacier change has a significant impact on hydrology in the northwest arid area, and glacier inflection points have appeared in some rivers. The melting water supply of the Tarim River basin makes up a large portion of water supplies (about 50%). In the future, the amount of surface water will probably remain at a high state of fluctuation.

Under CMIP5-RCP scenarios of RCP 2.6, RCP 4.5, and RCP 8.5, both air temperature and precipitation are showing increasing trends in the arid region of China, and warming trend is more significant. In 2046–2075, the temperature will increase by 1–2 °C in Southern Xinjiang and by more than 3 °C in Northern Xinjiang compared with that in the historical period of 1971–2000. Annual

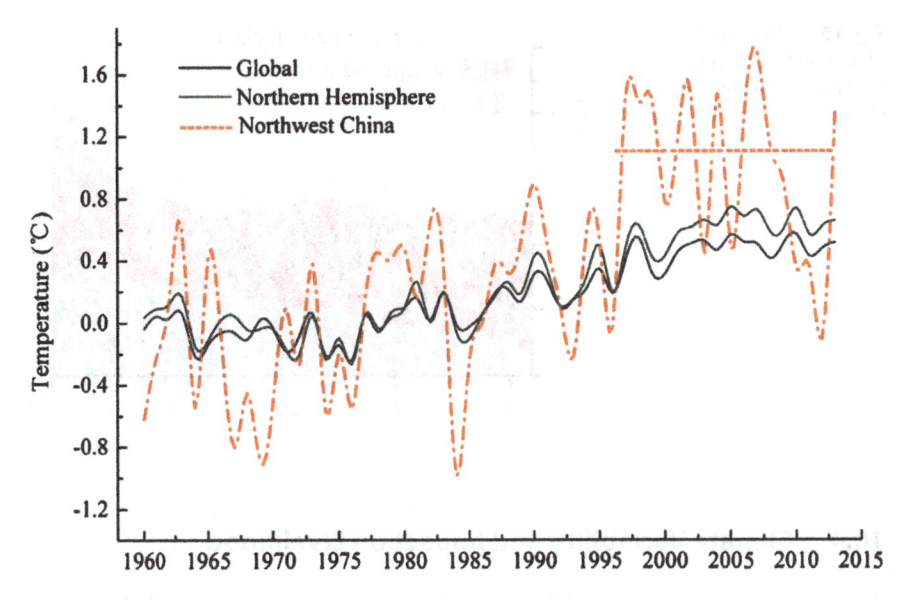

Fig. 10.4 Trends of temperature in the global, Northern Hemisphere, and arid region of Northwest China during 1960–2013 (Chen et al. 2015)

precipitation will also increase from 120 mm in the past (1971–2000) to 155–165 mm (Guo and Shen 2016).

The air temperature will increase in two terms of the twenty-first century in the whole arid region of China. The average annual temperature in the arid region of China was 6.1 °C during 1971–2000, which will increase to 7.9 °C, 8.0 °C, and 8.3 °C during 2016–2045 and 8.4 °C, 9.1 °C, and 10.3 °C during 2046–2075 under RCP2.6, RCP 4.5, and RCP 8.5 scenarios, respectively. Warming trend is more significant under the high emission scenario than under the low emission scenarios. Compared with the temperature in the historical period, the predicted temperature under the highest emission scenario in the middle term of the twenty-first century in the ARNWC increased by 67% (Guo and Shen 2016).

On a monthly scale, the temperature shows an increasing trend in all seasons, which is more significant in the summer and winter, most particularly in July. Under the highest emission scenario RCP 8.5, temperature will increase by more than 5 °C (Fig. 10.4). Spatially, the annual temperature shows a warming trend all over the arid region of China, which is more significant in the Northern and Eastern Xinjiang. In 2046–2075 under different emission scenarios, temperature will increase by 1–2 °C in Southern Xinjiang and by 3 °C in Northern Xinjiang and especially by more than 5 °C in some areas of the Tian Shan Mountains in Northern Xinjiang (Fig. 10.5).

Precipitation is one of the main water supplies to surface water resources and greatly affects the variation in water resources in the arid region of China. In a similar trend of temperature, annual precipitation will also increase in the two terms of the twenty-first century in the whole region. The increase trend is more significant

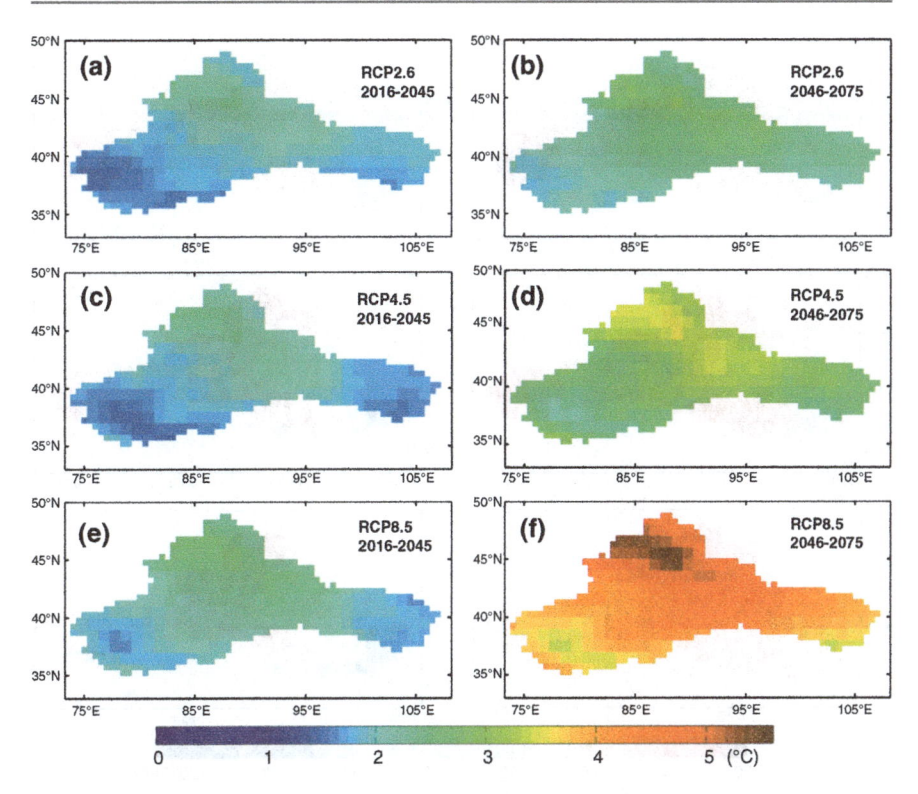

Fig. 10.5 Distributions of annual temperature change from 1971–2000 to the periods of 2016–2045 and 2046–2075 under three scenarios, respectively (Guo and Shen 2016)

under the high emission scenario than under the low emission scenarios and which is more significant during the period of 2046–2075 than that during 2016–2045. The average annual precipitation in the arid region of China is 120 mm during 1971–2000, which will increase to 141 mm, 145 mm, and 147 mm during 2016–2045 and 150 mm, 154 mm, and 165 mm during 2046–2075 under RCP 2.6, RCP 4.5, and RCP 8.5 scenarios, respectively.

From the seasonal variations, the precipitation shows an increasing trend in all seasons, which is more significant from January to August (Fig. 10.6). Spatially, monthly precipitation increases more significantly in the mountains, especially in the Tian Shan Mountains, the upstream of the Tarim River, and the Kunlun Mountains.

10.5 Recent Trends in Arid Lands and Future Scope

It is a commonly held belief that the rapid expansion of farmland (irrigated land, primarily cotton fields) and its subsequent abandonment or neglect would lead to increasing soil salinity and ultimately to desertification (Zhang et al. 2003; Ma et al. 2011).

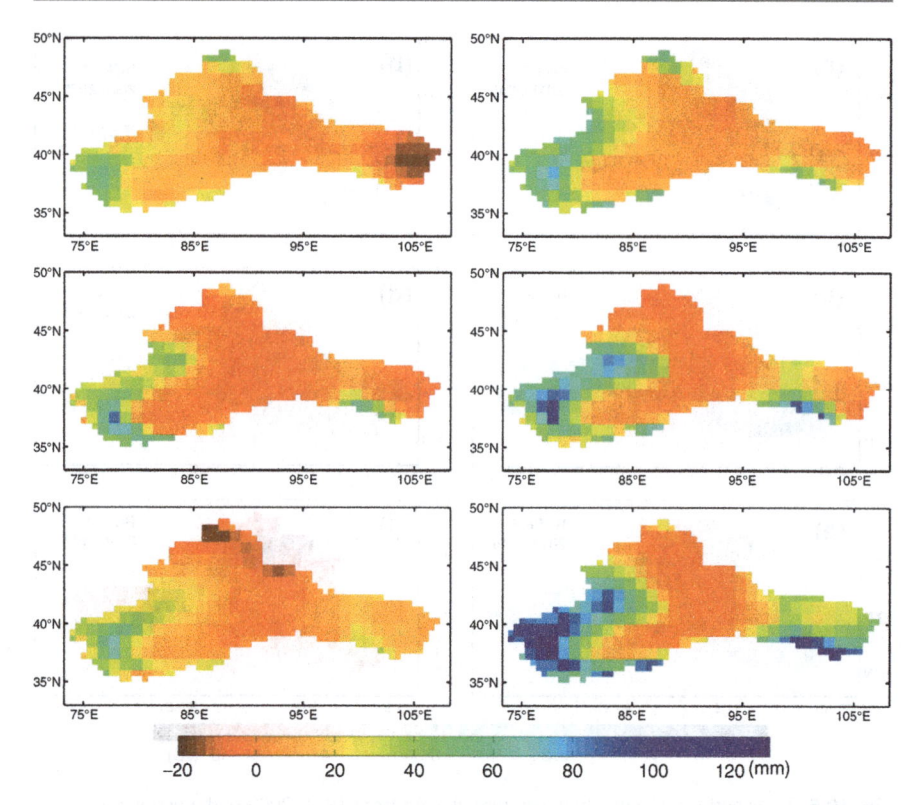

Fig. 10.6 Distributions of annual precipitation change from 1971–2000 to the periods of 2016–2045 and 2046–2075 under three scenarios, respectively (Guo and Shen 2016)

The mountain glaciers melt, and glacial retreat retreat has been accelerated by climate change, particularly by the impact of winter temperatures. In the northwest arid area, runoff is dependent on glaciers. Moreover, since changes in glaciers have a significant impact on water resources and their annual portion of water supplies, glacier inflection points have appeared in some rivers. Rivers with comparative development of glaciers and more glacial melt water supply may have runoff of high volatility for a long period of time. In addition, glacier inflection points will appear, glacial melt water will reduce, glaciers' swap function will decline, and runoff variability will increase due to the impact of precipitation anomalies. In the context of global climate change, the frequency and intensity of extreme hydrological events in the northwest arid region are increasing, which affect the safety of the water system and increase water resource vulnerability and uncertainty (Chen et al. 2015).

Agricultural land in the arid areas of China mainly located at the oasis with irrigation sourced from river water, which mainly originated from snow/glacier melting and summer rainfall in mountain areas. Climate change has large impacts on not only the quantity of river discharge but also river hydrographs in the northwestern arid basins in China. The latter is more important with regard to water supply.

In some river basins, even though the annual water resource can meet the irrigation requirement, there are huge deficits in some critical water-demanding periods, such as in April, May, and even June in some specific river, i.e., Hei River. This kind of seasonal water shortage reinforced the water stress in the region. On the other hand, the unprecedented enlargement of the crop-growing area has resulted in rapid increase in irrigation water demand and already caused ecological degradation in the lower reaches of the rivers (Shen and Chen 2010), such as the terminal lake shrinkage and desiccation of the floodplain in down stream reaches of Tarim River (Chen et al. 2012) and Hei River (Zhao et al. 2007). It is urgent to call for wise water management such as water price regulation, crop planting pattern adjustment, as well as water-saving technologies in this hyper-arid region.

Irrational irrigation patterns and imperfect water infrastructure resulted in the inefficient utilization of water resources, coupled with the rapid expansion of crop acreage, creating huge agricultural water consumption. The demand for water resources has further expanded in recent years with the development of society and the economy and a rapid increase in population. Water shortage situations are becoming increasingly serious. Meanwhile, this region is the most sensitive area to global climate changes. Climate change will further exacerbate the uncertainty of the water supply and increase extreme hydrological events in this region, such as floods and droughts. Under the combined effects of climate change and human activities, this region's water resource system, which is based on snow and ice meltwater, has become very fragile. Rising temperatures caused seasonal variations in meltwater runoff, thus reducing the stability of the water cycle system and the renewability of water resources and increasing the frequency and intensity of extreme climatic/hydrologic events. Climate change and extreme hydrological events have increasingly impacted the water supply system and increased the instability of oasis agricultural production. Increases in temperature and the uncertainty of precipitation directly affect the evaporation and transpiration of croplands, thus increasing crop irrigation water demand (Shen et al. 2013; Wang 2010).

Meanwhile, climate change has affected cropping systems and the layouts of crop varieties (Piao et al. 2010), thus impacting the irrigation water demand. In recent years, the contradiction between water supply and demand has become more severe, which has affected agricultural production, restricted industrial development, and impacted on domestic water use. Water shortages have led to more vulnerable ecosystems in the context of climate change. Water shortages have become a major limiting factor of socioeconomic development and environmental protection in the arid areas of China (Liu et al. 2016).

Future climate change will present greater challenges to the regional water supply and demand balance. Therefore, predicting future climate trends and their impact on future water supplies and quantitatively simulating agricultural water demand (the largest water consumption sector) under future climate change and socioeconomic development conditions is very important to grasp future agricultural water demand changes and to improve future water allocation and management. Meanwhile, proposing effective adaptation strategies according to possible

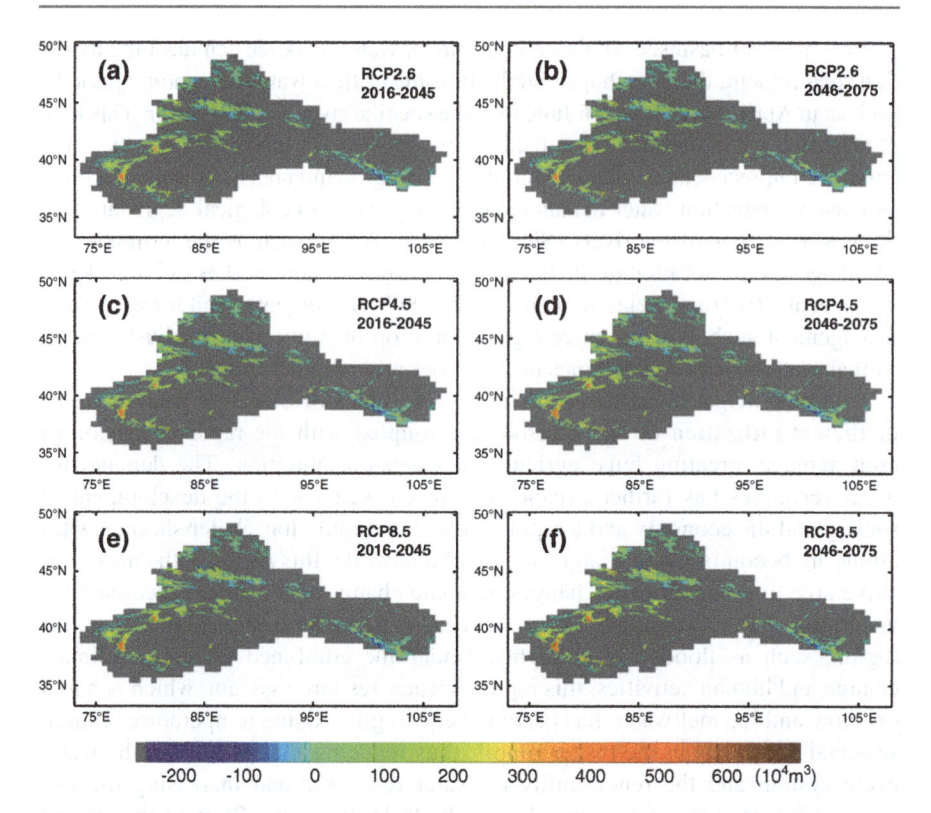

Fig. 10.7 Distributions of annual irrigation water demand (IWD) change from 1971–2000 to the periods of 2016–2045 and 2046–2075 under three scenarios, respectively (Guo and Shen 2016)

changes in the climate, water supply, and agricultural water demand has very important practical significance for the sustainable use and scientific management of water resources, sustainable socioeconomic development, and the restoration of the water and ecological environment in the arid areas of China.

The projections done by Guo and Shen (2016) showed that the irrigation water demand of five main crops (wheat, rice, maize, cotton, and oil crops) under future climate scenarios has an increasing trend and will reach 37.02–37.27 billion m^3 and 37.15–38.9 billion m^3 in 2016–2045 and 2046–2075, respectively. The amount increases by 4.27–6.15 billion m^3 compared with that in the past (1971–2000). Spatially, the largest IWD increases occur in Southern Xinjiang, the Ili River basin, and the Northern piedmont of Tian Shan Mountains (Fig. 10.7).

Annual runoff of the two future periods will only increase by 4.8–7.5 billion m^3 and 5.4–8.5 billion m^3, respectively, which is equivalent to or even less than the increased IWD of five main crops, and runoff in some river basins even shows declining trends. The potential changes of water supply and demand indicate that the water shortage situation will be more severe in the future. In some basins, the water scarcity index will be over 1, which indicates that the water resources in the basin will be unable to meet the water demand of the basin. The water crisis is also likely to bring about further deterioration of the ecological problem (Guo and Shen 2016).

Rational planning and effective use of water resources with adherence to "resource-saving" development is the only sustainable and basic ways for economic development and eco-environmental protection in these arid northwest regions of China. In this respect, there are many possibilities such as adoption of advanced irrigation techniques and development of water-saving agriculture. It has been proven that large-scale afforestation to fix shifting sand and the establishment of a protective forest network play a very important role in the control of sandy land desertification. The reversal of desertification in the middle reaches of the Tarim River basin and the Heihe River basin is a good example of these policies (Wang and Cheng 2000).

Land use policies are often designed with the intention of striking a balance between man and the natural environment and between economic development and ecological sustainability. In reality, the implementation of land use policies is affected by various economic, social, political, and environmental changes. Consequently, the outcomes of these policies may not be as intended, and the impact of land policies on environment can be either positive or adverse. Hence, monitoring and understanding the effects of land use policies on environmental issues are theoretically and practically important (Xu et al. 2016).

Climate change affects both the water resource system itself and the social, economic, and environmental changes that are caused by changes in water resource systems and human demand for water resources (Arnell 1999). Global warming will accelerate the process of atmospheric circulation and the hydrological cycle, subsequently affecting the spatial and temporal distribution of water resources and exacerbating water shortages, especially in arid regions. The irrigation water demand depends on evaporation and transpiration during crop growth periods, which are impacted by changes in temperatures and rainfall (Wang et al. 2012).

The expansion of agricultural oasis is a type of land use and cover change (LUCC), which greatly changes the vegetation distribution and surface biogeochemical and biophysical processes, leading to dramatic changes in the water cycle and ecosystem productivity. At the same time, water is the most limiting resource for sustaining crop production in agricultural oasis due to high water demands of the increased cropland area and population (Luo et al. 2003). To address this concern, the effects of expansion and associated management practices in agricultural oasis on regional carbon and water budgets have not been examined quantitatively (Bai et al. 2014).

Water, largely originated from the mountain areas, has been the most critical factor to drive the energy and mass circulation in this region, which responds sensitively to the global climate change; thus, it plays a crucial role in future sustainable development. With the increasing concern of global environmental and ecological degradation, there has been an urgent need to investigate the related water cycle changes.

References

N.W. Arnell, Climate change and global water resources. Glob. Environ. Chang. **9**(Supplement 1(0)), S31–S49 (1999). doi:http://dx.doi.org/10.1016/S0959-3780 (99) 00017-5

J. Bai, X. Chen, L. Li, G. Luo, Q. Yu, Quantifying the contributions of agricultural oasis expansion, management practices and climate change to net primary production and evapotranspiration in croplands in arid northwest China. J. Arid Environ. **2014**(100–101), 31–41 (2014)

Bureau of Xinjiang Water Resources, Xinjiang Water Resources Bulletin 2010, (2011), p. 48 (in Chinese)

Y. Chen, Q. Yang, Y. Luo, Y. Shen, Ponder on the issues of water resources in the arid region of northwest China. Arid Land Geograph. **01**, 1–9 (2012). (in Chinese with English abstract)

Y. Chen, Z. Li, Y. Fan, H. Wang, H. Deng, Progress and prospects of climate change impacts on hydrology in the arid region of northwest China. Environ. Res. **139**, 11–19 (2015)

Xi. Chin (Editor-in Chief). *Physical Geography of Arid Land in China*, (Science Press, Beijing, 2010). ISBN 978-7-03-027583-7. (in Chinese)

L. Ci, The impacts of global change on desertification in China. J. Nat. Resour. **9**, 290–302 (1994). (in Chinese)

Y. Guo, Y. Shen, Agricultural water supply/demand changes under projected future climate change in the arid region of Northwestern China. J. Hydrol. (2016). doi:http://dx.doi.org/10.1016/j.jhydrol.2016.06.033

Y. Hu, D. Li, Regional economic development strategy in Xinjiang. Finance Econ Xinjiang **6**, 16–21 (2010). (in Chinese)

S. Hu, Y. Shen, Y. Gan, J. Xang, Effect of saline water drip irrigation on soil salinity and cotton growth in an oasis field. Ecohydrology (2013). doi:10.1002/eco.1336

IPCC, Working Group Contribution to the IPCC Fifth Assessment Report. Clim. Change, The Physical Science Basis, Summary for Policymakers. (2013)

Q.H. Li, Y.N. Chen, Response of spatial and temporal distribution of NDVI to hydrothermal condition variation in arid regions of Northwest China during 1981–2006. J.Glaciol.Geocryol. **36**(2), 327–334 (2014)

H.-L. Liu, P. Willems, A.-M. Bao, L. Wang, X. Chen, Effect of climate change on the vulnerability of a socio-ecological system in an arid area. Glob. Planet. Chang. **137**, 1–9 (2016)

X. Luo, J. Yang, Researches on the questions and countermeasures of sustainable utilization of water resources in the northwest area of China. Areal Res. Dev. **01**, 73–76 (2003). (in Chinese with English abstract)

Y. Ma, Z. Wang, C. Pu, Analysis on the relationship between cotton production and land degradation in south of Xinjiang. Ecol. Econ. **3**, 148e151 (2011). (in Chinese)

S. Piao, J. Fang, H. Liu, B. Zhu, NDVI-indicated decline in desertification in China in the past two decades. Geophys. Res. Lett. **32**, L06402 (2005). doi:10.1029/2004GL021764

S.L. Piao et al., The impacts of climate change on water resources and agriculture in China. Nature **467**(7311), 43–51 (2010). doi:10.1038/Nature09364

Y. Shen, Y. Chen, Global perspective on hydrology, water balance, and water resources management in arid basins. Hydrol. Process. **24**(2), 129–135 (2010)

Y. Shen, S. Li, Y. Chen, Y. Qi, S. Zhang, Estimation of regional irrigation water requirement and water supply risk in the arid region of Northwestern China 1989–2010. Agric. Water Manag. **128**, 55–64 (2013)

B. Sun, Q. Zhou, Expressing the spatio-temporal pattern of farmland change in arid lands using landscape metrics. J. Arid Environ. **124**(2016), 118–127 (2016)

G. Wang, G. Cheng, The characteristics of water resources and the changes of the hydrological process and environment in the arid zone of northwest China. Environ. Geol. **39**(7), 783–790 (2000)

G.Q. Wang et al., Assessing water resources in China using PRECIS projections and a VIC model. Hydrol. Earth Syst. Sci. **16**(1), 231–240 (2012). doi:10.5194/hess-16-231-2012

X.Y. Wang, Irrigation water use efficiency of farmers and its determinants: Evidence from a survey in northwestern China. Agricultural Sciences in China **9**(9):1326–1337 (2010)

W.-L. Zhang, A.-g. Xu, R.-l. Zhang, H.-j. Ji, Review of soil classification and revision of china soil classification system. Sci. Agric. Sin. (2014). doi: 10.3864/j.issn.0578-1752.2014.16.009. (in Chinese with English abstract)

C. Xu, X. Yang, Y. Li, W. Wang, Changes of China agricultural climate resources under the background of climate change III. Spatiotemporal change characteristics of agricultural climate resources in Northwest Arid Area. Chin. J. Appl. Ecol. **03**, 763–772 (2011). (in Chinese with English abstract)

F. Xu, X.H. Helen, H.L. Bao, M.-P. Kwan, X. Huang, Land use policy and spatiotemporal changes in the water area of an arid region. Land Use Policy **54**, 366–377 (2016)

H. Zhang, J. Wu, Q. Zheng, Y. Yu, A preliminary study of oasis evolution in the Tarim Basin, Xinjiang, China. J. Arid Environ. **55**, 545–553 (2003). (in Chinese)

W. Zhao, X. Chang, Z. He, Z. Zhang, Study on vegetation ecological water requirement in Ejina Oasis. Sci. China Ser. D Earth Sci. **50**(1), 121–129 (2007)

X. Zhao, K. Tan, S. Zhao, et al., Changing climate affects vegetation growth in the arid region of the north western China. J. Arid Environ. **75**(10), 946–952 (2011). http://dx.doi.org/10.1016/j.jaridenv.2011.05.007

Q. Zhou, B. Li, A. Kurban, Trajectory analysis of land cover change in arid environments of China. Int. J. Remote Sens. **29**(4), 1093–1107 (2008)

Drylands of the Mediterranean Basin: Challenges, Problems and Prospects

11

Ahmed H. Mohamed and Victor R. Squires

11.1 Introduction: Overview and Context

The lands around the Mediterranean Sea (the Mediterranean Basin) form the largest of the five world regions subjected to a Mediterranean climate – long warm to hot dry summers and mild to cool, wet winters (Figs. 11.1 and 11.2). According to Gintzburger et al. (2006), the region considered stretches from the Turkish Anatolian plateau to the Red Sea in Southern Jordan over 1,200 km and from the Lebanese coastal zone through Syria and Iraq to the Iranian border. This is the cradle of ancient civilizations where pastoralism continues to be an essential agricultural activity (30% of agricultural income) supplying the local population with indispensable red meat and milk products.

Bioclimatically, the Mediterranean Basin (Le Houerou 2004) comprises a transition between southern desert (Saharan-Arabian deserts) and northern nondesert (European woodlands). Using UNEP's aridity classification, the political boundaries of all Mediterranean countries (Fig. 11.2) include the whole range of dryland types: from south to north, southern Mediterranean countries, which are closer to the Saharan-Arabian deserts than the northern Mediterranean countries, have hyperarid drylands (true deserts), semiarid drylands and dry subhumid drylands. Northern Mediterranean countries have semiarid drylands, dry subhumid drylands and non-dryland regions – humid areas.

The UNCCD does not regard hyperarid drylands as prone to desertification; hence, all Mediterranean countries have within their boundaries areas prone to desertification and areas not prone to desertification; in southern Mediterranean

---

A.H. Mohamed
Desert Research Center, Cairo, Egypt

V.R. Squires (✉)
Institute of Desertification Studies, Beijing, China
e-mail: dryland1812@internode.on.net

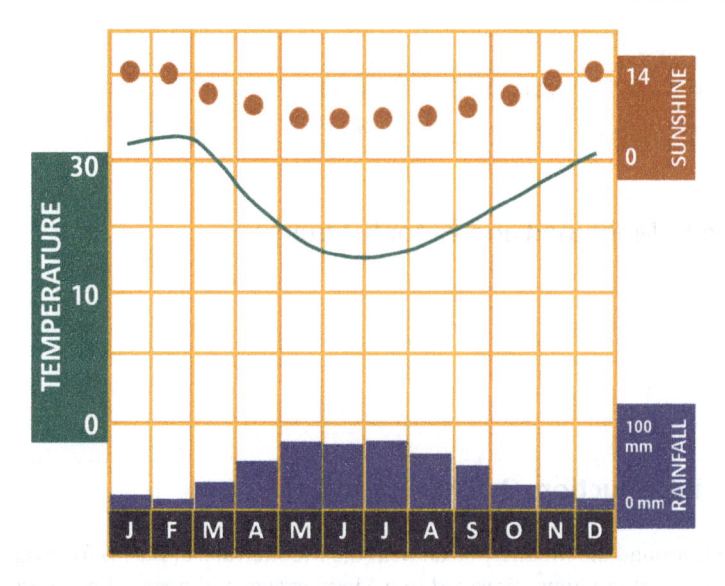

Fig. 11.1 A typical climatograph for a site within the Mediterranean Basin

Fig. 11.2 There are numerous countries in the Mediterranean Basin. The drier areas are shaded in darker colours

countries, areas not prone to desertification are the southernmost and driest regions, and in the northern Mediterranean countries, these are the northernmost and least dry regions. The European and/or the northern parts of the Mediterranean Basin do not have arid and hyperarid drylands. In the south-eastern parts of the basin, these dryland subtypes do appear and become widespread with increasing distance from the coast (the transition between semiarid and arid drylands occurs between 260 and 720 km due south of the Mediterranean coasts of North Africa whereas in the south-eastern Mediterranean arid drylands often appears close to the coastline).

The eastern Mediterranean countries – Israel, Lebanon and Syria combined – present the full gradient (southern-northern) of the global drylands. Thus, most drylands of the Mediterranean Basin are semiarid and dry subhumid drylands. Namely, drylands that are at least part of their extent are characterized by a degree of aridity that is most prone to desertification. Using UNEP's classification, the political boundaries of all Mediterranean countries include the whole range of dryland categories: from south to north, southern Mediterranean countries which are closer to the Saharan-Arabian deserts than the northern Mediterranean countries, have hyperarid drylands (true deserts), semiarid drylands and dry-subhumid drylands; northern Mediterranean countries have semiarid drylands, dry subhumid drylands and non-dryland regions – humid areas. These climatic conditions make the Mediterranean vegetation (woodlands, scrubs, steppes and grasslands) well adapted to survive dry conditions and to recover from droughts, floods and fires. Yet, land degradation and desertification do occur in croplands, rangelands and woodlands in the Mediterranean Basin. They have led to economic and social impacts such as land abandonment, migration to urban areas and reduction of the economic value and social function of the land. Population growth in both rural and urban areas increases food demand and leads to intensive cultivation which often exacerbates land degradation and desertification and causes declining productivity and poverty.

Desertification is the result of two factors operating either singly or in combination in arid Mediterranean zones. These are (1) periods of prolonged drought and (2) human exploitation of arid lands (Le Houerou 2005). The resulting degraded conditions threaten the long-term future of lands in the Mediterranean Basin. For the past century, conditions have deteriorated due to agricultural policies, such as governments providing subsidies on the costs of imported feed, and range mismanagement, such as overgrazing, risky rainfed cropping, fuel wood collection and increasing watering places by installing new boreholes combined with escalating use of trucks and tractors to cart livestock, feed and water to remote rangeland areas. Deteriorating conditions have a clear bearing on the lives of herders and nomads and may have future global impacts as impoverished rangeland populations leave the rangelands (see Sect. 11.2.1 below) (Table 11.1).

In Algeria, for example, rangelands are mostly located within a belt between 100 and 600 mm isohyets and cover 39% of the country. Rangelands support 38% of the country's human population and nearly 48% of the sheep units (SU). Livestock utilizing the rangelands contribute about half of the GDP of the country.

Many areas of the Mediterranean Basin are classified as drylands and hence are prone to desertification, but there are differences between the southern and the northern parts of the basin. In the southern part, most dryland areas are classified as arid and hyperarid, while in the northern parts, only few arid areas, but many dry subhumid and semiarid drylands, are found. The semiarid areas are currently the most affected by desertification, but the concern for dry subhumid areas is increasing.

Drylands that, for at least part of their areal extent, are characterized by a degree of aridity that is most prone to desertification, on account of both its moderate vulnerability impacted by moderate human population pressure. Indeed, the Mediterranean Basin has experienced desertification during its lengthy history of extensive and intensive land use (IDCC 2008, 2010, 2013). Experiences

Table 11.1 Economic indicators of North African rangelands

Economic indicators	Morocco	Algeria	Tunisia	Libya	Egypt
Contribution of livestock as proportion of agricultural GDP	260.	480.	0.30	(~0.15)[a]	
Rangelands as proportion of total area	420.	390.	250.	(~0.02)	010.
Human population on rangelands and desert areas as proportion of total population	380.	380.	280.	(~0.10)	
Rangeland contribution to yearly feed calendar	0.30	0.10	0.10	(~0.05)	0.05
Small ruminants (sheep equivalent[b]) as proportion of livestock in pastoral areas	0.75	0.48	0.20	(~0.90)	0.27

Adapted from Nefzaoui and El Mourid (2008)
[a]In brackets: estimate for Libya for 2011
[b]One sheep equivalent = one dry ewe (DE = 45 kg liveweight); one goat = 0.7 DE; one local cow = 5 DE; one dromedary = 7 DE

accumulated during the millennia of combating desertification and more recent advances in agricultural research have been and still are counterbalanced by population growth, though there are some very recent positive trends.

11.2 Land Degradation as a Response to Dryland Development

The notion that the very factor that has been urged upon developing countries now for about a century might be the principal root cause of dryland degradation will come as shock to many. Whole UN agencies (United Nations Development Programme, UNDP; International Fund for Agricultural Development, IFAD) have been set up to assist dryland countries (in particular) to develop. Bilateral donors such as the UK Department for International Development Agency (DFID), Canadian International Development Agency (CIDA) and Swedish International Development Cooperation Agency (SIDA) and the World Bank, Asian Development Bank, African Development Bank and Islamic Development Bank provide funds and technical assistance.

Land degradation in drylands (desertification) can be thought of as a response of dryland ecosystems to pressure of dryland development and is therefore what Safriel (2006) describes as a "manifestation of non-sustainable development". The pressure constitutes an attempt to overexploit the ecosystem service of biological productivity for generating subsistence crops or livestock. The pressure increases with reduced aridity of the drylands, whereas the vulnerability to desertification increases with aridity. These contrasting trends across the global aridity gradient place drylands of intermediate aridity, like those common within the Mediterranean Basin under the greatest risk of desertification.

11.2.1 The Interlinkages Between Development and Desertification

There is accumulating evidence that supports the description of *desertification* as a reaction of dryland *environment* (or ecosystem) to *development* (Kreutzmann 2012). Vulnerability and resilience to accelerated land degradation are both affected by development initiatives which created both the economic conditions for anthropogenic impacts on land and water resources (including climate change) and the social conditions that limit resilience capacity. This environmental response is driven by people striving to produce *essentials* for human livelihoods – food, fibres and forage – by exploiting a major service of dryland natural ecosystems, that is, land or biological productivity (Millennium Ecosystem Assessment 2003). Dryland development is characterized by the intensification of this exploitation, forcing the ecosystem to *increase* the provision of the service of biological productivity beyond its natural provision rate. However, this attempt often achieves in drylands the opposite result – the provision of biological productivity not only increases, but it *declines* to a level below that prevailing prior to the onset of development. This man-induced time-lagged reduced provision of the service of biological productivity, its causes and its repercussions jointly constitute the phenomenon of desertification. Thus, the first signs of desertification serve as an indicator that the threshold between sustainable and non-sustainable dryland development has been crossed (Safriel and Adeel 2008). Once their land becomes desertified, people abandon the land and migrate to new areas. As shown in Fig. 11.3, migration (*sens lat.*) can be at different scales (within boundary) and cross boundary of events that had led to the desertification of the land of their first choice. Only this time the downward spiral to desertification is much faster than the one experienced by the already desertified land. This is because the land of second choice is of biological productivity inherently lower than the one of first choice and hence more vulnerable. Thus, sooner than later, the people take a more dramatic, *cross-boundary* migration. Here, the term is used in its broadest sense too: migration through rural-urban boundary, through socio-economic boundaries and finally through ethnic, national and political boundaries. These cross-boundary migrations driven by the decline in provision of the dryland ecosystems' service of biological productivity make desertification to snowball from an environmental biophysical phenomenon to a societal security issue. These are often expressed in social, ethnic and political strife that further increase the demand for biological productivity (Fig. 11.3).

While there are development projects and practises that do not generate desertification (IDDC 2008, 2010, 2013) most desertification cases are tightly interlinked with dryland development – for example, exploiting fossil water to grow alfalfa in the deserts of Saudi Arabia or sedentarization of pastoralists, e.g. Bedouin in North Africa and the Gulf States or *Kuchi* in Afghanistan/Iran. In recent years, a renaissance of modernization theory-led development activities can be observed. Higher inputs from external funding, fencing of pastures and settlement of pastoralists in new townships are the vivid expression of "modern" pastoralism in urban contexts. The new modernization program incorporates resettlement and transformation of

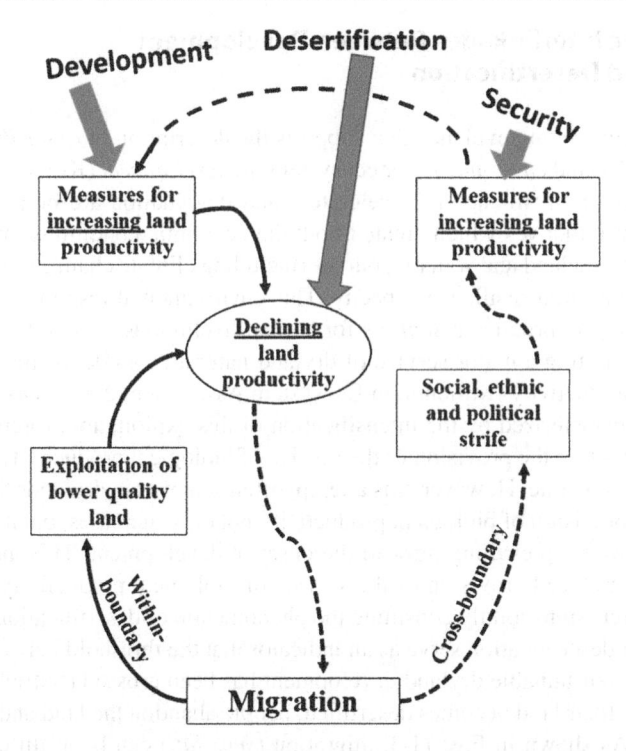

Fig. 11.3 The desertification vortex: development (*left-hand boxes*) driving desertification (*solid arrows, central circle*), which generates security issues (*dotted arrows, right-hand boxes*), which drive intensified development (*broken arrow*) and further desertification

lifestyles as to be justified by environmental pressure in order to reduce degradation in the age of climate change (Feng and Squires, 2017).

According to Kreutzmann (2012):

> In conventional views, pastoralism was classified as a stage of civilization that needed to be abolished and transcended in order to reach a higher level of development. In this context, global approaches to modernize a rural society have been ubiquitous phenomena independent of ideological contexts. The 20th century experienced a variety of concepts to settle mobile groups and to transfer their lifestyles to modern perceptions. Permanent settlements are the vivid expression of an ideology-driven approach. Modernization theory captured all walks of life and tried to optimize breeding techniques, pasture utilization, transport and processing concepts.

11.3 Climate Change and Its Implications

Data from many sites throughout the Mediterranean Basin show that warming and drying are widespread. The observed drying trend since 1950 was predominantly due to winter drying, with very little contribution from the summer. The historical region of Fertile Crescent (FC) was recently hit by an intense and prolonged drought episode during the two hydrological years 2007 and 2008. The impact of the 2007–2009 drought in vegetation was evaluated with normalized difference vegetation index (NDVI) obtained from VEGETATION instrument. It is shown that large sectors of south-eastern Turkey, Eastern Syria, Northern Iraq and Western Iran present up to 6 months of persistently stressed vegetation (negative NDVI anomalies) between January and June 2008. During the following dry years (2008–2009), dry areas are restricted to Northern Iraq with up to 5 months of stressed vegetation. The impacts on cereal production (wheat and barley) were severe in the major grain-growing countries in the area. Syria, Iraq and Iran were significantly affected by this drought, particularly in 2008. The economic impact was mostly due to the steep decline in agricultural productivity in the highly populated areas of the Euphrates and Tigris river basins.

The drought-affected region is located at the eastern end of the Mediterranean Basin, a region characterized by decreasing precipitation and river flow in the last few decades. The strong dependence of vegetation dynamics on water availability has been for long recognized throughout the Mediterranean Basin and especially in semiarid regions in North Africa (once a major source of cereal grains). In this environment, natural vegetation and nonirrigated crops are crucially dependent on soil moisture provided by seasonal rains or springtime snowmelt. This dependence leads to quite different vegetation activity levels on seasonal and inter-annual timescales. Consequently, opportunistic annual species may appear rapidly in response to humid condition of the soil, and their greenness is mainly related to recent precipitation (Zaitchik et al. 2005). On the other hand, winter crops and persistent vegetation are dependent on deeper reserves of soil moisture, and their vegetative cycle is the result of the combined effect of precipitation (over weeks and months), evaporation and, in some regions, of temperature. The vast majority of crops in these regions are nonirrigated and thus dependent on winter precipitation. Dry conditions during the planting period cause crop failure due to lack of water during the germination of the seeds. A combined effect of lack of precipitation over a certain period with other climatic anomalies, such as high temperature, high wind and low relative humidity over a particular area, may result in reduced green vegetation cover. When drought conditions end, recovery of vegetation may follow (Nicholson et al. 1998), but such recovery process may last for longer periods of time.

11.3.1 Drought Impact on Vegetation

The combined effects of this future precipitation decrease and the widely accepted future increment in the surface temperature on the Mediterranean (Christensen et al. 2007) will bear important changes in the region's hydrological water cycle. The strong seasonal and inter-annual variability of vegetation in most semiarid regions is a subject of particular interest due to the ecological and economic impacts. In particular, the high sensitivity of vegetation to climate forcing may result in rapid land use changes and high vulnerability to land degradation, as a result of human action. The changing land cover pattern reflects the precipitation regime in the area.

Over longer periods, changes may have a considerable impact on the viability of agricultural and pastoral systems.

Greenhouse gas emissions are rising throughout the basin. No climate model predicts that rising GHGs will cause the Mediterranean region to get wetter as a consequence of rising GHGs, so, instead of waiting for yet more model confirmation of what we already know, the time is ripe for Mediterranean countries to plan for the drier times ahead (e.g. Iglesias et al. 2007). Droughts are likely to be more frequent (Hoerling et al. 2012).

11.3.2 The Causes of Increasing Aridity

According to Seager et al. (2014), the hydroclimate consequences of warming and the associated thermodynamical adjustments of the hydrological cycle are well understood. The dynamical mechanisms for hydro-climate change – why the atmospheric mean and transient circulations change in the way they do – are not so well understood. It has been known for a while that the mid-latitude jets and storm tracks shift polewards under global warming, and the Hadley cell expands, causing expansion of subtropical dry zones. Wu et al. (2012, 2013) provide a review of proposed mechanisms for this and advance the case that this is a tropospheric response to stratospheric circulation adjustment with the signal propagating downwards via linear wave refraction.

11.4 Poverty, Land Degradation and Ecological Migration

Desertification and poverty are interlinked, culminating in land abandonment and trans-boundary, mostly South-North, migration. The current flow of immigration within the Mediterranean Basin is unidirectional – the geographical South-North migration from North Africa to Europe, as well as from the "socio-economic south" to Western Europe. At least in part, a substantial segment of this immigration is driven by loss of food and environmental security linked to desertification. This immigration then drives apparent or even tangible "soft" security concerns in the target countries, some of which are themselves desertification-affected countries. Attaining sustainability of dryland development, through relieving much of the pressure on the service of biological productivity of dryland ecosystems by replacing them with alternative dryland livelihoods, will reduce immigration and increase security within the Mediterranean Basin. The desertification-driven unidirectional migration will be replaced by mutual exchanges. Safriel (2006) suggests several South-North flows that can be envisaged. One is that of cash crops produced by intensive but low-impact agriculture (e.g. Egypt and Jordan export fresh fruit (grapes and melons) to Spain and other countries). Dryland aquaculture has considerable potential too. Dryland aquaculture is inherently advantageous on dryland agriculture. This is, paradoxically, due to the water scarcity or, more precisely, to the high evaporative losses that make an area a dryland. Although aquatic organisms live in water, they do not transpire water; hence, water losses from aquaculture can be analogous to evaporation, rather than to evapotranspiration. Furthermore, many more aquatic than terrestrial crop species are tolerant to salinity in water and even benefit from it. Thus, dryland aquaculture can thrive on fossil aquifers; though quite common in drylands, their salinity greatly curtails their usability by dryland agriculture. Both freshwater fish and species adapted to saline water (brine shrimp, etc.) can be used. Water that is no longer suitable for fish can be used to irrigate hardy species such as olive or grape. Another possible South-North flow is that of transported solar or wind energy generated in southern Mediterranean arid drylands that may provide a tangible segment of the North's energy demand. Wind power and solar energy "farms" generally require much land. Most drylands are sparsely populated and land is not scarce.

Generally, creation of better urban livelihoods can be based on the tourist and solar energy industries and better organized ways to capitalize on drylands' scenic and cultural attractions (eco-tourism and cultural tourism) and climatic attributes (abundant sunshine, mild winters, etc.). A combination of these developments has the potential to arrest the North-South Mediterranean migration and replace it with a northbound movement of cash crops and solar power and a southbound movement of tourists as well as finances linked to carbon trading, under the auspices of the Kyoto Protocol and the 2015 Paris Climate accord. Although "dryland development" and "rural development" are often synonymously used terms, dryland cities as an alternative to dryland villages may prove to be a sustainable option for settling more people in drylands[1] (Safriel 2006). Dryland urbanization may be advantageous, both over other land uses in drylands and over urbanization in the non-drylands. This is because dryland cities consume and hence impact fewer land resource than the drylands' farming and pastoral livelihoods and have a lower environmental impact than non-dryland cities. The prevailing notion of drylands is that of harsh climate and meagre livelihood opportunities; hence, the success of dryland cities depends on their ability to provide livelihoods as well as living conditions that are as advantageous as those provided by non-dryland cities.

Addressing land degradation, desertification and drought in the Mediterranean Basin includes policy reforms and enforcement, as well as research and extension that promote sustainable land use and restoration of already degraded lands. Reynolds et al. (2007) explain that the sustainable development for dryland ecological environment requires land user participation, and humans must be considered as part of the natural system. Sustainable development of agriculture *sens lat.* in the arid region must ensure the livelihoods of farmers, agro-pastoralists and herders in order to effectively implement sustainable development of the agro-ecological system. Climate change affects the stability of the ecosystem, and this adds to pressure as almost all land users seek to follow their own individual interests and to maximize profit. The challenge is to balance the interests of the government, livelihoods of land users and ecological benefits in the allocation and management of natural

---

[1] Tucson, and other locations in Arizona, have successfully established retirement settlements in the desert (see Squires, Chap. 8, this volume).

resources (Squires 2012). The focus on global drylands is shifting from an emphasis on negative images of desertification to a more forward-looking perspective concerning human livelihoods, based on interactions between and among human activities and natural-world processes.

In this millennium, global drylands face a myriad of problems that present tough research, management and policy challenges. Recent advances in dryland development, however, together with the integrative approaches of global change and sustainability science, suggest that concerns about land degradation, poverty, safeguarding biodiversity and protecting the culture of 2.5 billion people can be confronted with renewed optimism. The functioning of dryland ecosystems and the livelihood systems of their human residents is explained by Reynolds et al. 2007) in a new synthetic framework – the Drylands Development Paradigm (DDP). The DDP centres on the livelihoods of human populations in drylands, and their dependencies on these unique ecosystems, through the study of coupled human-environmental (H-E) systems (Qi et al. 2012). The DDP responds to recent research and policy trends that link ecosystem management with human livelihoods in order to best support the large, and rapidly expanding, populations of dryland dwellers. The DDP represents a convergence of insights and key advances drawn from a diverse array of research in desertification, vulnerability, poverty alleviation and community development. The DDP, supported by a growing and well-documented set of tools for policy and management action, helps navigate the inherent complexity of desertification and dryland development, identifying and synthesizing those factors important to research, management and policy communities.

Case Studies of Significant Subregions

1. North Africa

    This section draws on pasture profiles for Algeria (Nedjraoui 2001; Saidi and Gintzburger 2013), Morocco (Berkat and Tazi 2004) and Tunisia (Kayouli 2000). These North African countries have large areas of grazed land and many pastoral features in common and stretch from 13°E to 12°E and from 19°N to 37°19N; vast areas of their southern part are desert. The relief is in two broad categories, the Atlas and the Sahara. The Atlas is a group of ranges running southwest to northeast roughly adjacent and parallel to the Mediterranean coastline. South of the Atlas, a series of steppic plateaus descend to the Sahara, which is a great barrier between the Mediterranean zone and the tropics.

    The northern mountains capture most of the precipitation, and agriculture lands are concentrated in the north; the highest Atlas lands are forest and summer grazing. The climate is typically Mediterranean, with hot summers and rain occurring during the cool season. Temperature is governed both by altitude and the degree of continentality. The region has all the Mediterranean bioclimates, from per-humid to per-arid for bioclimatic levels and from cold to hot for temperatures.

Livestock are important throughout the zone, and in most farming systems, sheep are the most important and are the main livestock of the steppe, although small flocks are kept in most areas for domestic use; several local breeds are used according to regional adaptation. Cattle are mainly kept in the northern farming areas and are commonly fed on crop residues, by-products and concentrates; the traditional breeds were *Bos taurus* of the Atlas Brown type, but there are now many crosses with exotic dairy breeds, notable black and white ones.

2. The Near East (Often Lumped with North Africa and Called MENA)

    2.1   Syrian Arab Republic

The country is largely pastoral, with 45% of its land being grazing and pasture and 20% desert (Masri 2001). The climate is typically Mediterranean and precipitation is low, decreasing towards the interior. Most of the grazing land is in the semi-desert and desert (*badia*) with less than 200 mm rainfall; there is a little mountain grazing; plains with over 200 mm rainfall are now under rainfed crops. Cattle are kept in the agricultural areas but are absent from the main grazing lands.

From earliest times until the end of the Second World War, Syrian grazing lands were under tribal control, population density was low, and the herders were nomadic, moving seasonally with their flocks. Pastoral communities evolved codes of laws and customs and the organization of groups and subgroups based on family relationship. Each group used to maintain grazing rights on certain resources in its traditional land as *hema* (pasture reserved for use in drought or emergency) and negotiated when necessary with the other groups for movement of its livestock to areas of more favourable climatic conditions during periods of drought. The chief was the first among equals and unanimously obeyed and respected by members. The social structure of the pastoral groups was close to a cooperative organization.

Large areas of grazing land were only accessible once the autumn rains had fallen, and herds had to leave them once surface supplies of water ran out, so the pastures were rested for a long period of the year. There was no external feed source, so stock numbers were limited to what could be carried through the lean season on available pasture and water. There was no rainfed cropping on marginal lands.

After the war, the situation changed rapidly. The central authorities became much stronger and the tribal system was disintegrating. Motor transport, introduced during the war, allowed transport of goods, water and feed, making large areas of grazing accessible for much of the year. Grazing land was nationalized and became an open-access resource with no supervision over its use. Settlement of the Bedouin became an official policy; this greatly improved their access to medical care, education, water and other services and led on the one hand to a rapidly increasing population and on the other a great reduction in mobile herding. Cheap cereals allowed increasing numbers of stock to be kept through the lean season.

Marginal land was increasingly cleared for rainfed cropping. Yields are low and uncertain; if they seem too low or the crop is unlikely to mature because of drought, it will be grazed. Clearing of grazing land was encouraged by granting land rights to those who developed it.

Sheep and goat production is the main and sometimes only agricultural activity available to populations living on rangelands in the arid regions around the Mediterranean. Desertification threatens large areas of the Mediterranean, especially the North African arid rangelands but remains difficult to describe, quantify and accurately locate for management purposes (Saidi and Gintzburger 2013; Hirata et al. 2005). Sheep are the main grazing livestock; the only local breed, the *Awasii*, is a milch sheep that is well adapted to harsh desert conditions, and its fat tail provides a reserve of nutrients for periods of feed shortage. They graze in the *badia* from late autumn till late spring, with supplements, and then they migrate to the rainfed and irrigated areas, clearing all crop residues (cereal, cotton, beet and summer vegetables) before returning again to the *badia*. The main constraint to sheep production is degradation of grazing land which increases dependency on supplementary feed.

The subsidized state feed policy puts pressure on the already degraded pasture through an increase in sheep numbers, which get most of their food as concentrates but continue to eat any available herbage. The sheep population was 2.9 million in 1961; it rose to 5.5 million in 1971, 10.5 million by 1981 and peaked at 15.5 million in 1991; it then fell to just over 10 million and for the past 7 years has been just over 13 million (FAOSTAT 2004). Goats are the second most numerous livestock; their numbers rose from 439,000 in 1961 to 1 million in 1981 and have remained at that level. Two main types of goat are kept; the Shami goat is a milch breed, and they are kept around homesteads; the other is the mountain goat, which grazes in the mountain ranges.

Before the Second World War, the Syrian badia was in good condition; climax plants like *Salsola vermiculata, Atriplex leucoclada, Artemisia herba-alba* and *Stipa barbata* were widespread; and flocks of gazelle were present. Herders went to the *badia*[2] with the onset of rains in autumn and had to leave when the water supply dried up in late spring. Range livestock depended on grazing until 1958, when concentrate feeds were introduced. The rate of concentrate use increased to 25, 50 and 75% in the 1960s, 1970s and 1980s, respectively.

Jordan

About 90%, or 80,771 km², of the Kingdom is grazing land, 69,077 km² of which receives less than 100 mm of rainfall and 1,000 km² of marginal grazing with 100–200 mm annual rainfall. Natural and man-made forests cover 760 km², out of 1,300 km² registered as forests (Al-Jaloudy 2001). There are also about 500 km² of state-owned land used for grazing in mountainous areas.

The average altitude of the highlands ranges from 600 m in the north to 1,000 m in the middle and 1,500 m in the south. There is a semiarid zone (350–500 mm annual rainfall) with a small subhumid zone (over 500 mm annual rainfall). The arid

---

[2] Badia is a general term for semi-desert.

zone comprises the plains between the badia and the highlands. Rainfall ranges between 200 mm in the east and 350 mm in the west. Rainfed crops are mainly barley (in areas with 200–300 mm rainfall) wheat and fruit trees (in areas with between 300 and 350 mm rainfall). Badia (eastern desert), which covers about 8 million hectares – 90% of the Kingdom – has very sparse vegetation cover and an annual rainfall of less than 200 mm. In the past, it was only used for grazing. In the last two decades, however, 20,000 ha have been irrigated, using underground water.

Jordan is on the eastern margins of the Mediterranean climatic zone. This climate is characterized by hot, dry summers and cool, wet winters; more than 90% of the country receives less than 200 mm annual precipitation.

There are four bioclimatic zones.

- *Mediterranean*: This region is restricted to the highlands from 700 to 1,750 m above sea level. The rainfall ranges from 300 to 600 mm. The minimum annual temperature ranges from 5 to 10 °C.
- *Irano-Turanian*: A narrow strip that surrounds all the Mediterranean ecozone except in the north; it is treeless. The vegetation is mainly small shrubs and bushes such as *Artemisia herba-alba* and *Anabasis syriaca*. Altitudes range from 500 to 700 m, and rainfall ranges from 150 to 300 mm.
- *Saharo-Arabian*: This is the eastern desert or *badia* and comprises almost 80% of Jordan. It is flat except for a few hills or small volcanic mountains. Altitude ranges from 500 to 700 m. The mean annual rainfall ranges from 50 to 200 mm, and mean annual minimum temperatures range from 15 to 2 °C. Vegetation is dominated by small shrubs and small annuals in the wadi beds.
- *Sudanian*: It starts from the northern part of the Dead Sea and ends at the tip of the Gulf of Aqaba. The vegetation is characterized by a tropical tree element, such as *Acacia* spp. and *Ziziphus spina-christi*, in addition to some shrubs and annual herbs.

Badia (Semi-desert)

The most significant use of this zone is pastoralism. Sheep and goats graze the forage produced on the desert range in the short period following rainfall; precipitation is less than 100 mm/year, which falls off towards the east and the south till it reaches 50 mm or less. Most are state lands. *Artemisia herba-alba*, *Retama raetam*, *Achillea fragrantissima* and *Poa bulbosa* are common in the wadi beds, while the unpalatable *Anabasis* sp. is present in most places. Despite its deterioration, this is the main grazing land of Jordan. The average annual dry matter production is 40 kg/ha in normal years; this can rise to 150 kg/ha in protected areas and range reserves. Steppes were used generally for grazing, but it is estimated that about 90% of the steppe has been privatized and ploughed for barley.

There are 2,200,000 estimated heads of sheep and goats in Jordan. Nomadic grazing has declined to less than 10% of the sheep and goats, which belong to less than 5% of herders. The ratio of semi-settled herds has increased to more than 70% of sheep and goats. The remaining small ruminants (about 20%) follow a system that is mixed with agriculture, especially in the west of the Kingdom.

Small ruminant production systems developed gradually in the middle of the past century as a result of a number of changes: increasing settlement of the nomadic Bedouin in the marginal areas, a change to sheep and goat raising instead of camels, deterioration of traditional grazing systems (eastward and westward trips), widespread use of vehicles for movement of flocks and equipment and increased dependence on imported feed.

The *traditional mobile system* prevails in the arid to semiarid east and south. Herds move from one place to another, on foot or by truck, looking for grazing or water. The sheep depend on natural herbage as their main source of feed, in addition to the feed given in winter for a period that varies with availability of herbage.

In the *seminomadic system*, sheep depend partially on grazing and crop byproducts. They move to a land adjacent to the fields and spend the winter around the houses, where they survive on the feed given to them.

In the *settled (semi-extensive) system*, stocks are kept in fattening units but graze in the morning and return to their units in the afternoon. They feed on crop byproducts and the adjacent natural grazing. Supplementary feed is given as required.

In the *intensive system*, sheep are kept on permanent farms with modern facilities and equipment. They are given balanced feed, and health care is provided.

Existing statistics indicate that there are 2,200,000 small ruminants, which depend for half their food requirements on imported feed. Natural grazing supplies only 25–30% of their requirements, as its productivity has declined to half of its potential and the area has decreased. In the past, the availability of fodder and water, and the search for them, were the limiting factors for movement of herds. Nowadays, food and water are transported to herds wherever they are, and it is possible to quickly transport the herds themselves. In 1930, the sheep herd was 229,100 and remained at a similar level until 1950, by 1970 it had doubled, and by 1990 it had reached 1.5 million, where it currently stands. Goat numbers in 1930 were 289,500; they had risen in recent years to 479,000 in 1990 and 547,500 in 2003, but nothing like the extent of sheep, which are far better suited for intensive fattening.

Existing policies are not comprehensive and are incompatible with national needs and development plans. Feed subsidy policies from the 1980s to 1997 brought about the unusual increase in sheep and goats numbers and the deterioration in local production of feed. Also allocation of wide tracts of the best range to private ownership caused their deterioration and desertification.

Pastoral communities informally claim common tribal rights and enjoy free access and use of natural resources in their rangelands, but these claims are only recognized in settled areas. In all the unsettled areas, the state asserts ownership regardless of customary tribal claims. State claims over grazing lands changed the traditional welfare system, caused the breakdown of resource allocation mechanisms and transformed secured-access rights into secured-tenure rights. Consequently, customary management rules are often no longer being enforced. State appropriation did not deny local communities access to their traditional pasture but favoured a situation of open access to grazing and expansion of barley cultivation.

11.5 Summary and Conclusions

The rangelands of North Africa and the Middle East used to provide the population with fodder for their flocks and now mostly fuel wood for domestic use. The small livestock graze the steppe vegetation by the end of winter and in spring (March), collecting nearly all available annuals and ephemeroids on the rangelands. Between December and March, depending upon the precipitation, they are heavily dependent on hand-fed concentrates, agroindustrial by-products, cereal straw (mostly barley *T'ben*) and legume straw (chickpeas and lentils). In summer and autumn, livestock are now found grazing cereal stubble and irrigated residues in cropped areas, while in earlier days, they used to browse the chamaephytic vegetation (*Artemisia herba-alba*). This is due to the increase in both the human and small ruminant populations. In Syria, for example, the sheep population increased from some 3 million heads in 1961 to nearly 14 million in 2000 with a 4% average annual growth rate that continues – causing great concern.

According to Gintzburger et al. (2006) "the future of rangeland resources of the Middle East and Middle Asia, essential to local population, remains quite gloomy unless proper conservation measures are implemented by strong and relentless political will and enforced without further delay". Similar views have been expressed about the drylands in the north of the Mediterranean Sea (Safriel 2006). Gintzburger et al. (2006) in their excellent review paper do a forensic examination of the situation in the region. They dissect the various contributing factors and offer recommendations about ways to arrest and reverse land and water degradation.

Expanding grazing land in the MENA region is a limited option. The development in agriculture is increasingly driven by shifts in human diets towards livestock products, which in turn influences crop production decisions and patterns. Meat supply in MENA steadily increased for red meat from the early 1960s to 2010. During this period, the annual per capita (sheep and goats + bovine meat) supply moved from 7.7 + 2.5 kg in 1961, increasing to 10.8 + 2.5 kg in 2002 in Syria, 9.5 + 8.3 kg to 4.0 + 17.3 kg in Lebanon, 5.2 + 1.5 kg to 2.6 + 5.8 kg in Jordan, 1.1 + 7.7 to 1.3 + 22.2 kg in Israel, 9.0 + 3.6 kg to 4.7 + 4.7 kg in Turkey and 4.9 + 2.7 kg in 2002 for Palestine. Despite the shift from sheep and goats to bovine red meat, especially in rich countries, it still shows the importance of sheep and goat meat for the local population.

Government subsidies to provide for cheap animal feed and concentrates have contributed to the keeping of too many small livestock on degraded rangelands. In crop-livestock-rangeland rainfed mixed systems, livestock substitutes for natural and purchased inputs, in addition to producing meat and milk. Until the mid-1950s, about 65% of the small livestock population was based on the steppe where feed requirements were covered without any supplementary feed. In the mid-1990s, 75–80% of the flocks are steppe-range based. They however do not get most of their feed from the range but mostly as supplementary feed, complements and crop residues. The low current contribution of rangeland feeding is due to the tremendous increase in sheep numbers but also to the loss of traditional management tools and to the modification of land tenure arrangements and "use rights".

References

M. Al-Jaloudy, *Country Pasture/Forage Resource Profiles* (Ministry of Agriculture, Rome, 2001). See: www.fao.org/ag/AGP/AGPC/doc/Counprof/Jordan.htm (accessed July 2017)

O. Berkat, M. Tazi, Pasture profile for Morocco (2004), See: http://www.fao.org/ag/AGP/AGPC/doc/Counprof/Morocco/morocco.htm

A. El-Beltagy, M. Madkour, Impact of climate change on arid lands agriculture. Agric. Food Secur. **1**, 3–12 (2012)

J.H. Christensen, B. Hewitson, A. Busuioc, A. Chen, X. Gao, I. Held, R. Jones, R.K. Kolli, W.-T. Kwon, R. Laprise, V. Magaña Rueda, L. Mearns, C.G. Menéndez, J. Räisänen, A. Rinke, A. Sarr, P. Whetton, Regional climate projections, in *Climate Change 2007: The Physical Science Basis. Contribution of Working Group I to the Fourth Assessment Report of the Intergovernmental Panel on Climate Change*, ed. by S. Solomon, D. Qin, M. Manning, Z. Chen, M. Marquis, K. B. Averyt, M. Tignor, H. L. Miller (Cambridge University Press, Cambridge, 2007).

J. Evans, 21st century climate change in the Middle East. Clim. Chang. **92**, 417–432 (2009)

J.P. Evans, R.B. Smith, R.J. Oglesby, Middle East climate simulation and dominant precipitation processes. Int. J. Climatol. **24**, 1671–1694 (2004)

H. Y. Feng, V.R. Squires, Climate variability and impact on livelihoods in the cold arid Tibet Plateau, in *Climatic Variability Impacts on Land Use and Livelihoods*, ed. by M.K. Gaur, V.R. Squires (Springer, New York, 2017)

G. Gintzburger, H.N.L. Houérou, S. Saïdi, Near East-West Asia arid and semiarid rangelands. Sécheresse **17**(1-2), 152–168 (2006)

FAOSTAT, *Food and Agriculture Organization of the United Nations* (Rome, 2004)

M. Hirata, N. Koga, H. Shinjo, H. Fujita, G. Gintzburger, J. Ishida, A. Miyazaki, Measurement of above-ground plant biomass, forage availability and grazing impact by combining satellite image processing and field survey in a dry area of north-eastern Syria. Grass Forage Sci. **60**, 25–33 (2005)

M. Hoerling, J. Eischeid, J. Perlwitz, X. Quan, T. Zhang, P. Pegion, On the increased frequency of Mediterranean drought. J. Climate **25**, 2146–2161 (2012)

IDDC, Sustainable development in the drylands – meeting the challenge of global climate change. Proc. 9th International Conference on Dryland Development., Bibliotheca Alexandrina, Alexandria, Egypt, (2008)

IDDC, Meeting the challenge of sustainable development in drylands. Changing climate-moving from global to local. Proc. 10th International Conference on Development of Drylands. Cairo, Egypt, (2010)

IDDC, Global climate change and its impact on food & energy security in the drylands. Proc. 11th International Conference on Development of Drylands. Beijing, China, (2013)

A. Iglesias, L. Garrote, F. Flores, M. Moneo, Challenges to manage the risk of water scarcity and climate change in the Mediterranean. Water Resour. Manag. **21**(5), 227–288 (2007)

A. Karagöz, Pasture profile for Turkey (2001), See: http://www.fao.org/ag/AGP/AGPC/doc/Counprof/Turkey.htm

C. Kayouli, Pasture profile for Tunisia (2000), See: http://www.fao.org/ag/AGP/AGPC/doc/Counprof/TUNIS.htm

H. Kreutzmann (ed.), *Pastoral Practices in High Asia: Agency of 'Development' Effected by Modernization, Resettlement and Transformation* (Springer, Dordrecht, 2012)

H.N. Le Houerou, *Bioclimatology and Hytogeography of the World Arid and Semi-arid Isoclimatic Mediterranean Biomass* (Springer, Berlin, 2004)

A. Masri, Pasture profile for Syria (2001), See: http://www.fao.org/ag/AGP/AGPC/doc/Counprof/syria.htm

Millennium Ecosystem Assessment, *Ecosystems and Human Well-Being: Synthesis* (Island Press, Washington, DC, 2003)

D. Nedjraoui, Pasture profile for Algeria (2001), See: http://www.fao.org/ag/AGP/AGPC/doc/Counprof/Algeria.htm

A. Nefzaoui, M. El Mourid, *Rangeland Improvement and Management in Arid and Semi-arid Environments of West Asia and North Africa*. (ICARDA, Aleppo, 2008). Available at: http://icarda.org/wspublications//Regional_program_reports_archive/North_Africa/Rangeland_improvement.pdf

S.E. Nicholson, C.J. Tucker, M.B. Ba, Desertification, drought, and surface vegetation: an example from the West African Sahel. Bull. Am. Meteorol. Soc. **79**, 815–829 (1998)

J. Qi, J. Chen, S. Wan, L. Ai, Understanding the coupled natural and human systems in dryland East Asia. Environ. Res. Lett. **7**(015202), (7pp) (2012). doi:10.1088/1748-9326/7/1/015202

J.F. Reynolds, D.M. Stafford Smith, E.F. Lambin, B.L. Turner II, M. Mortimore, et al., Global desertification: building a science for dryland development. Science **316**, 847 (2007)

U.N. Safriel, Dryland development, desertification and security in the Mediterranean, in *Desertification in the Mediterranean Region*, ed. by D. A. Mouat, F. Pedrazzini, (Springer, Dordrecht, 2006)

U.N. Safriel, Z. Adeel, Development paths of drylands: thresholds and sustainability. Sustain. Sci. **3**, 117–123 (2008). doi:10.1007/s11625-007-0038-5

S. Saïdi, G. Gintzburger, A spatial desertification indicator for Mediterraneanarid rangelands: a case study in Algeria. Rangel. J. **35**, 47–62 (2013)

R. Seager, H. Liu, N. Henderson, I. Simpson, C. Kelley, T. Shaw, Y. Kushnir, M. Ting, Causes of increasing Aridification of the Mediterranean region in response to rising greenhouse gases. J. Clim. **12**, 4655 (2014)

V. Squires, *Rangeland Stewardship in Central Asia: Balancing Livelihoods, Biodiversity Conservation and Land Protection* (Springer, Dordrecht, 2012.) 482 p

V.R. Squires, L. Hua, D. Zhang, G. Li, *Towards Sustainable Use of Rangelands in North-West China* (Springer, Dordrecht, 2010.) 353 p

D.C.P. Thalen, *Ecology and Utilization of Desert Shrub Rangeland in Iraq* (W. Junk, The Hague, 1979)

Y. Wu, R. Seager, M. Ting, N. Naik, T. Shaw, Atmospheric circulation response to an instantaneous doubling of carbon dioxide. Part I: model experiments and transient thermal response in the troposphere. J. Clim. **25**, 2862–2879 (2012). doi:10.1175/JCLI-D-11-00284.1

Y. Wu, R. Seager, M. Ting, N. Naik, T. Shaw, Atmospheric circulation response to an instantaneous doubling of carbon dioxide. Part II: atmospheric transient adjustment and its dynamics. J. Clim. **26**, 918–935 (2013). doi:10.1175/JCLI-D-12-00104.1

B.F. Zaitchik, J. Evans, R.B. Smith, MODIS-derived boundary conditions for a mesoscale climate model: application to irrigated agriculture in the Euphrates basin. Mon. Weather Rev. **133**, 1727–1743 (2005)

Part IV

Southern Hemisphere Aridlands: Selected Examples

The southern hemisphere contains some large land masses, and there are extensive regions on three continents (South America, Africa and Australia) that deserve attention here. Most of the aridlands in the southern hemisphere are sparsely populated, in sharp contrast to aridlands in China and India where population densities far exceed the UN guideline.

Klaus Kellner and his team of contributors (*Graham von Maltitz, Mary Seely, Julius Atlhopheng* and *Lehman Lindeque*) provide a comprehensive overview of the aridlands of Southern Africa (Botswana, Zimbabwe, South Africa, Namibia). They review problems and prospects and conclude, that in the arid and semiarid areas of Southern Africa, many land users still depend on natural resources, such as soil, vegetation and water, for sustainable livelihoods. Due to higher population densities and more intensive land use practices, the semiarid areas in particular are prone to man-made or accelerated land degradation such as loss of vegetative cover, increase in alien and invader species, bush encroachment, change in species composition and eventually loss of habitats. All of these problems will be magnified if climate change is as bad as predicted.

Carlos Busso and *Osvaldo Fernandez* present an analysis of the situation on the arid and semiarid rangelands of Argentina. They also draw attention to the impact (both actual and projected) of climate change. Vegetation is being affected by changing land use, and problems are exacerbated by climate variability especially manifested as the migration of plant communities to higher elevations as climates warm up.

Australia has a large part of its land territory classified as dryland. There are several true deserts and a large tract of semiarid and dry subhumid land. Because of its location, Australia also has monsoonal tropics. **David Eldridge** and **Genevieve Beecham** have analysed the relevant data and model predictions about climate variability and its likely consequences for Australia's drylands. They conclude that climate change will reduce ground-storey plant cover, reduce livestock and crop yields, place a greater burden on a declining pastoral land and increase land degradation and soil nutrient loss. Increasing climatic variability will affect local communities by intensifying the displacement of people from rural areas, reducing community size and the provision of infrastructure and health services and placing an increasing social burden on communities. Overall, the adaptive capacity of rangeland managers will depend on their financial and social resilience and their ability to innovate and access new information. Overcoming policy and social constraints to adaptation will need to be a high priority of government.

Southern African Arid Lands: Current Status and Future Prospects

12

Klaus Kellner, Graham von Maltitz, Mary Seely, Julius Atlhopheng, and Lehman Lindeque

12.1 Climate and Environment

Most of the southern African region south of 15° S can be considered as hyperarid, arid or semiarid when using an aridity index based on the ratio of mean annual precipitation to potential evaporation (Fig. 12.1). This ranges from the hyperarid Namib Desert in the west through a diversity of arid vegetation types, including the Succulent Karoo, dwarf shrubland of the greater Karoo, some areas of the Fynbos, as well as large areas of grassland and savannas across the southern parts of the African subcontinent. Southern Angola, southern Zambia, all of Namibia and Botswana, most of Zimbabwe, most of southern Mozambique and most of South Africa (excluding some of the eastern section) can be considered as arid or semiarid. Current climate change predictions are that in general it is the arid regions which are likely to experience reduced rainfall as a consequence of climate change, whilst the moister eastern side of the subcontinent may become slightly wetter. Given that the interior is warming at greater rate than the global average, this is likely to deepen the

---

K. Kellner (✉)
Unit for Environmental Sciences and Management, North-West University, Potchefstroom, South Africa
e-mail: Klaus.Kellner@nwu.ac.za

G. von Maltitz
Global Change and Ecosystems Dynamics: Natural Resources and the Environment, CSIR, Meiring Naude Road, Brummeria, Pretoria, South Africa

M. Seely
Desert Research Foundation of Namibia (DRFN), Windhoek, Namibia

J. Atlhopheng
Geomorphology and Climate Change, P/Bag UB0704, Faculty of Science, University of Botswana, Gaborone, Botswana

L. Lindeque
United Nations Development Programme (UNDP) Pretoria, Pretoria, South Africa

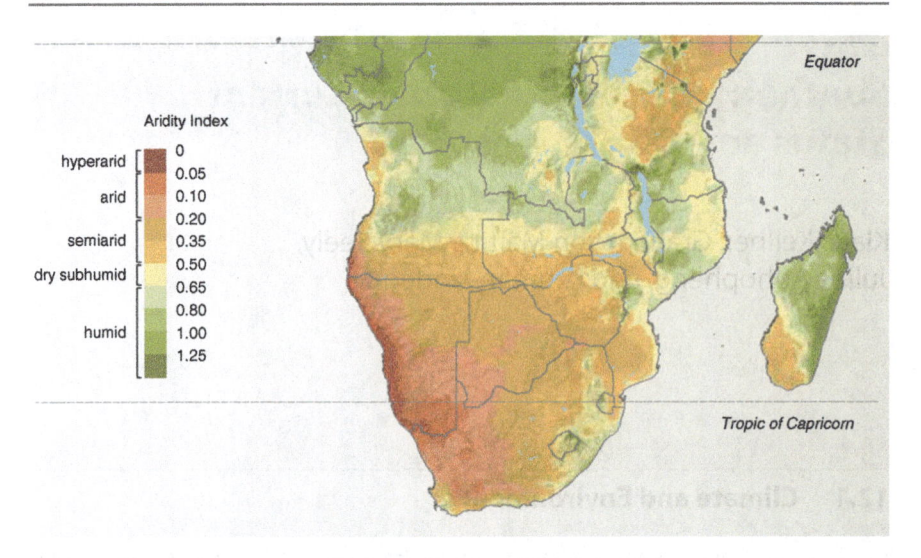

Fig. 12.1 Aridity map of southern Africa (Sebastian 2014. Reproduced with permission from the International Food Policy Research Institute)

levels of aridity on the western half of the subcontinent in the future. This will increase the vulnerability of habitats, biodiversity and people living in these regions.

There is a clear rainfall gradient from the most arid areas on the western coastal region in Namibia which receives less than 20 mm rainfall per year, to the moister eastern coastline in Mozambique which can get over 1,000 mm rain per year (Fig. 12.3a). In addition, a few of the mountainous area in Southern Africa also attract higher rainfall. Most of the region has a strong summer rainfall with dry winters. Rainfall mostly occurs between the months of October and March/April, which is the summer season of the southern hemisphere. These rain-fed semiarid areas in southern Africa are characterised by very seasonal and highly variable rainfall events (inter-annually and intra-seasonally). Rainfall is often unpredictable, and people living from the land, especially for agricultural purposes, have to rely on groundwater resources. These can also deplete very easily, due to consecutive low-rainfall seasons and drought conditions. The exception being the south and western regions of South Africa which gets winter rainfall, which mainly includes the Fynbos and Succulent Karoo vegetation types (Fig. 12.2). Although many ecoregions occur over the gradient from west to east in the African subcontinent, it is the vast areas of arid savanna and woodlands including the Kalahari region that stretch from Angola through Botswana to the northern parts of South Africa that dominate the drylands. This region will be the main focus of this chapter.

Though there is extensive variation in the vegetation due to geographic, soil and climatic differences, these savannas are often dominated by *Acacia* trees (now called *Senegalia* or *Vachellia*), and though arid, a variety of herbaceous species occur in this region. The composition will mainly depend on the climatic, land use and management strategies; this vegetation type is very important to the livestock

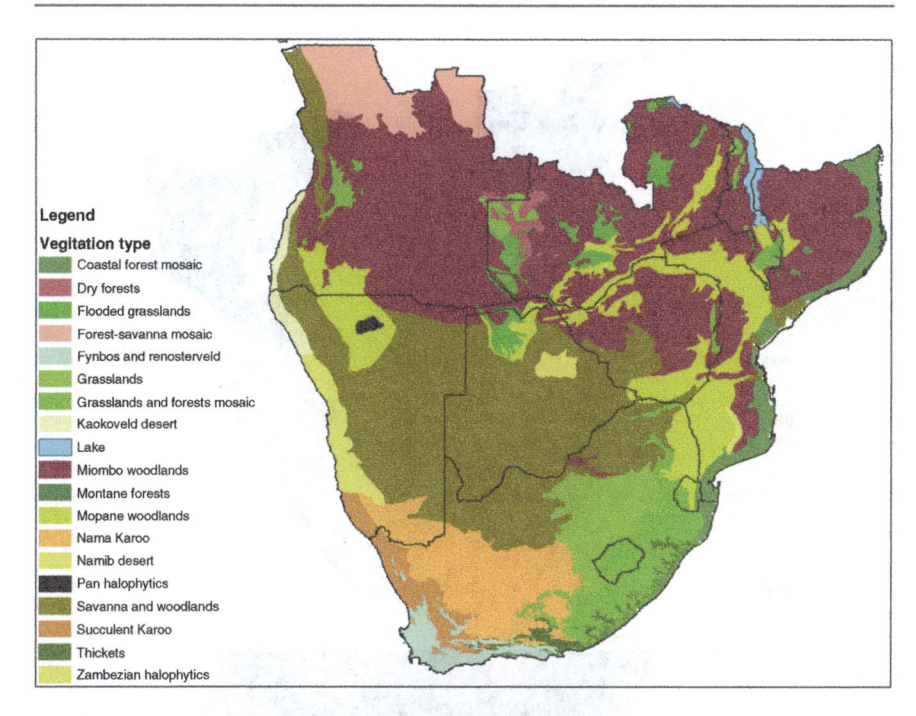

Fig. 12.2 Broad vegetation types (Derived from Olson et al. 2001)

and/or game industry. Although a variety of soil types occur in this dryland region, it is mainly dominated by the deep Kalahari sands. Cold winters (Fig. 12.3b) separate the true grasslands from the savanna regions. To the north and east, moister versions of the savannas are found; these are dominated by trees of the genus *Baikiaea* and locally referred to as the Miombo woodlands. The Miombo woodlands, though moist, have very low soil fertility and provide poor grazing and browsing from the predominantly "sour" grasses and high tannin content in trees. Within many of river systems in this hot and dry region, a vegetation type dominated by the species *Colophospermum mopane* gives rise to the Mopani veld (Huntley 1984; Musina and Rutherford 2006; Olson et al. 2001), with dwarf shrubs, grasses and succulent plants adapted to the more arid conditions. The relatively moister eastern side is dominated by more mesic savannas.

Much of the data reported here are from the ASSAR (Adaptation at Scale in Semiarid Regions) project, i.e. "Vulnerability and Adaptation to Climate Change in the Semi-Arid Regions of Southern Africa". The ASSAR series of working papers are funded by Canada's International Development Research Centre (IDRC) and the UK's Department for International Development (DFID) through the Collaborative Adaptation Research Initiative in Africa and Asia (CARIAA). CARIAA aims to build the resilience of vulnerable populations and their livelihoods in three climate change hot spots in Africa and Asia. The programme supports collaborative research

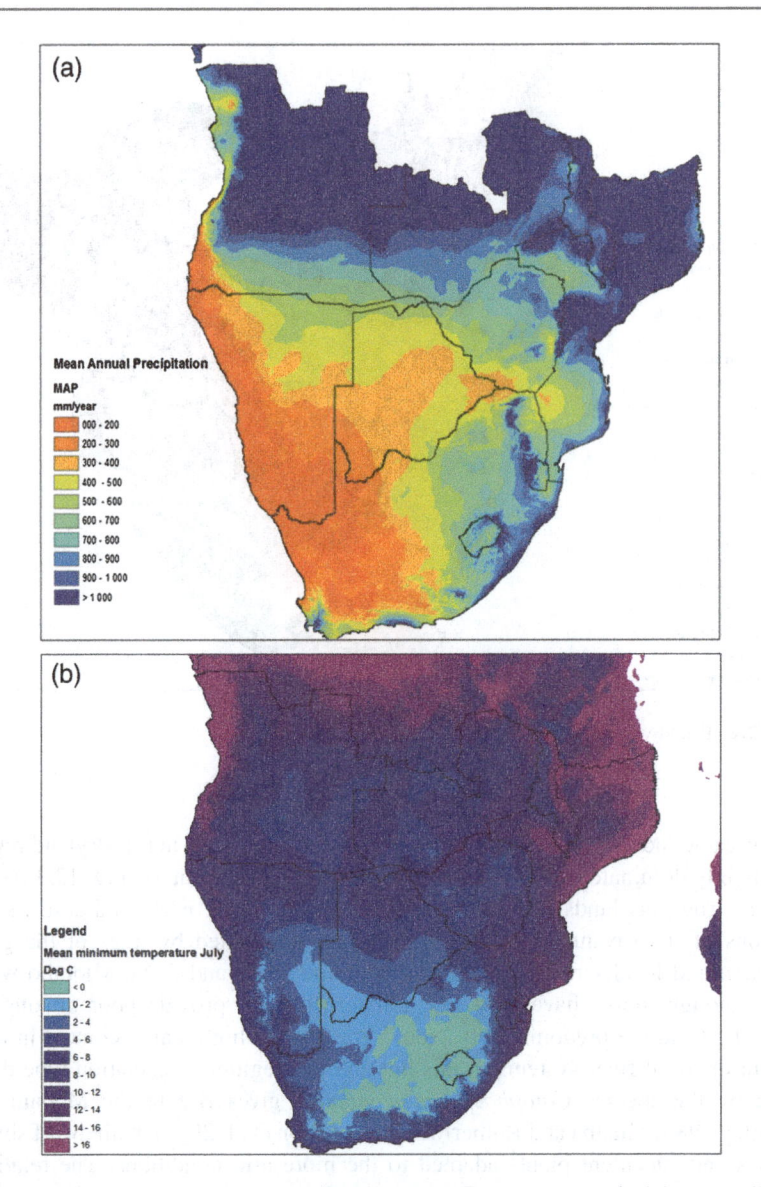

Fig. 12.3 (**a**) Mean annual rainfall and (**b**) mean minimum temperature in July (normally the coldest month) (From World climate data)

to inform adaptation policy and practice. In the report areas covered, role and phases of the regional diagnostic studies (RDS) used in the ASSAR project can be seen (Spear et al. 2015). It would appear sustainability in rangelands calls for "integrated landscape-scale planning of land use" so as to retain the biodiversity and ecological value of the southern Kalahari (Dougill et al. 2016).

12.2 The Socio-economic Context

12.2.1 Land Tenure

The southern African countries have a diversity of land tenure regimes, which have profound impacts on land management practices. In essence, there are three main tenure types found across the region, but with country-specific nuances to each.

(a) Private and leasehold tenure

In South Africa, about 80% of land is held in private tenure in large commercial-managed areas. These farms are owned and managed by private farmers, companies or government institutions. Within the drylands, this land is predominantly used as rangeland, traditionally for livestock grazing, but more recently with large amounts of the area being converted to either just game or mixed game and livestock management. Farming areas are typically large, especially in the arid and semiarid regions where the grazing and browsing capacity per hectare is very low. These large areas are necessary to make it financially viable for the land user and justify inputs to the livestock/game industry. In the moister areas, extensive agriculture for cash crops, such as maize, wheat and alfalfa, is practised, including crops and grasses for fodder production by planted pastures. In Namibia, large commercial farms are also common. Botswana has relatively few large commercial farms though there are a few such as in the Ghanzi and Tuli block regions. In addition, there are large hunting and photography concession areas. Livestock and/or game farming dominates in both Namibia and Botswana. Historically, Zimbabwe had extensive areas of large commercial farms, but land reform in the country has seen some of these reverting to communal ownership or small-scale farmers. Large-scale commercial farming activities have ceased in many cases due to changes in ownership. In Mozambique, commercial farms are on a leasehold rather than private ownership basis, and in many cases, the large commercial operations were disrupted by the civil war and farming practices have not resumed.

(b) Communal land

In all the southern African countries, a large amount of land is under some form of communal management, often linked to traditional land management authorities. This land is used predominantly for subsistence-type livelihoods, often including some type of communal management structure. Most of these dryland agricultural areas contributing largely to subsistence livelihoods are mainly used for livestock ranching. In South Africa, this communal land has its legacy in the so-called Bantustans or homelands. It formed about 13% of the land area and has a relatively high population density. Land reform has in effect increased the extent of this land from a practical perspective even though from a legal perspective, much of the land reform has created private land owned by a group of individuals. Recently in South Africa, land reform farms belong to the state and beneficiaries of the land reform programme lease the land from the government. Namibia has a similar past, and it is in the far north, where the largest areas of communal land are found. Namibia also has a number of large conservancies in the communal land-managed areas. In Botswana, all land belongs to the state, but the majority of the country in used communally, accounting for about 70% of the country (Land Policy 2011). Zimbabwe has large areas referred to historically as tribal trust land that are managed communally. In addition, there are large areas of redistributed land that are allocated to smallholders. Most of Mozambique is either formally or informally used as communal land. In all countries, there are degrees to which individuals can have privately owned croplands, though most of the rangeland is used communally.

(c) Conservation land

In all countries, there are extensive areas of privately or state-owned conservation land. This ranges from about 8% in South Africa up to about 31% in Botswana. This land is predominantly used for conservation, hunting and ecotourism activities in wildlife national parks and games reserves. Private conservation areas, the transboundary conservation initiatives between the countries in the region, conservancies and biosphere reserves in effect increase the extent of this conservation land, e.g. upwards of 50% in Namibia if combined livestock and game farming are included.

12.2.2 Land Use Practices

The land use practices and human population density differ with more rangeland management and low population densities found to the west of the southern African continent. Some activities, such as ecotourism, are practised over the whole gradient. As mentioned above, the rain-fed semiarid areas in the western regions of southern Africa, where the seasonal rainfall events are highly variable, people live from the land for agricultural purposes, mainly practising livestock and wildlife management. The areas are culturally and ecologically diverse with a high proportion of poor people with limited access to services such as water, sanitation, markets and infrastructure (Spear et al. 2015). The latter is especially true for the more remote areas. The employment rate is low and depending on the country; most people rely on mining, agriculture, tourism, social grants and remittances for their daily livelihoods. After agriculture, the mining sector offers one of the largest employment opportunities, but this is localised. Most tourism activities are practised in conservation areas, either privately or state-owned, whereas the agricultural lands are mainly used for extensive livestock or wildlife grazing dominated by a mix of privately owned commercial and subsistence agriculture and state-owned land (the ratio of these differing by country). Unskilled workers often have to find their income from agricultural and conservation enterprises offering low salaries. Although the energy generation, communication, health and education systems have improved over the past quarter century, it is still very poor in the rural environments. The latter contributes to the low level of literacy, especially where youth are far from centres offering better and more advanced schooling and skills development. As the unemployment rate for women is usually lower than for men, government initiatives and programmes focus on the capacity building of youth and women to increase the livelihoods and well-being. The government encourages the development of small business enterprises to address the large unemployment rate. Due to the higher unemployment in rural areas, urbanisation is on the increase, and many small towns and cities have an influx of people seeking job opportunities. The irony is that many of these small towns are, themselves, collapsing economically due to a decline in agricultural and because wealthier farmers tend to take their business to larger cities as this is where they are sending their children to school. People that do not get employment in the urban areas go back to the land and try and make a living from the natural resource. A lot of money is transferring between rural and urban centres (including cross border transfers between SA and other countries) in terms of remittances. This in part supports rural areas and can also create novel pressures, e.g. a person in town invests in cattle that are placed on the commonage in the rangelands further away in the rural areas. This creates in increasing pressure on the land, especially in areas with a high population growth and is exacerbated by high-income earners purchasing livestock and sending them back to the rural areas to be managed on their behalf. Higher population growth coupled with one of the globally highest rates of HIV and AIDS rates, places increasing pressure on health systems.

One of the main challenges for livestock and/or game-producing land users is the limited access to markets, which have a negative effect on the socio-economics, especially in communally managed regions. Livestock and game farmers in Botswana and Namibia however have preferential access to European markets, for high quality products. Due to the pressure on land, livestock are often moved (where possible) from the unproductive, low-rainfall areas to more productive land, especially in below-rainfall seasons. Many rural people consider livestock their wealth and hold back sales when conditions are poor, and market rates are low, often losing many of their livestock. Local-level monitoring is one approach to prepare for reducing livestock in an acceptable manner (Matamabo and Seely 2011). People, often backed by non-government organisations (NGOs), try to increase their income by selling or exchanging of medicinal plants and veld (rangeland) products, increasing the pressure on the natural resource. The same market challenges (market for the poor) affect new ventures that attempt to diversify livelihood strategies such as in smallholder irrigation, threatening their sustainability (Mutambara et al. 2015).

Land tenure remains an important challenge throughout the region, and despite most countries embarking on some level of tenure reform, numerous problems remain. In communal areas, customary tenure dominates, with individuals not having title deeds to land. This hinders investment, can lead to tenure insecurity and may mitigate against long term investment in the land. Further there are complexities relating to the communal management of shared resources. Historic imbalances in land ownership, with whites owning much of the commercial land, is also problematic. Redistribution programs in all countries have been slow, problematic and often not achieving the desired rural upliftment objectives.

In a study by Shackleton et al. (2008), to establish the links between ecosystem services (ES) and human well-being, and to concentrate especially on the opportunities for poverty alleviation through the provision and management of ecosystem services, the following were found:

- For the *Provisioning services*, which included natural products, fuelwood, fodder and water, the poor households use provisioning services for income generation and to address poverty-related issues. Provisioning services were seen to account for between one-quarter and one-third of household income, although this was variable, being over 50% for some households and less than 10% for others. Many case studies demonstrate that the poorer sectors of society make the greatest use of provisioning services, both for home consumption and income generation. Provisioning services also provide a "safety net" or can be used as "insurance" during times of unexpected shock or added stress. Poorer households turn to the natural resources during such times, which may lead to increased land degradation and misuse of the natural resources.
- The *Regulating services*, which included soil fertility and water, cannot be sold to generate income and therefore are less important for poverty alleviation than the provisioning services. Studies are under way to investigate possible mechanisms for the payment of ecosystem services (PES) to counteract the over- or misuse of the services gained from the ecosystem, such as water and other resources. Services gained from the ecosystem that regulate the functioning that would increase the production in local livelihoods are less direct, although many case studies show that the population depend on them. Communities are susceptible in situations where regulating services are diminished, for example, through flooding, drought, poor air quality, areas with higher disease incidence and degraded or exhausted soils. If these incidents occur, people move away (migrate) to areas where they are less serious and are least affected. If they have enough resources available, they often are able to buffer these negative impacts, e.g. buying fertiliser if the soil condition is decreased.
- Shackleton et al. (2008) could find little evidence regarding how *Cultural services*, which included the cultural and spiritual values of the ecosystem, are used in poverty alleviation. It is however well known that people have local rituals and respect for the environment and that they are concerned if these services are decreased. The tourism potential is high, as it attracts a growing number of international and domestic tourists for cultural and ecotourism experiences, which contribute to the poverty alleviation of the rural people. Although ecotourism is within the top contributors to national GDP, it is important that the revenue gained through this enterprise reaches the poor households.
- In the study by Shackleton et al. 2008, biodiversity was classified as a *Supporting service*, because the fauna, flora and landscape in southern Africa is the basis of the ecotourism industry. As such, biodiversity is therefore one of the main services to support the poor. The conservation and sustainable management of the habitat is therefore needed to protect one of the main resources required by rural people. It, for example, supports the people by providing key resource areas and species during times of stress, such as during periods of droughts or disease. Sustainably managed lands also support the community by decreasing food

insecurity and can provide revenue for established small business enterprises. Biodiversity resources can be used for commercial gain, either in the form of intellectual property rights or cultivation agreements. The consequences, and the need for conservation of genetic diversity, are especially critical during the threat of climate change, i.e. the use of genetically modified crops and the need to build resilience and adaptability. Nonetheless, elements of biodiversity are being seriously targeted by international poachers reducing potential for income to poor people.

In the concluding remarks of the study by Shackleton et al. (2008), it was mentioned that whilst all ecosystem services are important to a greater or lesser extent, water is a particularly important ecosystem service in arid and semiarid areas. According to predictions, these areas will be severely water stressed within two or three decades and will affect the poor to a greater extent. A second aspect that is highlighted by Shackleton et al. (2008) is that there is an inadequate understanding and appreciation of the importance and value of ES by planners, bureaucrats and policymakers, resulting in many avoidable negative trade-offs, hence there is an urgent need for better research and particularly communication of ES aspects to these agencies. Based on their extensive research in the Kalahari rangelands, Reed and Stringer (2015) have concluded that to address land degradation, critical natural capital needs to be preserved and livelihoods need to be based on a wider range of ES. They also emphasised the need for economic mechanisms to move from land degradation to sustainable land management, that the monetary evaluation of ecosystem services was not enough – and that benefits should accrue to both rich and poor, without degrading the resource base.

12.3 Climate Change, Trends and Projections

Climate change is one of the major challenges of our time and adds considerable stress to the societies and environment in the southern African continent. From shifting weather patterns that threaten food production to rising sea levels that increase the risk of catastrophic flooding, the impacts of climate change are global in scope and unprecedented in scale. Without drastic action today, adapting to these impacts in the future will be more difficult and costly (UNEP 2010).

As for the other regions, climate in the semiarid areas is also strongly affected by the El Niño Southern Oscillation (ENSO) patterns, which characterise a cyclical variation in the surface temperature of the tropical eastern Pacific Ocean. The effects of El Niño and La Niña events have a great influence on the seasonal trends in this region, with strong El Niño events being associated with droughts. It is a feature of the arid areas that the impacts of drought are negatively correlated with annual rainfall, as such the driest areas are also the areas with the greatest variance in rainfall, and these areas are most strongly impacted by drought. Although La Niña patterns generally characterising higher rainfall events occur after the low-rainfall events of

El Niño, these patterns are less predictable, and land users cannot depend on these cycles in the agricultural business.

There is a growing evidence that the Kalahari region has historically warmed faster than the global mean and is projected to warm at almost twice the global mean into the future (Engelbrecht et al. 2015). Predictions of precipitation changes in the region are less clear, but as a general trend, it appears that many of the more arid western areas will experience a reduction in rainfall, whilst the eastern side may gain slightly. The combined impact of this is that the arid areas are likely to become more arid. This will be offset slightly by the raised CO_2 which has the effect of making plants a bit more drought resistant. In all regions, there is likely to be fewer rainfall events, but with individual events having greater intensity.

Climate change and land degradation are closely interlinked and populations living in resource-dependent ecoregions are mostly affected (Reed and Stringer 2015). It is essential to understand these interlinkages if the targets of the Sustainable Development Goals (SDG) are to be addressed. These include the reduction of poverty and improving the well-being of the rural poor. Climate change causing drought and leading to land degradation operates differently in different habitats and under different forms of land management, which makes this interlinkage very complex, especially in the arid and semiarid regions of Southern Africa. It is important that policymakers respond in a timely and effective way to introduce effective adaptation strategies (Reed and Stringer 2015). Nevertheless, to date, this type of response has not been realised.

Due to the specific climatic conditions in the arid and semiarid regions of Southern Africa, the vulnerability of the people to climate change in these areas is high. Climate change increases food insecurity, either through reduced local agricultural production for home consumption or through impacts on livelihood activities such as livestock ranching. This is due to the low and unpredictable rainfall patterns and predictions of increased variability in precipitation at the local and regional scale. There will be more droughts and the prediction is that the temperatures will increase between 1 and 4 °C by 2050. Fodder production may be directly impacted, and livestock and wildlife may also be impacted on by heat and increased diseases.

Impacts from climate change are likely to be more pronounced in areas where people are poor and depend on the natural resource for food, fibre, construction and water. Women are particularly vulnerable to the impacts of climate change as in many of these areas they are responsible for the provision of food, water and firewood. These commodities become difficult to obtain during times of drought (Spear et al. 2015).

Land users in the semiarid regions of southern Africa often have limited institutional support (Seely and Montgomery 2011), which includes agricultural extension services and appropriate and integrated policy, especially knowledge of the latest legislation and developments by national and regional government. This includes information about climate change and adaptations by national and regional programmes regarding initiatives that are implemented to support the people in these areas. Unemployment, especially in the agricultural sector, will be increased due to climate change, as the productivity decreases and land users cannot afford workers.

This will increase poverty and have a negative effect on the socio-economic outlook of the people, with aspects such as health, education, infrastructure, access to markets and biodiversity also decreasing. The loss in biodiversity and changing environments may lead to negative impacts on the tourism industry due to changing weather patterns caused by climate change, ultimately affecting the ES of the rural poor. As mentioned, people seek more job opportunities in the urban areas, increasing the population density in villages and cities. However, if people do not get jobs in the city, they often rely on the natural resources in the rural areas, increasing the pressure on the land. If people have to depend on the land for an income, aspects such as land degradation, caused by overgrazing, pollution and deforestation will increase by the overpopulation of the rural areas. Higher population growth rates and agricultural activities also lead to increased pressure on groundwater resources as the groundwater recharge is reduced.

The increase in the density of woody shrubs and trees, also known as "bush encroachment", is regarded as one of the main causes of land degradation in the arid and semiarid savannas in southern Africa. Although the main cause of bush encroachment is often regarded as an impact of overutilisation or overgrazing of the herbaceous layer, giving the woody seeds and saplings a competitive advantage for soil moisture in these rangelands; other factors that are identified as causes of bush encroachment include the absence of fire due to deliberate suppression or as a consequence of heavy utilisation by livestock removing sufficient biomass to prevent fires. The elevation of atmospheric CO_2, together with changing rainfall and temperature patterns, may also be a direct cause of increasing woody densities due to C3 trees gaining a competitive advantage over C4 grasses in a high CO_2 environment (Bond et al. 2003; O'Connor et al. 2014). Bush encroachment is not an event but a complex dynamic change of vegetation composition, structure and density over time. To understand this phenomenon, a historic perspective is required which may vary in different ecoregions of southern Africa (O'Connor et al. 2014). Understanding the degree to which CO_2 fertilisation caused by climate change is a contributing factor to bush encroachment is important because its impacts are often outside of the control of the land manager, whereas grazing intensity and fire management can be controlled. Though sound rangeland management may reduce the occurrence of bush, to many livestock and game farmers in South Africa, the only option is to manage the increase in the density of woody shrubs (bush). It is estimated that already >13 million ha in South Africa and 26 million ha in Namibia (De Klerk 2004) are subject to bush encroachment. The woody plant density can be as high as 5,000 TE/ha (1 TE tree equivalent is a single stemmed tree 1.5 m high, Teague et al. 1981). Although the type of species causing bush encroachment differs in the ecoregions, *Senegalia mellifera* (previously known as *Acacia mellifera*) is the most dominant indigenous woody species causing bush thickening. At a density of 1,000 trees/ha, *Senegalia mellifera* uses about 65 l/day of water, which not only depletes the above- and underground water resources but also competes with the grass species in the upper soil layers. Based on a comparison between an area with almost no trees and an area with 1,500 TE/ha, the grass forage production decreased from 2,066 kg/ha in un-encroached areas to 410 kg/ha in encroached areas. The grazing capacity in encroached areas is 93.6–48.7 ha/LSU, whereas the grazing capacity in un-encroached areas was 10–12 ha/LSU. An area with 1,500 TE/ha suffered a decrease of 81.5% in forage production (Harmse 2014). The thickening of woody plants is also caused by the invasion of the alien species such as *Prosopis* spp., which have a negative effect on the biodiversity, water retension, forage production and ecosystem services in some areas. *Prosopis* spp. mainly occur in high densities in the arid and semiarid areas in the riparian areas in river beds and on river banks. In a study by Van den Berg et al. (2013), remote sensing techniques were used to quantify the change in distribution and density as well as the spatial dynamics of *Prosopis* from 1974 to 2007 in the Northern Cape of South Africa. The total area infected increased from 127,821 ha in 1974 to nearly 1,474 million ha in 2007, and the total percentage of land covered increased from 0.35% in 1974 to 4.06% in 2007.

Land users have implemented many different bush control technologies, either by chemical, manual, mechanical or a combination of applications, depending on the density and type of species encroaching (Barac 2003; De Klerk 2004). The control or combating of bush encroachment normally has positive results, especially for the livestock farmer, as it increases the grass sward for better fodder production. The latter will depend on the soil type and rainfall pattern and application implemented. However, the economics and impacts on the ground water of bush control are questionable (Harmse et al. 2016) unless value is gained from the harvested wood material. In this regard, a large charcoal producing industry has been established and there are investigations underway for use of the bush for power generation and other value adding products.

Trends in climate change also increase the migration rate of people seeking better livelihoods in more productive areas which offer greater job opportunities. Migrations occur within regions of one country or between countries. Unfortunately, the migration also leads to higher urbanisation rates creating increased pressures on municipal services.

An assessment by the Department of Environment Affairs (DEA) in South Africa (DEA 2015) showed that climate change projections will lead to significant changes across the ecosystems in the biomes through the alteration of existing habitats, seasonal rainfall, species distribution and impacts on the ecosystems. The DEA developed the climate change adaptation action plans in collaboration with the provincial and local government and civil society. All stakeholders identify and prioritise the actions that have to be considered on a landscape, ecosystem and species scale. The ecosystem services, such as grazing, water supply, sanitation, security and tourism opportunities provided by the ecosystems in the environment and the various cultural and economic activities, are considered. This programme will be implemented soon and hopefully adopted by the local communities.

The climate change "threats" and adaptive actions that were identified, such as rising temperatures, erratic rainfall patterns over seasons with reduced water availability, increased diseases, a reduction in the crop and fodder for humans and livestock and rising atmospheric concentrations of carbon dioxide (including rise in greenhouse gasses), as well as secondary "threats", such as an increase in the severity of fires and changes in land use, can be addressed by the following actions (Shackleton et al. 2008):

- Better governance, spatial planning and human demographic approaches (include social networks).
- Better land use management approaches which adjust the way in which the land uses are executed, e.g. conservation of biodiversity and better species management).
- Reducing the stresses (e.g. overgrazing, pollution, overpopulation) that might impact on ecosystems and improve the socio-economic status of the people.
- Restoring damaged ecosystems and ecosystem functioning and services.
- Expand protected areas on private and state-owned land to conserve biodiversity.
- Promote sustainable land management.

The factors above occur to a higher or lesser degree in all the arid land biomes in Southern Africa, such as the desert, savanna, grassland and Karoo biomes. Adaptation plans and processes should however be implemented to different degrees across the regions. The challenge is that the climate change adaptation action plan is sensitive to the spatial heterogeneity of social and environmental conditions and has a comprehensive understanding of the driving forces and accounts for major environmental and climatic feedbacks. A multi-sector process that addresses several factors with proactive responses that can exploit opportunities is followed (Reynolds et al. 2007; Verstraete et al. 2009). Aspects such as awareness, training, education, outreach and extension, which are critical for land users and policymakers, must be included. More strategic approaches that are required include adaptation actions, especially during periods of drought. Land management and policy decisions need to be guided by local scale, evidence-based, information on the impacts of climate change. All government interventions, such as the "Working for…" programmes in the Natural Resource Monitoring (NRM) programme (in South Africa), should include monitoring and evaluation (M&E) plans in order to adjust proactively to climate change and other unforeseen changes (Midgley et al. 2012). In a report prepared by the DEA (2015) regarding the development of climate change adaptation action plans for the different biomes in South Africa, adaptation actions were prioritised which apply to the broad South African landscape. This includes the ecosystems within which they occur, the species which make up those ecosystems and the ecosystem services such as grazing, water supply and tourism opportunities provided by the landscapes, as well as the various cultural and economic activities which take place in and depend on those landscapes. The main threats and adaptation options that were identified for the biomes in South Africa include extreme temperature events in the savanna, Nama Karoo, succulent Karoo, and desert biomes of South Africa as well as increased variable rainfall with extreme events and intensities, sea level rise and a rising CO_2 level in the grassland and savanna biomes (DEA 2015). Potential adaptation options for the arid and semiarid Karoo, savanna and desert biomes include the development of new viable land uses, the restoration of degraded land, adjustment of stocking rates and changes in cultivars or species, as well as actions to improve education, outreach and extension. It therefore seems that the Department of Environmental Affairs in South Africa has recognised the challenges that might occur with climate change and is contributing to the actions to address these problems. Although these challenges may be recognised and actions proposed, some barriers regarding the implementation of the adaptation plans have to be overcome. Based on Spear et al. (2015) and Muller and Shackleton (2013) these include:

- At national and regional level, proper governance, including better coordination, understanding of situation, technical capacity and availability of resources
- At the local level, better access to information, natural and financial resources, technical know-how and incentives
- At the community level, better knowledge on, for example, the impacts of climate change and needs during drought, measures of adaptation, appropriate crop varieties and other adaptation options, costing of adaptation measures and socio-economic impacts and the effectiveness of existing policy and practice with regard to the implementation of adaptation measures

The ASSAP report (Spear et al. 2015) further emphasises that an adaptation plan and programme to address climate change scenarios can only be achieved if the following additional aspects are included:

- Information on options are made readily available to practitioners.
- Alternative livelihood options created at the local level which require policies that enable and promote new and diverse livelihood options, e.g. increasing markets.
- Goals need to be better integrated, and if a common goal is selected, it should be applied across different sectors in the policy and practice process to achieve widespread and effective implementation and adaptation.
- Communities need an evidence base, as well as demonstrations of the benefits, especially with regard to financial and social aspects.
- The adaptation plan should be monitored and progress measured so that timely changes can be made.

12.4 Scenarios of Change

The drivers of change causing a decline in the ecosystem services and well-being of the populations are mainly local and often due to historical consequences. Some of the main drivers that exist across different spatio-temporal scales include (Shackleton et al. 2008):

- Climate change factors, mainly changes in rainfall and temperature
- Land transformation
- Health and well-being (e.g. HIV/AIDS)

- Overuse of resources and land degradation (harvesting, grazing, increase in woody density also called "bush encroachment", abstraction)
- Land tenure policies and weakened governance
- National and international policies and processes
- Urbanisation and expansion in peri-urban areas

A number of social, political and climatic factors are likely to impact on the dynamics of the drylands into the future. These areas have a high population growth rate, and this is likely to increase pressure on a resource that is simultaneously dwindling as a consequence of climate change. This will likely increase the existing trend of rural-urban migration and is very likely to increase the levels of poverty within especially the communal areas. In the commercial areas, profitability of operations may decline, and the trend of consolidating land holdings into larger unit is likely to continue. It is also likely that the trend from livestock to game will continue due to game having better adaptive capacity to the changing environment, especially the increase in woody densities ("bush encroachment") in the savanna regions.

Land reform, especially in South Africa and Namibia, is likely to see an increase of land under some form of group ownership. This will be driven by political priorities to see changes in land ownership. However, unless the current models of land reform are fundamentally changed, this might result in lowered land productivity and may not contribute to improved livelihoods.

Long- and short-term droughts can have serious ecological and socio-economic consequences in the arid lands of southern Africa (Vetter 2009). Droughts can even lead to economic collapse which in turn leads to large migrations. This is especially true for people living from natural resources based on rangeland management systems. Through the resilience thinking process, land users and government officials try to understand the forces that lead to change and how to adapt to shifting ecological systems affecting the socio-economic outcomes of the communities and how to manage the land more sustainably if changes or disturbances (such as drought) occur. It is important to try and understand the resilience of a system and how to overcome some thresholds before applying better or more appropriate management programmes. Management decisions, developments and initiatives should therefore be based on resilience thinking theory, especially in the light of current and future climate change and variability (Dixon and Stringer 2015). Dixon and Stringer (2015) further propose six recommendations for assessing the climate resilience of smallholder farming systems in sub-Saharan Africa (SSA) in order to enhance the linkages between resilience theory and practice. These are (1) better integrate vulnerability and resilience; (2) recognise that resilience does not equal development or poverty reduction; (3) recognise the benefits and limitations of adopting flexible, participatory approaches; (4) integrate issues of power into assessment tools; (5) target specific systems; and (6) encourage knowledge sharing, empirical studies and critical evaluation.

Another strategy that is becoming important to mitigate the impacts of climate change in southern Africa is the adoption of conservation agriculture (CA). CA is a farm land management strategy that focuses on the preservation of soil moisture and prevention of soil erosion through reduced tillage, permanent organic soil cover and the use of crop rotation or intercropping. CA is widely promoted as a climate-smart agriculture (CSA) which pursues development, climate mitigation and adaptation goals simultaneously. As such, the CA research directly contribute to the wider theme of climate compatible development (CCD).

12.5 Conclusion

In the arid and semiarid areas of southern Africa, many land users still depend on natural resources, such as soil, vegetation and water for sustainable livelihoods. Due to higher population densities and more intensive land use practices, the semiarid areas in particular are prone to man-made or accelerated land degradation including: loss of vegetative cover, increase in alien and invader species, bush encroachment, change in species composition and eventually loss of habitats (Lindeque and Koegelenberg 2015). There are many direct causes for land degradation in the area mostly related to the inability of socioecological systems to adapt to a changing environment and the inability to restore resilience in degraded ecosystems through rehabilitation and restoration. Currently, farmers are responding to land degradation challenges by implementing sustainable land management (SLM) practices to some extent, but the effectiveness and maintenance of these measures are in many cases slow and depend on the rate and degree of the degradation, as well as long-term policy contributions. Considering that a number of social, political and climatic factors are likely to impact on the dynamics of the drylands in Southern Arica in future, it emphasises the importance to create an enabling environment to allow land users to adopt climate-smart land management best practices to enable the future sustainable development and use of the drylands.

For any successful application for adaptations and change needed to take place and to ensure that social, economic and environmental goals are reached, the emphasis should be on participatory environmental management based on community-based natural resource management (CBNRM) principles (Dyer et al. 2014). Communities have to be empowered and fully participate in project developments and design.

References

A.S. Barac, EcoRestore: a decision support system for the restoration of degraded rangelands in southern Africa. M.Sc. Thesis, Potchefstroom: North West University, (2003), 323 p

W.J. Bond, G.F. Midgley, F.I. Woodward, What controls South African vegetation – climate or fire? S. Afr. J. Bot. **69**, 1–13 (2003)

N. De Klerk, *Bush Encroachment in Namibia: Report on Phase 1 of the Bush Encroachment Research, Monitoring and Management Project* (Ministry of Environment and Tourism, Government of the Republic of Namibia, Windhoek, 2004), p. 255

DEA – Department of Environmental Affairs, Climate Change Adaptation Frameworks for South African Biomes, in *Pretoria*, ed. by C. Davis, D. Dziba, R. Scholes, E. van Garderen, G. von Maltitz, D. Le Maitre, S. Archibald, D. Lotter, H. van Deventer, G. Midgely, T. Hoffman. (Department of Environmental Affairs, Pretoria, 2015)

J.L. Dixon, L.C. Stringer, Towards a theoretical grounding of climate resilience assessments for smallholder farming systems in Sub-Saharan Africa. Resources **4**, 128–154 (2015)

A.J. Dougill, L. Akanyang, J.S. Perkins, F.D. Eckardt, L.C. Stringer, F. Favretto, J. Atlhopheng, K. Mulale, Land use, rangeland degradation and ecological changes in the southern Kalahari, Botswana. Afr. J. Ecol. **54**, 59–67 (2016)

J. Dyer, L.C. Stringer, A.J. Dougill, J. Leventon, M. Nshimbi, F. Chama, A. Kafwifwi, J.I. Muledi, J.-M.K. Kaumbu, M. Falcao, S. Muhorro, F. Munyemba, G.M. Kalaba, S. Syampungani, Assessing participatory practices in community-based natural resource management: experiences in community engagement from southern Africa. J. Environ. Manag. **137**, 137–145 (2014)

C.J. Engelbrecht, W.A. Landman, F.A. Engelbrecht, J. Malherbe, A synoptic decomposition of rainfall over the Cape south coast of South Africa. Clim. Dyn. **44**, 2589–2607 (2015)

C.J. Harmse, *Evaluation of Restoration and Management Actions in the Molopo Savanna of South Africa: An Integrative Perspective*. MSc thesis. (North-West University, Potchefstroom, 2014)

C.J. Harmse, K. Kellner, N. Dreber, Restoring productive rangelands: A comparative assessment of selective and non-selective chemical bush control in a semi-arid Kalahari savanna. J. Arid Environ. **135**, 39–49 (2016)

B.J. Huntley, Characteristics of South African biomes, in *Ecological Effects of Fire in South African Ecosystems*, ed. by P. de V Booysen, N. M. Tainton, (Springer, Berlin, 1984), pp. 1–17

Land Policy, *Ministry of Land and Housing. Government Paper of June 2011* (Government Printing and Publishing Services, Gaborone, 2011)

G.H.L. Lindeque, F.A. Koegelenberg, *Perceptions on Land Degradation and Current Responses to Land Degradation Problems in South Africa: Local Municipal Fact Sheet Series* (Department of Agriculture, Forestry and Fisheries, Pretoria, 2015)

S. Matamabo, M. Seely, World Resources Report Case Study. Namibia: Combating Desertification with Tools for Local Level Decision Making (World Resources Report, Washington DC, 2011). Available online at http://www.worldresourcesreport.org

G. Midgley, S. Marais, M. Barnett, K. Wågsæther, (2012), Biodiversity, climate change and sustainable development–harnessing synergies and celebrating successes. Final technical report. Available at: http://cap.org.za/downloads/Biodiversity Climate Change and Sustainable Development Final Technical Report April 2012_final.pdf

C. Muller, S.E. Shackleton, Perceptions of climate change and barriers to adaptation amongst commonage and commercial livestock farmers in the semi-arid Eastern Cape Karoo, Afr. J. Range Forage Sci. (2013). online version http://dx.doi.org/10.2989/10220119.2013.845606

L. Musina, M.C. Rutherford (ed.), The Vegetation of South Africa, Lesotho and Swaziland. Strelitzia 19 (South African National Biodiversity Institute, Pretoria, South Africa, 2006), (808 pp with CD GIS-database)

S. Mutambara, M.D.K. Darkoh, J.R. Atlhopheng, Making markets work for the poor (M4P) approach and smallholder irrigation farming. Irrig. Drain. Syst. Eng. **4**, 1 (2015). ISSN: 2168-9768 IDSE, an open access journal

T.G. O'Connor, J.R. Puttick, M.T. Hoffman, Bush encroachment in southern Africa: changes and causes. Afr. J. Range Forage Sci. **31**(2), 67–88 (2014)

D.M. Olson, E. Dinerstein, E.D. Wikramanayake, N.D. Burgess, G.V.N. Powell, E.C. Underwood, J.A. D'Amico, I. Itoua, H.E. Strand, J.C. Morrison, C.J. Loucks, T.F. Allnutt, T.H. Ricketts, Y. Kura, J.F. Lamoreux, W.W. Wettengel, P. Hedao, K.R. Kassem, Terrestrial ecoregions of the world: a new map of life on earth. Bioscience **51**(11), 933–938 (2001)

M.S. Reed, L.C. Stringer, (2015), Impulse Report Climate change and desertification: anticipating, assessing & adapting to future change in drylands. UNCCD 3rd Scientific Conference, 9–12 Mar 2015, Cancún, Mexico. http://www.unccd.int/

J.F. Reynolds, D.M.S. Smith, E.F. Lambin, B. Turner, M. Mortimore, S.P. Batterbury, T.E. Downing, H. Dowlatabadi, R.J. Fernández, J.E. Herrick, Global desertification: building a science for dryland development. Science **316**(5826), 847–851 (2007)

K. Sebastian (ed.), *Atlas of African Agriculture Research & Development: Revealing Agriculture's Place in Africa. Map 1, "Aridity Index"* (International Food Policy Research Institute (IFPRI), Washington, DC, 2014). https://www.ifpri.org/publication/atlas-african-agriculture-research-development

M. Seely, M. S. Montgomery, S. Proud of our deserts: combating desertification – An NGO perspective on a National Programme to Combat Desertification, ISBN: 978-99945-73-27-1, Desert Research Foundation of Namibia (DRFN) 2011, Windhoek, Namibia

C.E. Shackleton, S. Shackleton, J. Gambiza, E. Nel, K. Rowntree, P. Urquhart, (2008), Links between ecosystem services and poverty alleviation: Situation analysis for arid and semi-arid lands in southern Africa. Consortium on Ecosystems and Poverty in Sub-Saharan Africa (CEPSA). Submitted to: Ecosystem Services and Poverty Reduction Research Programme: DFID, NERC, ESRC, p. 200

D. Spear, E. Haimbili, M. Angula, M. Baudoin, S. Hegga, M. Zaroug, A. Okeyo, *Vulnerability and Adaptation to Climate Change in Semi-Arid Areas in Southern Africa* (Collaborative Adaptation Research Initiative in Africa and Asia c/o International Development Research Centre, Ottawa, 2015). Email: cariaa@idrc.ca

W.R. Teague, W.S.W. Trollope, A.J. Aucamp, Veld management in the semi-arid bush-grass communities of the Eastern Cape. Proc. Ann. Congresses Grassl.Soc. S. Afr. **16**, 23–28 (1981)

UNEP, Assessing the Environmental Impacts of Consumption and Production: Priority Products and Materials, A Report of the Working Group on the Environmental Impacts of Products and Materials to the International Panel for Sustainable Resource Management. Hertwich, E., van der Voet, E., Suh, S., Tukker, A., Huijbregts M., Kazmierczyk, P., Lenzen, M., McNeely, J., Moriguchi, Y. (2010)

E.C. Van den Berg, I. Kotze, H. Beukes, Detection, quantification and monitoring of *Prosopis* in the Northern Cape Province of South Africa using remote sensing and GIS. S. Afr. J. Geomatics **2**(2), 68–81 (2013)

M.M. Verstraete, R.J. Scholes, M.S. Smith, Climate and desertification: looking at an old problem through new lenses. Front. Ecol. Environ. **7**(8), 421–428 (2009)

S. Vetter, Drought, change and resilience in South Africa's arid and semi-arid rangelands South African. J. Sci. **105**, 29–33 (2009)

Arid and Semiarid Rangelands of Argentina

C.A. Busso and Osvaldo A. Fernández

13.1 Introduction

The continental surface area of Argentina is 2,780,400 km^2. It extends from latitude 22°S to 55°S. As a result, its length is approximately 4,000 km between the northern extreme in La Quiaca and the southern limit in Ushuaia. Its maximum width is about 1,200 km. Topography varies from 7,000 m above sea level (a.s.l.) in the Cordillera de los Andes, which crosses the country from north to south at the western extreme to sea level on the extensive shores of the Atlantic Ocean.

Two thirds of the almost 3,000,000 km^2 continental surface area are arid and semiarid territories (Fig. 13.1). Such areas, not available for cropping because of climate and/or topography constraints, are what we call rangelands. Its major economic use should be livestock production, utilizing the natural vegetation for animal feeding. The phytogeographic regions contained in these extensive surfaces present an outstanding variety in plant community structure (Cabrera 1976). This is because of the development from north to south through various thousands of kilometers and elevations which vary from sea level to the east to the tallest mountains in all the Americas to the west. Climate variations, which go from subtropical to subantarctic depending on latitude, together with the soil substrate and the biota translate into well-differentiated biomes, each with its own ecological characteristics (Barbosa et al. 2015).

Briefly, and making reference to the arid and semiarid territories from north to south, we can distinguish the following (see Fig. 13.2 below):

1. Puna, with its high plains to elevations above 3,000 m a.s.l.
2. Mesophytic and xerophytic forests of the Chaco Occidental

C.A. Busso (✉) • O.A. Fernández
Departamento de Agronomía-CERZOS (National Council for Scientific and Technological Research of Argentina: CONICET), Universidad Nacional del Sur,
8000 Bahía Blanca, Provincia de Buenos Aires, Argentina
e-mail: carlosbusso1@gmail.com; cebusso@uns.edu.ar

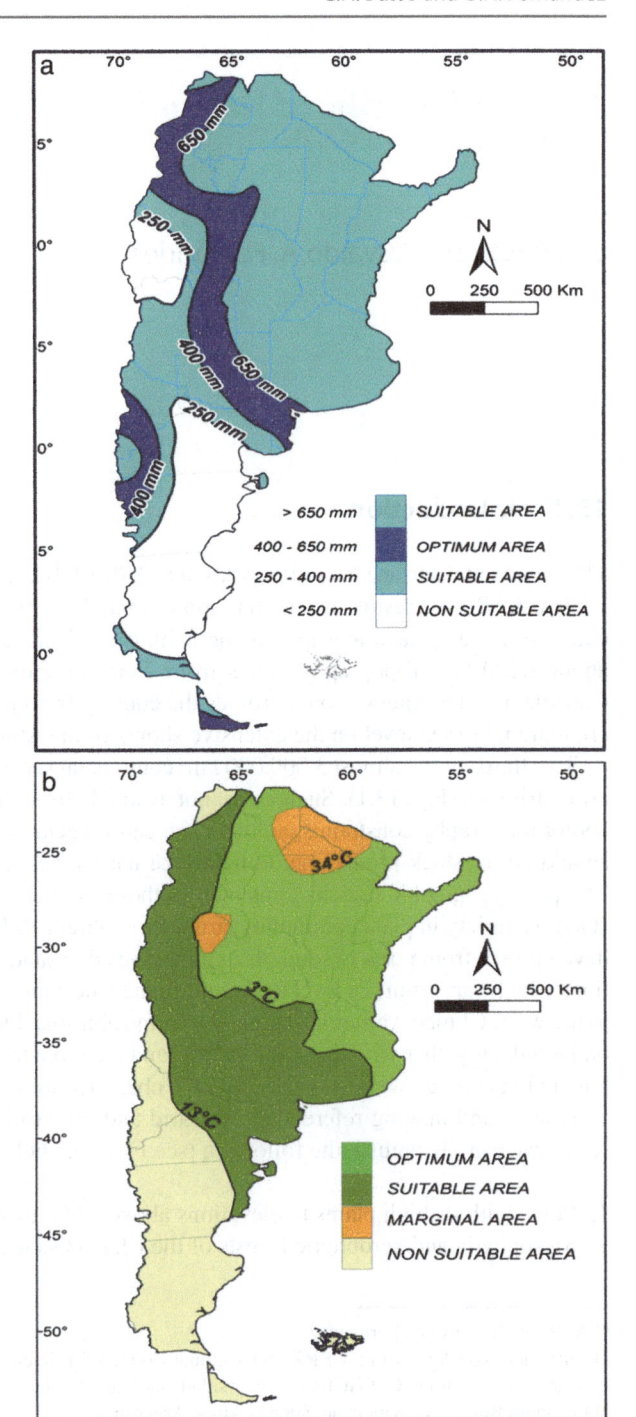

Fig. 13.1 (**a, b**) Arid and semiarid territories

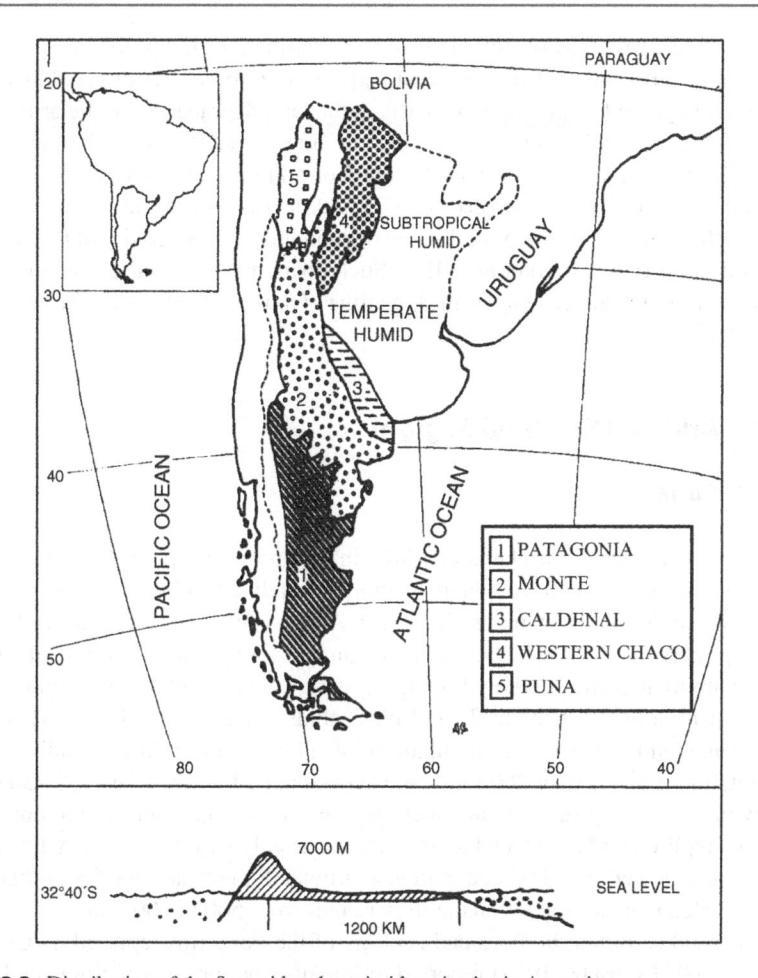

Fig. 13.2 Distribution of the five arid and semiarid territories in Argentina

3. Arid and semiarid territories to the west surrounding the Andes, characterized by an homogeneous shrubby vegetation as the dominant community on about 50 million hectares (Mha) recognized as Monte
4. Prairie rich on high forage value grasses with rather isolated shrubs and trees of the Caldenal at the borders of the cropped territories to the east
5. Subantarctic regions at the south of the country known as Patagonia, which covers an estimated surface area of 60 Mha, and constitutes one of the few cool deserts in the world

This chapter draws on the works of Fernández and Busso (1999) and Fernández et al. (2016). It includes a synthesis of research conducted during the past 20 years. Overall, it presents a (1) brief description of the climate, soil, and vegetation of those territories, which are the focus of this review; (2) discussion of the effects of their past and current utilization; and (3) an assessment of the problems of desertification. The term "desertification" identifies a series of processes directed by nature and anthropogenic forces. We consider it as the land degradation of the arid, semiarid and dry, and subhumid zones because of climate variations and human activities, following the definition given by the Convention of the United Nations for Controlling Desertification (Holtz 2003). Land degradation or desertification also refers to the decrease or loss of biological or economical productivity of the arid and semiarid territories (Ulrich et al. 2014). Such phenomena are currently among the major environmental challenges in Argentina (Quiroga et al. 2013; Torres et al. 2014, 2015).

13.2 Arid and Semiarid Regions

13.2.1 Puna

The arid Puna extends on plains and hills in the Andes at 3,200–4,000 m a.s.l. It goes from the border with Bolivia up to approximately 27° S latitude. This region is one of the coldest ecosystems in the world. Additionally, it is characterized by its landscape heterogeneity in association with an interior mountain system, which has basins with their corresponding drainage systems, plains, valleys, and cattle paths. Predominant soils are haplargids and plaeorthids (Etchevehere 1971). There are often dunes and extensive accumulations of alluvial and colluvial, saline soils. Precipitation is lower than 200 mm and is concentrated in a wet season. Radiation is very intense during summer, and there is a very wide, maximum-minimum daily, thermic amplitude. Mean annual temperature is usually less than 10 °C, while daily temperatures lower than 0 °C occur all year around. Snowy days are fewer than five per year. Mean annual evapotranspiration ranges from 500 to 600 mm.

A detailed study on the flora and ecology of the Puna was reported by Cabrera (1957, 1968). Ruthsatz (1974) reported on the distribution, morphology, and phenology of 115 species in the study region, also including the mountain belt of the pre-Puna. Vegetation in the Puna is intrinsically related with the Patagonic flora. Much of its dominant genera (i.e., *Junellia, Fabiana, Chuquiraga, Nardophyllum, Adesmia*, and *Mulinum*) are frequent in both territories (Cabrera 1976). Predominant life forms are nanophanerophytes from 0.30 to 1.00 m height. The climax community is represented by an extensive steppe of shrubby species with extensive unvegetated, bare sites between the plants. Dominant shrubs are *Fabiana densa, Adesmia horriduscula*, and *Baccharis boliviensis*. Other frequent species include *Adesmia spinosissima, Junellia seriphioides, Baccharis incarum, Senecio viridis, Acantholippia hastulata, Ephedra breana*, and *Tetraglochin cristatus*. Some cacti such as *Opuntia soehrendsii* and *Tephrocactus atacamensis* are also common. There is a sparse herbaceous cover which can only partially cover the soil. Representative grass species are *Stipa cespitosa* and *S. leptostachya*. Other herb species, particularly associated with the wet season, are *Mutisia friesiana, M. hamata, Paraonichia cabrerae, Mitracarpus brevis, Hoffmannseggia gracilis, Conyza deserticola*,

Portulaca perennis, P. rotundifolia, and *Dichondra argentea*. While *Atriplex microphylla* dominates in the salty plains, *Baccharis caespitosa, Distichlis humilis, Festuca hypsophila*, and *Triglochin maritima* are the most common species at their borders (Cabrera 1976).

The major economic activity in the Puna is raising of goats, llamas (i.e., *Lama glama*), donkeys, and cattle. Small surface areas are cropped with maize and quinoa (i.e., *Chenopodium quinoa*). Several shrub species with good forage value (e.g., *Ephedra breana, E. rupestris, Krameria iluca, Buddleja hieronymi*, and *Acantholippia hastulat*) are exposed to intensive browsing pressure (Ruthsatz 1974). Vegetation overuse by domestic livestock, as a result of a lack of an appropriate management, has had a negative impact on the ecosystem. Extensive areas are deteriorated because of the loss of both plant cover and desirable (i.e., palatable) species. In addition, soil erosion is serious. The loss of plant diversity in the native flora affects its potential utilization for other purposes. This is because various species of the Puna are recognized for its therapeutic value, such as *Pellaea nivea, Chenopodium graveolens, Artemisia copa, Ephedra americana, E. breana, Azorella compacta*, and *Haplopappus rigidus* (Cabrera 1957; Ruthsatz 1974).

13.2.2 Chaco Occidental

The phytogeographic region known as the Gran Chaco is a low-elevation plain of about 65 Mha constituted mostly of sediments derived from the east of the Andes. It includes the north of Argentina, west of Paraguay, and southeast of Bolivia and agrees in general with the "Provincia Chaqueña" described by Cabrera and Willink (1980). Only the Argentinian portion of the Gran Chaco is included in Fig. 13.1, which corresponds to the driest area of this region. Precipitations increase from west to east with increasing distance from the Andes. The borders of this subregion were reported by Morello (1968) as the "Chaco Leñoso," which mostly overlaps with the "Parque Chaqueño Occidental" cited by Ragonese and Castiglioni (1970) or the "Distrito Chaqueño Occidental" described by Cabrera (1970).

Hot and humid summers and dry and temperate winters characterize the climate of the Chaco Occidental. The absolute maximum and minimum temperatures can reach 48 °C and −8 °C, respectively. The reduction of precipitation from the forest area to the east, with values of 800 mm per annum, to the shrubby region of the arid Monte to the west, with 320 mm annually, results in a floristic change and impoverishment of species (Cabido et al. 1993).

Predominant soils in this region, which have a dark epipedon because of the presence of organic matter, have coarse texture without a silty horizon. Soils of alluvial origin occur on river flood plains and support typical riparian plant communities. The hydric regime of the Chaco has been classified as aridic and hyperthermic or thermic for the north or south areas, respectively.

Vegetation consists of a medium to a low forest of mesophytic or xerophytic trees with a dense lower shrub cover of the Bromeliaceae and Cactaceae families. In non-utilized or slightly utilized zones, a rich prairie of good forage species is present (i.e., *Leptochloa virgata, Paspalum inaequivalve, P. unispicatum, Melica argyrea, Setaria gracilis, S. argentina*, and *Trichloris crinita*) associated with the shrubby vegetation.

The climax vegetation of the system is represented by forests of *Schinopsis lorentzii* (Quebracho colorado, Anacardiaceae) and *Aspidosperma quebrachoblanco* (Quebracho blanco, Apocynaceae). These are trees which can reach 15–20 m height. In addition, *S. lorentzii* is used for tannin extraction. *Bulnesia sarmientoi* (Zygophyllaceae) is a species, the wood of which is much appreciated because of its perfume and greeny color. Trees pertaining to the leguminous family are locally abundant. This is why *Prosopis alba* and *P. nigra* are considered among the most useful species in the Chaco Occidental. Their pods are important for feeding of human beings and livestock; in addition, they provide firewood. *Caesalpinia paraguariensis* is another tree of particular interest because of its various practical applications (Aronson and Saravia Toledo 1992). Other legumes include *Acacia aroma, A. caven, Prosopis chilensis, P. flexuosa, P. ruscifolia, P. kuntzei,* and *Geoffroea decorticans*. Frequent representatives of the dense, shrubby vegetation are *Bulnesia foliosa, B. bonariensis, Bougainvillea praecox, B. infesta, Castela coccinea, Ruprechtia apelata, R. triflora, Schinus piliferus, S. sinuatus, Mimosa detinens, Acacia furcatispina,* and *Larrea divaricata*.

These lands are exploited intensively for coal production, forestry, and grazing. The livestock industry is characterized by an excessive stocking rate on the grass and herbaceous prairie. Also, the excessive advance of the agricultural boundary (conversion to cropland to grow soy beans) in the last 20 years has forced the displacement of the livestock activities to marginal areas and to the loss and degradation of various natural ecosystems (Paruelo et al. 2005; Zarrilli 2010). In this context, forests have been the most adversely affected ecosystems in this process of resettlement of the agricultural-livestock activities. The secretary of Ambient and Sustainable Development of Argentina has reported that 2.5 Mha have been cleared in the last 10 years. Such a loss is alarming. Very little is known about the status of the remaining forests. Montenegro et al. (2007) estimated that the loss of native forests in the Chaco Park has been 805,260 ha between 1998 and 2002. This forest region has been greatly influenced by the advance of the agricultural frontier. In this region, Montenegro et al. (2005) determined that only 7 % of the forests were at a pristine stage, while the remaining 93 % showed signs of anthropic action. If this expansionist policy continues, the rich biological diversity that currently includes important forest areas will be seriously compromised in the medium term. Decades of abusive management have degraded the ecosystems. Productivity has declined, and plant species composition has changed in comparison with the pristine situation (Morello and Saravia Toledo 1959).

While the grasslands either tend to decrease or disappear as a result of an excessive utilization, cattle increases their grazing pressure on shrubs and the new seedlings of trees. Cattle are replaced by goats on heavily degraded, overstocked surface areas. Goats are able to browse on almost any kind of available vegetation. This sequence of rangeland use translates into the disappearance of the forest, which was representative of the "climax" vegetation and of the herbaceous prairie cover over wide areas. The result is a dense chaparral of spiny shrub species which have little or no value as forestry or for livestock. The shrub *Prosopis ruscifolia*, often known as "vinal" and considered an important weed, is typical in that chaparral, which constitutes a new ecosystem in the north of this territory. The colonization mechanism of the "vinal" can be associated with environmental changes imposed by human activities. This is because of previous to the introduction of domestic livestock, the natural ecosystem was always exposed to periodic wildfire and floods, factors which regulated the invasive potential of that species (Morello 1970; Morello et al. 1971). The elimination of the occurrence of these two natural disturbances determined the explosive irruption of *P. ruscifolia*. Other genera with similar characteristics include the dense shrub communities of *Acacia and Celtis* and other *Prosopis* species.

The western Chaco has been less damaged than other semiarid forest regions in the world, some of which have reached a point beyond threshold levels. However, continuation of an abusive use of the ecological system in this region would determine a progressive advance toward a greater degradation of plant community structure and soil erosion (Romero et al. 2014). Fortunately, there are some good examples of agroforestry management guidelines in this region, based on a rational land use which would allow sustainability of their productive systems.

In most of the native forests of Argentina, livestock activities of different intensity and with various levels of planning have developed. These activities go from extensive and communal to intensive models which translate into the transformation of forests to woodlands and savannas. The law number 26331 constitutes the current normative framework for any intervention in the native forests. In such sense, general guidelines and minimum contents have been established to secure both forest persistence and maintenance of both their original structure and ecological functions after anthropogenic disturbances (i.e., their resilience). Among these sustainable management strategies are the agroforestry systems: woody species (shrubs and trees) of economic or environmental importance (e.g., wood, fruits, forage) interact with herbaceous forage species (native or cultivated) grazed by animals within an integrated management system (Manghi et al. 2013).

Since 2010, those Argentinian provinces which approved by law and established the territorial ordering of their native forests have participated of the National Fund for the Enrichment and Conservation of the Native Forests. About 74 % of the forest management plans (FMP) correspond to the agroforestry type (AT), which represent 80 % in terms of surface area (Manghi et al. 2013). According to these authors, the total surface occupied by the FMP-AT was 1,000,000 ha until 2013. The application of productive techniques of sustainable anthropic intervention, such as the use of select native forage species in management plans, is one of the most important strategies to preserve the productive potential of the native forests and its biological diversity. The use of native forage species in restoration and domestication plans requires long-term programs of collection, preservation, and characterization, trying to maximize the initial plant genetic diversity. This is the base for the selection of promising genotypes to be introduced into the AT systems which is the collection of native forage germplasm with the maximum genetic variability

(Roggers et al. 2005). Despite the importance of our native forage resources, the germplasm banks which are dedicated to collect and preserve native forage species are scarce in Argentina. Only collections of the genus *Bromus* (Bank of INTA Pergamino, Argentina) and *Paspalum* (Institute of Botany of the Northeast (IBONE) of the Faculty of Agricultural Sciences, National University of the Northeast) can be highlighted.

13.2.3 Monte

The Monte desert extends from north to south in the central and western regions of the country. It is an extensive and almost uninterrupted, uniform area of shrub species which reach about 50 Mha. A typical landscape of depressions between mountains, valleys, and plains belonging to the Pampean mountains characterize its northern portion; rivers are intermittent in general, and there are extensive salty areas. The central region, however, has a landscape of undulating plains of fluvial lacustrine and eolian origin from the quaternary. The southern portion is a landscape of plains characterized by wide ecotones in the north of Patagonia. The Monte constitutes one of the most arid territories of Argentina.

It has a dry climate when it is considered in all its extension, being warmer at the north and turning gradually colder further south. Aridity at the north portion is related with its position between the Andes to the west and the Pampean mountains to the east; mountains in both extreme regions intercept the wet winds coming from the Pacific and Atlantic Oceans, respectively. Rainfall occurs mostly during summer time, ranging from 80 to 200 mm annually. Annual evapotranspiration decreases from 1,000 mm at the west to 700 mm at the east. Mean annual temperature ranges from 15 to 19 °C. The intermediate, central region has a continental climate influenced by the dry, warm winds coming from the west; summers tend to be very hot with maximum values between 40 and 45 °C; absolute minimum temperatures can be as low as −15° to −20 °C. Precipitations vary between 250 and 500 mm annually, with an annual evapotranspiration of about 800 mm. The most southerly portion has a mean temperature of 12–14 °C; its annual precipitation is from 200 to 300 mm, which is concentrated during winter and spring. Finally, its evapotranspiration is analogous to that mentioned for the north and intermediate regions (see above).

Pioneering studies on the vegetation at the Monte include those of Hauman-Merck (1913), Morello (1958), Roig (1970), Boucher et al. (1972), Ruiz Leal (1972), Cabrera (1976), and Balmaceda (1979). Vegetation at the Monte is associated with a shrubby steppe where xerophytic and microphyllus shrubs 1–3 m height are dominant. It has a relatively uniform physiognomy and floristic composition despite its large surface area and variability in soil and climate characteristics.

The characteristic plant community on extensive surface areas of the Monte is the "jarillal" (Zygophyllaceae), including *Larrea divaricata*, *L. cuneifolia*, and *L. nitida* as the most frequent species. These evergreen shrubs have various strategies which allow them to survive severe droughts in the region (Ezcurra et al. 1991). Other evergreen shrubs are *Atamisquea emarginata* and *Zuccagnia punctata*. Shrubs with deciduous foliage comprise *Plectocarpa rougesii, Prosopidastrum globosum, Prosopis alpataco, P. flexuosa, Lycium chilense,* and *Geoffroea decorticans.* The Cactaceae family is well represented with the genera *Opuntia* and *Cereus,* the abundance of which increases toward the north in the ecotone with the forest Chaco. Aphylly is frequently present in some species such as *Bulnesia retama, Neosparton aphyllum, Cassia aphylla,* and *Monttea aphylla*; photosynthesis mostly occurs in stems and branches in these species. The herbaceous stratum is partly represented by *Cottea pappophoroides, Munroa argentina, Pappophorum mucronulatum, Aristida adscensionis, Bouteloua aristidoides, B. barbata, Euphorbia serpens, Boerhavia paniculata, Pectis sessiliflora, Trichloris crinita, Tribulus terrestris,* and *Eragrostis argentina.* Some of these species are short-lived, ephemeral species which abundance is strictly dependent on the seasonal rainfalls.

Guevara et al. (1997) have conducted studies on productivity at the Monte. They reported that vegetation in this region has been degraded as a result of overgrazing in the last 100 years. This has been the result of various factors: lack of appropriate fencing; scarcity of water sources for livestock; abusive, intensive exploitation for making firewood (e.g., use of *Prosopis spp., Acacia spp.,* and *C. microphylla*); and the frequent existence of uncontrolled fires. In general, the local disappearance of the most desirable, palatable, forage perennial grasses; the increase in the abundance of undesirable (i.e., unpalatable) perennial grasses, shrubs, and herbaceous, weedy dicots; and the increase of unvegetated, bare soil surface areas have been the net result of this poor management (Cerda et al. 2012). Because of this, the existence of desirable, perennial grasses is limited to populations located under the protection of spiny shrubs (e.g., *Prosopis sp., Condalia microphylla*); even shrubs considered as desirable (i.e., often grazed by livestock) are sparse (e.g., *Brachyclados lycioides, L. chilense, Atriplex lampa, Acantholippia seriphioides,* and *Bredemeyera microphylla*).

Changes in plant species composition as a result of an abusive renewable natural resource use have determined shifts in the type of domestic herbivore which feeds on natural vegetation. Cattle and sheep have been replaced by goats. On the other hand, rotational grazing practices in well-managed experimental farms have demonstrated a sustainable stocking rate of 7.8 ha/animal unit, which represents a 54 % increase on the regional mean stocking rate (Giorgetti et al. 2006). An animal unit is defined as the annual average dry forage requirement of a 400 kg cow that goes through gestation and subsequent nursing of a calf, until the 160 kg, 6-month-old calf is weaned, including the forage consumed by the calf (Giorgetti et al. 2006).

13.2.4 Caldenal

This is a territory located as an ecotone between the more arid region of the Monte to the west, with annual precipitation equal to or lower than 200 mm, and the humid prairie to the east, suitable for various crop species. It is found in the southern portion of the phytogeographical province of the "Espinal," and it is known as the "Caldén" District (Cabrera 1976) or "Pampeano" Forest (Parodi

1964). They are approximately 10 Mha which include part of the center east of the province of La Pampa, south of San Luis, and south of the province of Buenos Aires. Various recent, scientific works on the flora, soil, and management of the livestock industry in the Caldenal, and its impact on the natural ecosystem, are summarized in these reviews (Bóo and Peláez 1991; Busso 1997; Fernández and Busso 1999; Fernández 2003; Peláez et al. 2003; Fernández et al. 2008, 2009; Bentivegna and Fernández 2010; Laborde et al. 2013; Busso et al. 2013; González-Roglich et al. 2015).

The Caldenal is a slightly undulated plain covered with sandy loessic sediments of eolian origin. There are no rivers, but salty zones and sweetwater lagoons (which are the result of rainfall on sandy dunes) are present. Its dominant soils are mollisols, also including entisols and aridisols (INTA 1980). A characteristic is the presence of a calcareous, solid horizon known as "tosca" to a depth from 0.5 to 1.0 m from the soil surface, which sometimes is exposed at the soil surface area. Climate is semi-arid, and aridity increases to the west and south. Annual precipitation, which mostly occurs in autumn and spring, varies from 350 to 550 mm. Annual evapotranspiration is 800 mm. Severe droughts occur at the end of spring and during summer. Mean annual temperature is 15.3 °C, while absolute maximum and minimum temperatures are 42.6 °C and −12.8 °C, respectively. The free-frost period is 188 days, and mean relative humidity is 60 % with minimum and maximum values in December (summer) and June (winter), respectively. Strong winds often occur during spring.

This territory is physiognomically characterized by the Caldén, a typical and almost exclusive tree species which gives the name to the area. The Caldén is an endemic, xerophytic species that taxonomically is identified as *Prosopis caldenia*. Its presence tends to decrease toward the west and south (i.e., in the Monte). This species can reach up to 10–12 m in height. In wet years, the soil underneath its canopy turns to a yellow color; this is because of the presence of pods, which are a good food for livestock. Fernández (2003) reported that counting of rings in a tree of great diameter indicated an age of approximately 720 years.

One characteristic of the region is its flora rich in native grasses of good forage value which are locally known as "flechillas." This makes the Caldenal one of the richest rangeland areas in the country, and likely in the world. This is due to a coincidence of favorable factors of soil and climate, with a precipitation that might appear high for arid territories, but that is insufficient for crop production. Perennial grasses are abundant such as *Nassella tenuis*, *Piptochaetium napostaense*, *S. speciosa*, *S. gynerioides*, *P. lanuginosa*, *Digitaria californica*, *Bromus brevis*, *Aristida subulata*, *Setaria mendocina*, *Trichloris crinita*, *Nassella clarazii*, and *Poa ligularis*. Two introduced, annual species, *Medicago minima* and *Erodium cicutarium*, have good forage value (Cano 1988; Fresnillo Fedorenko et al. 1991). However, its productivity can be limited by either a short growing season or lack of rains (Fresnillo Fedorenko et al. 1991). *Medicago minima*, a legume native of the Mediterranean, has been already reported by Hauman-Merck in 1913 as an adventitious species of the Argentinian flora. Its persistence in the Caldenal is ensured by its high potential of colonizing open, overgrazed areas and growing and developing in association with perennial grasses (Fresnillo Fedorenko et al. 1991, 1994, 1995).

The shrub stratum has a high species richness. There are caducifolius species, such as *Prosopis alpataco*, *P. flexuosa*, *Condalia microphylla*, *Geofafroea decorticans*, *Prosopidastrum globosum*, *Lycium chilense*, and *L. gilliesianum*, and perennial, evergreen species including *Larrea divaricata*, *Baccharis ulicina*, and *Capparis atamisquea*. Other species almost with no leaves are *Ephedra triandra*, *E. ochreata*, and *Senna aphylla*. Saline habitats where halophytic species like *Cyclolepis genistoides*, *Atriplex undulata*, *A. lampa*, and *Salicornia ambigua* are dominant. Other species on saline soils include the grasses *Distichlis scoparia* and *D. spicata*. Covas (1971) reported an inventory of the woody species grazed by cattle and sheep. *Lycium chilense* and *Ephedra triandra* are, possibly, the most palatable shrubs in the territory. The production of pods by *P. caldenia*, *P. alpataco* and *P. flexuosa* are a valuable food resource for livestock such as cattle and sheep. The fruits of *Geoffroea decorticans* are also eaten by livestock.

At the pristine condition of the ecological system, or in areas that are almost never utilized, the landscape was composed of large, spaced trees of Caldén, shrubs, which were always present but at a low density, and a steppe of grasses as the lower stratum.

The only planted pasture with a margin of success, in small areas of 50–200 ha, is the Pasto Llorón (*Eragrostis curvula*). This forage, perennial grass can produce forage to livestock during many years after being established. Its reduced nutritive value during the summer months, time when natural pastures are at their minimum nutritious stage, only provides levels for livestock maintenance. An extensive publication on the biology and utilization of this species was edited by Fernández et al. (1991).

The importance of fire in the Caldenal has been a subject investigated by many authors (Bóo 1980, 1990; Busso et al. 1993; Distel and Bóo 1995; Bóo et al. 1996, 1997; Peláez et al. 2003, 2012). Both grazing and fire history influence the dominance of a particular species group in a given area (Distel and Bóo 1995). On the basis of historic information, fire frequency was considered a natural phenomenon in the region associated with the vegetation dynamics. Previous to the introduction of livestock, fires were estimated to occur every 5 years. From that time onward, fire frequency has diminished as a result of the lack of fine fuel (because of overgrazing), and its prevention by the land owners. Currently, fires occur either naturally or by accident, but sometimes they are promoted. Such fires occur mostly during the warmest months in the summer and extend out of control covering thousands of hectares. The exception are those fires which are controlled and used as a management strategy. Appropriate use of fire combined with a controlled grazing can result in an adequate strategy for vegetation recovery.

The major economic activity on the Caldenal ecosystem is livestock production using cattle, some sheep, and occasionally goats. Natural vegetation is used as the only food resource, primarily grasses and secondarily some shrubs. From the anthropogenic view point, the ecological degradation of the Caldenal was initiated from the time when livestock production was intensified at the end of the eighteenth century, with the arrival of the European settlers. During the first few decades of the last century, cattle raising and extraction of wood for fuel were the major causes of changes on vegetation structure in the region. Human population has remained the same, and it is still very low (0.05–1.3 inhabitants Km^2), and rangeland properties (ranches) have usually a surface area between 1,000 and 10,000 ha (Morris and Ubici 1996). Ranchers brought their domestic animals (which did not exist previously in this region) to graze the natural system. Settlers fenced the land and stocked it at high animal density. At that time, they thought that the resource had no end. However, a few decades of abusive use of rangelands were enough to demonstrate the fragility of these systems, where the plant community was easily deteriorated and often unable to be restored. Management guidelines for conserving the natural ecological system have been almost nonexistent at the time of first utilizing the Caldenal rangelands. Even today, there are very few ranchers who apply a management regime that preserves the most desirable natural species for animal production and which can persist at a stocking rate of about five to seven cattle per hectare (Distel and Bóo 2002).

There are almost no areas in the Caldenal which have escaped the abusive use of livestock production or exploitation of the Caldén. As a result, extensive areas are progressing toward desertification. The most significant impacts are related to the breakdown of the ecological productive integrity with associated changes in community structure and soil erosion (Fernández et al. 2008). Each of the phenomena will be briefly discussed below.

A significant alteration was the replacement of palatable, perennial grasses by unpalatable species (i.e., having either low or no preference by livestock). Frequent and intensive herbivory has been the rule since the introduction of cattle. Under these conditions, the palatable perennial grasses of high animal preference gave a competitive advantage to the perennial grasses of low palatability, which often appeared as subordinate species (Flemmer et al. 2003). The potential of a species to persist in the community was compromised because of being constantly, selectively grazed. Flowering, seed set, and dispersal of their fruits (*pilous anthecia* + awn) were reduced as was their presence in the soil seed bank. Because of this, many of the best forage grasses almost disappeared from the landscape. This long-term process finally determined a floristic change in the community: *Nassella clarazii* and *Poa ligularis*, two of the grazing-tolerant perennial grasses originally dominant, but highly selected by the grazing animals, and *Nassella tenuis* and *Piptochaetium napostaense* (two species of good forage value) were replaced by a community where currently the dominant unpalatable perennial grasses known as "pajas" (e.g., *Amelichloa ambigua*, *Stipa speciosa*, and *Stipa ichu*) dominate. This is the current situation over large areas, where the vegetation, originally characterized by good forage value species, has been transformed to another with high lignin content and low nutritive value and low palatability to animals. The livestock productive potential of this new ecological system has been greatly diminished. Even though this new ecological equilibrium is dominated by perennial grasses of low or no animal preference, soil cover by these grasses ensures that the system is relatively protected: soil erosion is, prevented, and the nutrient, water, and energy cycles as well as the animal biodiversity are reasonably favored. This new type of ecological equilibrium, however, can be a catastrophe for the land owners from an economical viewpoint.

Another important change in the vegetation has been the shrubbiness. Woody species initially present in lowland areas or limited surfaces have currently advanced occupying extensive surfaces, where the proportion of low nutritive value perennial grasses is either scarce or absent (Cano 1975). This change in community structure is due to a combination of a continued, intensive grazing and an alteration of fire frequency (Distel and Bóo 1995). The reduction of the herbaceous stratum as a result of the abusive grazing and uncontrolled fires diminishes its competitive potential, favoring at the same time the establishment of shrub seedlings after timely rains (Peláez et al. 1992). With time, sites dominated by the palatable perennial grasses previously mentioned have been transformed on sites where shrub density is high and species such as *Prosopis flexuosa*, *Condalia microphylla*, *Larrea divaricata*, *Chuquiraga erinacea*, and *Geoffroea decorticans* are dominant. There are some woody species which are preferred as forage by livestock such as *Lycium chilense* and "efedra" (*Ephedra ocreata*). However, they are often grazed to the soil surface level. Aboveground, herbaceous net primary productivity below the shrubby stratum is always very low. When a situation like this has consolidated, any restoration attempt toward having productive perennial grasses implies the elimination of shrubs. This does not occur because it means to either reduce or eliminate the stocking rate. Controlling both fire and post-fire grazing might be a tool for restoration.

A more advanced step in the perception of degradation is the damage that can be caused to the soil because of a wrong use of the natural system. It frequently results in the presence of wide spaces of bare ground (i.e., without plant cover). These surfaces tend to get loose because of the rain drops, animal trampling, and a very limited superficial root system, all of which contribute to soil erosion. Total lack of soil plant cover always translate into the loss of the superficial soil horizon. This is because of the effects of wind and laminar hydric erosion by rainfall water, which runs on a bare surface area. The presence of plants with their root system exposed to sight is a proof of the erosion of the superficial soil layer. The existence of hydric erosion in an area of low precipitations might catch our attention. However, a few intense rains to variable intervals of months or years (Giorgetti et al. 1997) are enough for transport soil from unprotected soil to other areas.

Research conducted at the Center of Renewable Natural Resources of the Semiarid Region [CERZOS (CONICET)] and the Agronomy Department of the National University of the South indicate that the plant community can return from a stage of degradation to the preponderance of the most desirable and productive species (Distel and Bóo 1995). This requires an appropriate management on areas where the availability of desirable species or their disseminules is still possible. Studies focused on obtaining the basic information necessary for a better understanding of the Caldenal ecosystem functioning under the current conditions of livestock production. The lack of this kind of primary information would make more difficult the development of guidelines for a conservative management regime for ecological system. The results obtained present a contrasting situation between the degraded rangelands and the ecophysiological characteristics of the most palatable species to livestock: (1) the desirable grasses are better biologically provided to behave as dominant species; they are more tolerant to drought than those undesirable in their critical stages of regeneration (e.g., germination, seedling establishment, root growth dynamics), (2) the floristic reconstruction of the natural system through studies of soil phytoliths postulated the dominance of the desirable perennial grass species before the introduction of domestic animals in the region. If this is not the situation to date, the result has been the shrubbiness of the ecological system. Shrub invasion is because of a lack of an appropriate vegetation management by man, which includes the abusive livestock grazing and a lack of natural fires with a sufficient frequency. The undesirable grasses and woody species have been less tolerant to fire than the desirable grasses (Bóo et al. 1996, 1997). A management that controls the frequency and intensity of grazing, combined with a controlled use of fire can influence the floristic composition of rangelands, favoring the most desirable grass species.

Minimizing the adverse effect caused by yearly, continuous grazing over the same surface area in the Caldenal is one condition to improve the deteriorated condition of the renewable natural resources. This can be obtained by introducing a discontinuous feeding system (i.e., dividing the whole grazing surface area into subunits over which rotational grazing is practiced). This allows the desirable species to have vigorous growth and a better opportunity for reproducing and establishing. This management would contribute to restore the ecological system toward a stable, conservative system with greater meat production.

13.2.5 Patagonia

This territory extends between latitudes 40 and 55° S and from the Atlantic Ocean to the spurs of the "Cordillera de los Andes," a mountain system, to the west in the border with Chile. It occupies 780,000 km^2 of Monte and Patagonian steppe ecosystems (Fernández et al. 2002; Morello et al. 2012). It is identified as the Patagonic province by Cabrera (1976). Its landscape physiognomy consists of a plateau and flattened hills system. A highlighting characteristic is the presence of the named "Tehuelches Patagónicos," or hills of glaciofluvial origin, of eroded stones and rocks. These systems form a desertified pavement because of the strong winds which do not allow the deposition of new materials on its surface. Soriano (1983) published a synthetic report on the geology and soils, climate, and vegetation of the Patagonia; Schulze et al. (1996) reported the existence of water at greater depths. One of the most notable characteristics of the climate in Patagonia is the predominance of strong winds from the west during the whole year. These winds lose their humidity on the Andes at the side of the Pacific Ocean. Because of this, rains only reach from 100 to 200 annual mm in the Patagonian rangelands, except at the border of the oceanic coast where precipitations can reach less than 300 mm annually. Soil moisture is directly influenced by the annual recharge via precipitations (Coronato and Bertiller 1996). Rains occur mostly during autumn and winter, and the highest soil moisture levels occur at the end of winter after snow melts. Mean annual temperature range from 6 to 14 °C, and the absolute maximum and minimum values alternate from 30 to 40 °C and −15 to −20 °C, respectively. Annual evaporation fluctuates between 550 and 750 mm. All these values tend to decrease from the northeast to the southeast. Frost occurrence is at least 60 days annually, where 5–20 days correspond to snowy days mostly in the west and south of the territory. Drought in Patagonia results from the combination of low precipitation, high temperatures during the summer months, and strong winds all year around which determine high evaporation values.

Recuperation of degraded areas in arid and semiarid zones, like those in Patagonia, might require centuries of human intervention. This situation is the result of cumulative processes of desertification as a result of various disturbance histories (Bainbridge 2007; Dalmasso 2010; Granados-Sánchez et al. 2013). Despite the fact that Argentina is part of the agreement of the United Nations of fight against desertification, and in particular Patagonia has been included in various intervention programs, the problem of desertification remains unsolved (Morales 2012; Abraham et al. 2014). In some provinces of Patagonia (e.g., Neuquén), desertification problems reach 92 % of the territory (Del Valle et al. 1998). In Neuquén, not only long-term overgrazing but also petroleum exploitation (since 1916: Turic and Ferrari 2000) are the major reasons for desertification (Ciano 2010). Urgent measures are needed to restore severely damaged areas. In Patagonia, several rehabilitation projects have been developed using native species coming from nurseries, evaluating different amendments, and facilitation among species (Ciano et al. 1998, 2000; Dalmasso et al. 2002; Pérez et al. 2010; González et al. 2013; Farinaccio et al. 2013). Even though seedling survival has been high enough to justify the use of those projects, their cost is very high (Valladares et al. 2011; Abella et al. 2012). Under this scenario, other techniques for restoration are currently being studied such as topsoil transplants (0–10 cm soil surface layer) or seed-enriched top soil transplants (Walker and Del Moral 2003; Banerjee et al. 2006; De Falco et al. 2010).

The variation among the plant communities is considerable in Patagonia. Soriano (1983) has published a review of the Patagonia's vegetation and reports the existence of six floristic districts. Other authors have described the vegetation for the central portion of Patagonia (Bertiller et al. 1977, 1981a, b) or the ecotone vegetation between the north and the shrubby formations characteristics of the Monte (Soriano 1949, 1950; Ragonese and Piccinini 1969; Ruiz Leal 1972). Physiognomy of the Patagonic vegetation is a low shrubby steppe which alternates with undesirable perennial grasses. Floristic composition of the system is dominated by the grass families, and the genus *Stipa* is dominant, including *S. humilis*, *S. speciosa*, *S. ibari*, *S. neaei*, *S. psylantha*, and *S. subplumosa*. Other significant components of the grass family are *Poa ligularis*, *P. lanuginosa*, *Festuca argentina*, *F. pallescens*, *F. gracillima*, and *Bromus setifolius*. Extensive surfaces are characterized in its physiognomy by the presence of shrubs of rounded aspect and of less than 1 m height; the most frequent species include *Chuquiraga avellanedae*, *Colliguaja integerrima*, *Mulinum spinosum*, *Senecio filagiroides*, *Verbena tridens*, *Pseudoabutilon bicolor*, *Berberis heterophylla*, *B. cuneata*, *Baccharis darwinii*, *Anarthrophyllum rigidum*, *Nassauvia glomerulosa*, *Lycium chilense*, and *Trevoa patagonica*. Total soil cover varies notably from 15 % to 60 %, depending on the site and particularly of the land use history (Larregy et al. 2014; Wiesmeier 2014). Lower areas frequently present halophytic vegetation characterized by communities of *Frankenia patagonica*, *Atriplex lampa*, and *A. sagittifolia*. *Schinus polygamus* can reach a plant height of up to 3 m being one of the greatest shrubs of the austral Patagonia; it has been almost extinguished on extensive surfaces because of its properties as fuel.

Factors which affect plant species composition and diversity in Patagonia are associated with the presence of various topographies (i.e., plains versus sierras and mountains), which have a direct effect on the heterogeneity of the abiotic environmental factors (Campanella and Bertiller 2013). The composition of the plant communities would appear to depend on vegetation functional groups because of the distribution of shrubs, and grasses would respond to environmental gradients (Jobbágy et al. 1996). Slope areas oriented to the east would be protected from the strong winds coming from the west, which are stronger during summer, giving a more favorable thermic balance comparatively to the high, plain territories (Coronato and Bertiller 1996).

Aguiar et al. (1996) emphasized that plant species in Patagonia can be grouped in three growth forms: shrubs, grasses, and herbaceous species. Shrubs include deciduous and evergreen species with a height greater than 0.5 m and in general without the presence of a well-developed main stem. Grasses have a caespitose growth form with green foliage during the whole summer and a C_3 photosynthetic pathway. Herbaceous species include in general either annual or perennial or deciduous or evergreen dicots. Shrubs and grasses differ in their strategies for obtaining soil water, one of the greatest limitations for plant growth in Patagonia. Shrubs obtain water mostly from the lower soil horizons, and initially depend on the deep recharge of water during winter. Soil water uptake of grasses is mostly from the superficial soil horizons (Soriano and Sala 1983; Soriano et al. 1987; Fernández and Paruelo 1988).

The formation of vegetation patches by shrubs and grasses is an important factor in the control of plant community dynamics. This subject has been studied by various authors (Soriano and Sala 1986; Sala et al. 1989; Aguiar et al. 1992; Aguiar and Sala 1994, 1997; Mazzarino et al. 1996; Busso and Bonvissuto 2009a, b; Busso et al. 2012; Bonvissuto and Busso 2013). Patch formation rules ecosystem functioning regulating the spatial models of soil organic matter and texture and the dynamics of water and nutrients (Concostrina-Zubiri et al. 2014; Tabeni et al. 2014). As a result of different vegetation structures corresponding to shrubs and grasses, particular microsites are created which allow the germination and subsequent growth of new seedlings (Busso et al. 2012). In addition, the possibility of desirable grass seedling establishment and growth in the unvegetated patches of vegetation is strongly controlled by summer droughts, wind, and herbivory (Defossé et al. 1997; Busso et al. 2012).

Rangelands of Patagonia have been utilized for sheep production since the end of the 1800s (Morrison 1917). Animal stocking rate varies from 400 to 1,500 heads per 2,500 ha. Animals are often maintained on rangelands all year round, and they might be managed separately on summer and winter pastures. The pressure from herbivory has resulted in an advanced desertification process over large areas. Loss of grass cover associated with the advance of shrubbiness and the increase of unvegetated surface areas has resulted in a reduction of plant biomass of prairies with good forage value for livestock. Persistence of the original vegetation structure is minimal, or it has disappeared. The reduction of plant cover because of grazing has determined an increase of nutrient concentration in the either remaining or recent vegetation patches, leaving wide spaces of bare soil with limiting nutritive resources, and where plant recolonization is problematic (Mazzarino et al. 1996; Busso et al. 2012). Even more, the existence of strong winds from the west is a major additional factor associated with a generalized phenomenon of soil erosion (Rostagno and del Valle 1988).

Currently, the need of using conservationist management guidelines that secure the productive ecological sustainability of the Patagonic natural system is in the conscience of government entities and their research stations, and others like the National Agricultural Technology Institute (i.e., INTA), the Research Centers of the National Council for Scientific and Technological Research of Argentina (i.e., CONICET), and the Universities.

A detailed map of the utilization of arid and semiarid territories (i.e., rangelands) of Argentina shows that only inaccessible areas are free from anthropogenic perturbation. Such territories are associated with a high biodiversity and provide a wide range of ecological services (i.e., carbon sequestration, water quality, plant products, erosion control, livestock industry systems, intimately linked with the productivity of such surfaces). Its environmental preservation and the maintenance of its ecological structure are essential to secure an unlimited provision of benefits through time and the persistence of biodiversity. However, this situation of privilege is currently under threat, particularly because of its abusive utilization and a deficient management of the natural system.

Because of its nature, arid and semiarid ecosystems have in common, that is, a fragile ecological stability, with a marked susceptibility to desertification. Economically, they also have a limited threshold of productivity. From the beginning of its utilization, there has been an uncoupling in the balance between the demand and the supply of services that this type of ecosystems can provide. Interests in these potential services have often made difficult to diagnose an adequate combination of conservationist management strategies. This is an essential objective to secure their sustainability.

The process of environmental deterioration of the ecological systems on each of the studied regions is the result of a succession of degradation steps, which mostly were either not perceived or ignored at a regional level. Either ignorance or indifference to the progressive subtle changes on the quality of the system is an attitude which has prevailed on the uncultivated, arid, and semiarid territories of the country. It is possible that ranchers and technicians have ignored the slow process of deterioration of the desirable flora and the subsequent loss of the most valuable forage species, thus affecting the livestock production potential of the land. Under these circumstances, it is understandable that they think that the current state of the ecological system is what it should be, rather than the result of something arising from more than a hundred years of mismanagement. The identification of the progressive stages which have determined the desertification of the system is a major subject in the agenda of scientific and technological investigations. This is essential because beyond a limit of accumulated losses a threshold will be reached which did not allow the reestablishment of the utilization potential of the renewable natural resources.

From the anthropogenic point of view, the modification of the ecological systems of the arid and semiarid territories of Argentina was initiated at the time when livestock production and other extractive uses like lumber and firewood were intensified from the end of the fifteenth century with the arrival of the Europeans. Human populations have remained in such territories at a very low density. Colonizers introduced a foreign herbivore to the natural system, with new food habits, in territories where large herbivores were scarce from the Pleistocene until their occupation of the virgin land. As a result, the model of the natural, original equilibrium started to change toward a new structure when the first domestic ruminant was integrated into the community, and it grazed selectively on the available vegetation composed by a mixture of grasses, forbs, and woody species. The preferential diet of an herbivore for any given vegetation type is influenced by its palatability, which is a complex phenomenon integrated by odor, taste, texture, and the post-digestive effects of nutrients and secondary compounds. Initially, it would appear to early settlers that the resource was infinite, inexhaustible. At that time, there was no concern about what we call today "environmental impact," currently an everyday subject in newspapers.

The term "environmental impact" refers to what happen to a natural ecosystem (plants, animals, soil, water, etc.) which is exposed to biotic (e.g., grazing) and/or abiotic (e.g., drought, waterlogging) disturbances. The main characteristics (intensity, frequency, duration, etc.) exceed the limit of those to which the ecological system should be exposed normally in response to the natural evolution process. Following the historical perspective for any disturbance, there would be a time when a new, different history begins, initiated by the arrival of the new anthropogenic impacts, in particular of livestock. The previous stage, which will start to transform with disappearing species, corresponds to the pristine situation of equilibrium, possibly in a climax situation under the natural, biotic, and abiotic, conditions of its environment. This era closed with the introduction of the first sheep and cows. This second history is very recent when viewed at an appropriate time scale. It begins since approximately something more than a hundred years ago. In ecological, evolutionary terms, this is an insignificant period, is nothing, like meaning "yesterday." In a short time, domestic animals and the exploitation of other resources, upon specific to a region, dominated the scene. In the utilization of land, conservation management guidelines of the natural system oriented to preserve the most desirable plant species were either very scarce or nonexistent. Questions as simple as what type of plants should predominate? What are those that I do not want to predominate or invade my land? What can I do to give priority to the first question and not the second? What is going on with the soil? These questions were not made for a long time, despite the rancher knowing each plant by its aspect, shape, and degree of animal preference; he is an ecologist by his own nature. The researcher in turn, already too late, when the system had already begun the way toward its impoverishment or evident trajectory toward its disappearance as productively useful, began to ask questions and suggest responses that can then be tested.

All people involved in agriculture and cattle production systems, such as professionals, technicians, and ranchers, know about the difficulty of studying the principles which rule the stability and sustainability of a natural ecological system. This is because of the multiplicity of environmental, biological, and management factors which influence them. This makes it difficult to assemble an integral model of "ecological functioning" of the system, or at least some of its more significant constituents (e.g., primary productivity; soil; nutrient, water, and energy cycles; biological cycles; biodiversity, etc.).

To date, the devastation of territories, and subsequent natural utilization of the arid and semiarid regions of Argentina, does not reach levels of environmental degradation like in other parts of the world where desertification is an irreversible process. The ecological degradation of arid and semiarid lands of Argentina is associated with overgrazing, lumber and firewood extraction, modification of the frequency of occurrence of fires, and, in general, a poor management of the ecosystem. The domestic livestock will feed not only of grasses but also of shrubby, woody species, and tree seedlings. Even more, animals will try to reach any accessible space where they can obtain some kind of food. The impoverishment of rangelands has led to a change in the kind of livestock in various areas. Regions originally able to sustain cattle have changed to the production of goats. These are able to survive with the rough remnants of any possible food. It is a complex task when trying to explain the negative impact of the domestic livestock on vegetation or the ecosystem as a whole. An appropriate management of cattle requires the use of specific methodologies under the supervision of specialists. Pristine plant associations which persist to date are very few.

The successive stages of vegetation type replacement can be exemplified with what happened in the Monte:

1. Uncontrolled overgrazing has been the usual vegetation management type.
2. As a result, the pristine, climax-like situation of original equilibrium has disappeared.
3. Because of the scarcity or disappearance of the herbaceous cover, cattle grazed on shrubs and tree seedlings originated from seeds.
4. In the most affected areas, cattle was later replaced by goats, which are able to feed on any kind of vegetation type.
5. This sequence ends up with the presence of a new type of ecosystem where most of the original vegetation was replaced by a low-productive, woody, spiny chaparral.
6. If this scenario is continued, severe grazing under drought conditions translates into soil erosion because of the wind and water.

Plant species or livestock use might well end up being different for each of the mentioned regions. However, the final result might be similar because of an irrational use of the natural system; like this, sheep replaces cattle in most part of Patagonia, and the final result is similar: bare, unvegetated soil, loss of productive potential, and desertification.

The lack of public policies directed to achieve an adequate sheep management in Patagonia increased the migratory process from the rural zones to the urban centers. In general, rural people who emigrate to cities do so as unskilled labor. Other problems associated with this process are the fragmentation of the family and the growth of peripheral settlements. Alternatively, it has been postulated the need of recognizing the social function of the ranchers in Patagonia as guards of a renewable natural resource that needs to be preserved for the future generations and also as economic agents in a rural space where there exists few replacement alternatives. However, there is no objective information to budget for the cost of applying a policy of subsidies oriented toward the figure of environmental guard and contrast it with the social, economic, and environmental costs of doing nothing. Future research should focus on determining the environmental, economic, and social trajectory of the productive systems of small ranchers of Patagonia identifying the costs associated with the current productive model. Public policies to that sector should also be evaluated as well as their impact to it. It is necessary to design a proposal based on the recognition of the function of environmental guard and sustainable rancher. The economic supports external to the system that should be done for recognizing these functions, maintaining and improving the natural resources that sustain this activity, should be objectively evaluated. All these needs imply the work of an interdisciplinary team: specialists in sheep production, sociology, natural resources, satellite images, economy, and rural development.

The loss of plant species richness and biodiversity directly affect other possible uses of the rich flora of deserts. There exists numerous examples which constitute a valuable material for either human use such as the production of resins and waxes, products used as foods, pharmaceutics, and medicine, or reestablishment in open spaces of land. Parodi (1934) has reported many woody species of arid and semiarid territories of the country as ornamentals of parks, gardens, and streets of any city. Even more, the disappearance of wild species represents a definite loss of genetic material; this is a poorly considered subject from the scientific and research viewpoint. In relation to this subject, Hunziker et al. (1986) reported their concern on the exhaustive exploitation of *Prosopis* spp. and the urgent need of the genetic conservation of the most valuable and promising species. They emphasize that the country should take a more active role to safeguard the valuable germplasm that these species represent. This is because Argentina is the center of greatest diversity of these species in America, Africa, and Asia Occidental. Such species can provide food to the human being and animals, gums and tannins, and materials for the construction and industry of furniture; they are equally useful as alcohol producers, and in the melliferous flora, and contribute additionally to the control of soil erosion.

One of the most important concerns about the process of desertification is related with the damage done to the soil. There are various studies on the effects of various disturbances and their combinations on vegetation deterioration (Busso and Richards 1989; Busso and Richards 1995; Busso et al. 2004), but very few regarding their effects on soil erosive processes. Most of these regions present a dry soil surface horizon, and a density of vegetation which is not adequate for an intensive or abusive economic use. A soil surface covered by vegetation represents the continuation of important functions in the community: it improves soil structure, facilitates infiltration and water retention, protects the soil surface horizon from erosion, improves sun energy capture, produces organic debris which eventually are incorporated into the soil as organic matter, and increases the soil seed bank (Busso and Bonvissuto 2009a, b). Accentuated deterioration of plant biomass prevents soil protection and exposes it to an erosive process as a result. The ways soils degrade are associated with wind erosion and transport of it by superficial, runoff water, the impact of rain drops, and the continuous, frequently intense animal trampling. This situation affects the sites for germination and establishment of new seedlings (Busso et al. 2012). Under the conditions of a bare, unvegetated soil and an either incipient or advanced process of soil erosion, one of the most significant, limiting factors of the dry regions is affected: the water cycle. This is because water will not get throughout one of the most critical destinies, such is the system of primary productivity; in turn, biogeochemical and energy cycles will be altered. The final result is the loss of fertility and physical, structural degradation of the surface soil layer. Recuperation of the system may take many years, when a change in vegetation structure has occurred. The ecosystem can change to an irreversible stage of desertification when the soil surface horizon has disappeared, with the loss of its abiotic (e.g., fertility, organic matter, structure, etc.) and biotic (e.g., microorganisms). A clear evidence of the superficial soil loss is the abundance of plant growing on pedestals of 10–20 cm height, which shows that the superficial soil horizon has disappeared from its location. The phenomenon of degradation can be considered as one of the most serious environmental problems of the rangeland territories of Argentina (Busso and Bonvissuto 2009a; Busso et al. 2012).

A side effect of an inadequate livestock industry management regime and exploitation of woody species has been their impact on the native fauna (from small animals such as insects to large mammals) and the soil microflora and microfauna (Ambrosino et al. 2013a, b, c). Various species of these organisms tend to disappear (become locally extinct), some animals have economical possibilities. With a poor flora, conditions for the existence of many animals and soil microorganisms have been modified. This is due to the loss of their niches and food, and fewer locations for their reproduction. Several decades ago, the guanacos (*Lama guanicoe*) and avestruces (*Struthio camelus*) were common representatives of the native fauna; contrarily, today, they are scarce or absent actors of the natural landscape, or they are shut away in inaccessible areas, national parks, and occasionally private ranches. A common fact has been that frequently some representatives of the native fauna have been considered as predatory species or competitive forages of the domestic livestock, with detrimental economic consequences for the livestock industry. This is that zorros (*Pseudophilautus zorro*), pumas (*Puma concolor*), guanacos, and avestruces, for example, have been the target of uncertain measures of "pest control." Indiscriminate control methods such as poisons or traps have damaged other nontarget species such as birds, rodents, armadillos, etc. Some of these species which were efficient in supporting the equilibrium or control of others have diminished their effectiveness. Because of this, other species like weasels have become pests. Fortunately, the human perception of the importance of preserving the wondrously rich fauna of these regions is rapidly increasing; the leadership of research projects toward its preservation is substantial.

Argentinians have been always amazed by the farming systems of the Pampa prairie, with deep, fertile soils, a temperate climate, and abundant rainfalls which allow a great diversity of crops of high economical returns. As a result, we have been somehow careless with respect to the rest of the country, in which arid and semiarid territories represent approximately 70 % (Fernández and Busso 1999). Since overgrazing and droughts characterize these territories, with subsequent production reductions in general, we should have already learnt that there are limits on the actions we take on the renewable natural resources and that exceeding those limits will cause them to degrade.

Nevertheless, it is necessary to distinguish on which we consider as "not degraded" or "degraded" systems. The situation of "degradation" is assigned, for example, to the replacement of desirable, palatable grasses by undesirable, unpalatable ones to livestock, or to the local disappearance of grasses from the system, as it has been emphasized in the Caldenal. However, the meaning of degradation can have different interpretations depending on whom or how it is looked by. The term "good health" of an ecological, natural system (rangeland in this case) is frequently used in relation to its characteristics and properties which allow that it is properly used by the human being. When it does not provide us with a service or any good that we desire (e.g., meat production, water quality, wood, etc.), we consider that it is "degraded." This is correct, since the world human population, in constant growth, depends of that stage of "good health" for its survival. However, all that has occurred is that the ecosystem has changed to a new one, which neither does not like nor it is useful to us (different plants, less soil, etc.); it has not disappeared by any point of view. Under this viewpoint, the concept of "degradation" because of, for example, the replacement of desirable by undesirable grasses is strictly anthropogenic. Nature does not recognize plants as "desirable" or "undesirable," or "good" or "bad."

The desertification process is currently one of the most serious environmental problems of the country. This process of degradation must end immediately. Otherwise, within one or two generations, large territories will be transformed in unproductive areas. People in Argentina understand that this serious process of land degradation is on the way to affect about 70 % of its surface area because of ignorance, indifference, or lack of vision. A process of recuperation might mean many years of corrective works since the time of its introduction.

One of the most significant, current achievements is having taken public consciousness of the problem, including ranchers, livestock associations, and national and province agencies. This condition is strictly necessary in a country where the natural land is mostly private property, with the exception of areas limited as national or provincial parks. There are many entities of higher learning which have included these subjects in their pre-graduate and postgraduate programs. This has also been done by entities like the (1) National Agricultural Technology Institute (INTA) and (2) Research Centers of the National Council for Scientific and

Technological Research of Argentina (CONICET) such as the Center of Natural Renewable Resources of the Semiarid Region (CERZOS), Argentinian Institute for Research in Arid Zones (IADIZA), Patagonian National Center (CENPAT), Austral Center of Scientific Research (CADIC), Research Institute of Agricultural Physiology and Ecology (IFEVA), and many others. We also need to include provincial agencies which have special programs for research and conservation-oriented management. The national government subscribed the Convention of the United Nations for fighting against desertification in 1994. The secretary of Ambient and Sustainable Development of Argentina is involved in such subject through the direction of soil conservation and fight against desertification.

To date, the devastation of territories, and subsequent natural utilization of the arid and semiarid regions of Argentina, does not reach levels of environmental degradation like in other parts of the world where desertification is an irreversible process. Even with various degrees of degradation, the ecological system and the species still exist over extensive surface areas. We can make an improvement on land use which will allow a sustainable production of a magnitude depending on the degree of degradation of the ecological system. Of course this will demand changes in the utilization of its renewable natural resources.

The question which rises immediately is as follows: have we realized (1) the challenges that we will need to confront in the near future to solve the social and political consequences of continuing the current aggressive modes and exploitation of renewable natural resources and (2) specific actions that we have to undertake for eliminating the great distance that exists between the current "parasitic" exploitation of nature and instead embrace land stewardship and adopt of an ecological ethic which allows a sustainable economic management to secure a viable world for our descendants?

13.3 Rangeland and Climate Change

Climatic change can adversely impact the rangeland ecosystems in Argentina and their economic potential and ecological sustainability (Gitay et al. 2001). The direct effects of climate and climate change on rangeland ecosystem processes have relatively short response times and are somewhat easier to predict than the indirect effects. The direct effects of climate change on rangeland ecosystems from alterations in precipitation regimes, temperature, and atmospheric concentrations of CO_2 include (1) changes in decomposition rates; (2) changes in aboveground net primary production (ANPP); (3) shifts in C3/C4 species mix in grasslands/rangelands; (4) changes in fluxes of greenhouse gases such as carbon dioxide (CO_2), methane (CH_4), ammonia (NH_3), nitric oxide (NO), and nitrous oxide (N_2O); (5) changes in evapotranspiration and runoff; (6) changes in forage quality; (7) changes in grassland storage of soil carbon; (8) changes in soil quality; (9) alteration in plant cover; (10) changes to ecosystem structure; (11) changes in water storage; (12) alteration in biodiversity; and (13) changes in habitat suitability (Ojima et al. 1993; Baker and Vigliozzi 1998; Breymeyer et al. 1996; IPCC 1996; Letcher 2016). Barros et al.

(2014) contributed an excellent review on trends, projections, impacts, and adaptation to climate change in Argentina. Even more, they also report on climate projections for the short term (2016–2035) and the long term (2081–2100) relative to the reference period (1986–2005) in Argentina.

The indirect effects are less well understood. Because of lags in the response time of the system and the complexity of the feedbacks that are involved, the impacts of climate change take longer to identify. Also human activities, including burning, cropping, and management of grazing animals, have the potential to modify indirect effects by either accelerating or slowing these processes. Thus predicting the direction and magnitude of the impact is more difficult. Examples of the indirect effects of climate change on rangeland ecosystems include (1) changes in the structure of vegetation communities; (2) changes in vegetation cover that could alter surface albedo, humidity, and ground level wind patterns; (3) changes in the C:N ratios or lignification of vegetation that alter litter quality and affect soil nutrients; (4) alterations in soil characteristics; (5) changes in biogeochemical cycles that lead to the invasion of exotic species of plants; (6) possible influence on frequency of wildfires in some areas; and (7) changes in species diversity (Baker and Viglizzo 1998; Fischlin et al. 2007).

During the twentieth century, global atmospheric temperatures rose at an average rate of 0.06 °C per decade, although for the period from 1956 to 2005, the warming rate was nearly 0.13 °C per decade (Sherman et al. 2012). Rainfall trends and patterns are more difficult to determine, and significant regional differences are evident. Hoffman and Vogel (2008) reported that temperature, rainfall, and atmospheric CO_2 concentration interact with grazing and land cover change to influence rangeland quality and composition. Increased temperature, for example, not only increases drought stress in plants but also increases lignification of their tissues, which affects both its digestibility as well as its rate of decomposition. Increased temperature and lower rainfall also increases vegetation flammability, resulting in a shift in species composition as a result of an increased fire frequency.

Rapid expansion of invasive species can be attributed to ongoing perturbations resulting from elevated CO_2 and N-deposition, past and present land uses, and the direct and indirect effects of climate change.

The most significant invasion in the region is the expansion and dominance of exotic annual grasses and in mid- to low-elevation woodlands and shrublands (Fresnillo Fedorenko et al. 1991; Fernández and Busso 1999). Numerous other noxious species are rapidly spreading through the region's shrublands and woodlands. Many of these invaders are capable of displacing native plant communities and altering watershed functions (Busso et al. 2013).

Climate change will impact rangelands in Argentina in a number of ways. Higher concentrations of CO_2 will generally increase the productivity of rangelands, alter vegetation composition, particularly the balance between woody plants and the herbaceous layer, and decrease forage quality (Wilson 1969; Wigley et al. 2010; Donohue et al. 2013). Increasing temperature will affect the length of the growing season, plant productivity, and animal production through both reduced cold stress in temperate climates and increased heat loads in summer. Livestock diseases are also likely to be affected by climate change through changes in pathogen behavior, host vulnerability, distribution of insect and other vectors, and epidemiology of diseases.

Rainfall projections are highly uncertain, though there is a general trend for midlatitudes to become drier while higher subtropical regions are more likely to experience increasing rainfall (IPCC 2007). A combined impact of increasing temperature and even slightly decreasing precipitation in the early growing season may increase dryness, reducing plant productivity in mid-latitudes. Indeed, even though projected declines in mean rainfall may seem small compared with interannual variability, they have the potential to greatly reduce forage production.

References

S.R. Abella, D.J. Craig, S.D. Smith, A.C. Newton, Identifying native vegetation for reducing exotic species during the restoration of desert ecosystems. Restor. Ecol. **20**(6), 781–787 (2012)
E.M. Abraham, H. Matallo, J.R. De Lima, Ciencia y desertificación en América Latina. Zonas Áridas **15**(2), 349–360 (2014)
M.R. Aguiar, O.E. Sala, Competition, facilitation, seed distribution and the origin of patches in a Patagonian steppe. Oikos **70**, 26–34 (1994)
M.R. Aguiar, O.E. Sala, Seed distribution constrains the dynamics of the Patagonian Steppe. Ecology **78**, 93–100 (1997)
M.R. Aguiar, A. Soriano, O.E. Sala, Competition and facilitation in the recruitment of seedlings in the Patagonian steppe. Funct. Ecol. **6**, 66–70 (1992)
M.R. Aguiar, J.M. Paruelo, O.E. Sala, W.K. Lauenroth, Ecosystem responses to changes in plant functional type composition: an example from the Patagonian steppe. J. Veg. Sci. **7**, 381–390 (1996)
M. Ambrosino, M. Montechia, C.A. Busso, D. Cardillo, Y. Torres, L. Ithurrart, O. Montenegro, D. Ponce, H. Giorgetti, Rodrìguez, Actividad microbiana del suelo debajo del follaje de Nasella tenuis, *Poa ligularis* y *Amelichloa ambigua* expuestas a defoliaciòn. II Taller regional sobre rehabilitación y restauración en la diagonal árida de la Argentina. Mendoza, 23, 24 y 25 de Octubre. Disponible en CD (2013a)
M. Ambrosino, M. Montechia, C.A. Busso, D. Cardillo, Y. Torres, O. Montenegro, L. Ithurrart, D. Ponce, H. Giorgetti, G. Rodrìguez, Análisis de las comunidades microbianas de suelos de gramíneas perennes expuestas a defoliación. IX Reunión Nacional Cientìfico-Técnica de Biología de Suelos y I Congreso Nacional de Biología Molecular de Suelos. Santiago del Estero 4, 5 y 6 de Setiembre, disponible en CD (2013b)
M. Ambrosino, M. Cabello, M. Velázquez, C.A. Busso, D. Cardillo, Y. Torres, D. Ponce, L. Ithurrart, O. Montenegro, H. Giorgetti and G. Rodríguez, Especies de hongos formadores de micorrizas arbusculares en gramíneas perennes expuestas a defoliación. XXXIV Jornadas Argentinas de Botánica, 2 al 6 de Setiembre 2013, La Plata, Buenos Aires, 2013c, p. 205
J. Aronson, C. Saravia Toledo, *Caesalpinia paraguariensis* (Fabaceae): forage tree for all seasons. Econ. Bot. **46**, 121–132 (1992)
D. Bainbridge, *A Guide for Desert and Dryland Restoration: A New Hope for Arid Lands. Society for Ecological Restoration International* (Island Press, Washington, DC, 2007), p. 416
B. Baker, E.F. Viglizzo, in *Handbook on Methods for Climate Change Impact Assessment and Adaptation Strategies*, ed by J.F. Feenstra, I. Burton, J.B. Smith, S.J. Richard. Rangeland and livestock (UNEP/IVM Handbook, Ecosystems Research International, United Nations Environment Programme, Institute for Environmental Studies, Amsterdam, 1998), p. 34
N.A. Balmaceda, Vegetación, in *Estudio de clima, geomorfología, suelos, vegetación y erosión*. Bol. Ministerio de Agricultura, Ganadería y Minería, Rio Negro, Argentina (1979), pp. 74–93

M.J. Banerjee, V.J. Gerhart, E.P. Glenn, Native plant regeneration on abandoned desert farmland: effects of irrigation, soil preparation, and amendments on seedling establishment. Restor. Ecol. **14**(3), 339–348 (2006)

H.A. Barbosa, T.V. Lakshmi Kumar, L.R.M. Silva, Recent trends in vegetation dynamics in the South America and their relationship to rainfall. Nat. Hazards **77**, 883–899 (2015)

V.R. Barros, J.A. Boninsegna, I.A. Camilloni, M. Chidiak, G.O. Magrin, M. Rusticucci, Climate change in Argentina: trends, projections, impacts and adaptation. WIREs Clim. Chang. (2014). doi:10.1002/wcc.316

D.J. Bentivegna, O.A. Fernández, Malezas invasoras: estrategias para una determinación y manejo apropiados. EdiUNS **13**, 5–7 (2010)

M.B. Bertiller, A.M. Beeskow, M.P. Irisarri, Caracteres fisonómicos y florísticos de las unidades de vegetación del Chubut. IDIA **35**, 247–296 (1977)

M.B. Bertiller, A.M. Beeskow, M.P. Irisarri, Caracteres fisonómicos y florísticos de la vegetación del Chubut. 1. Sierra de San Bernardo, llanura y valle aluvial del río Senguer, Pampa de Maria Santísima, Valle Hermoso y Pampa del Castillo. Contribución No. 40. Centro Nac. Patagónico, Puerto Madryn, Argentina (1981a)

M.B. Bertiller, A.M. Beeskow, M.P. Irisarri, Caracteres fisonómicos y florísticos de la vegetación del Chubut. 2. La Peninsula Valdés y el Istmo Ameghino. Contribución No. 41. Centro Nac. Patagónico, Puerto Madryn, Argentina (1981b)

G.L. Bonvissuto, C.A. Busso, En: Restauracón Ecológica en la Diagonal Arida de la Argentina, ed by R.D. Pérez, A.D. Rovere, M.A. Rodríguez Araujo. Establecimiento de plántulas en microambientes del Monte Austral Neuquino (UNCo, Buenos Aires, 2013), p. 518

R.M. Bóo, El fuego en los pastizales. Ecología Argent. **4**, 13–17 (1980)

R.M. Bóo, Algunos aspectos a considerar en el empleo del fuego. Rev. Fac. Agron. Univ. Nac La Pampa **5**, 63–80 (1990)

R.M. Bóo, D.V. Peláez, Ordenamiento y clasificación de la vegetación en un area del sur del Distrito del Caldén. Bol. Soc. Argent. Bot. **27**, 135–141 (1991)

R.M. Bóo, D.V. Peláez, S.C. Bunting, O.R. Elia, M.D. Mayor, Effect of fire on grasses in central semi-arid Argentina. J. Arid Environ. **32**, 259–269 (1996)

R.M. Bóo, D.V. Peláez, S.C. Bunting, M.D. Mayor, O.R. Elía, Effect of fire on woody species in central semi-arid Argentina. J. Arid Environ. **35**, 87–94 (1997)

T.W. Boucher, J.P. Hjerting, K. Rahn, Botanical studies in the Atuel valley area, Mendoza Province, Argentina. Part 111. Dan. Bot. Ark. **22**, 195–358 (1972)

A. I. Breymeyer, D. O. Hall, J. M. Melillo, G. I. Ågren (eds.), *Global Change: Effects on Coniferous Forests and Grasslands. Scientific Committee on Problems of the Environment 56* (Wiley, Chichester, 1996)

C.A. Busso, Towards an increased and sustainable production in semi-arid rangelands of central Argentina: two decades of research. J. Arid Environ. **36**, 197–210 (1997)

C.A. Busso, J.H. Richards, Drought and clipping effects on tiller demography andgrowth of two tussock grasses in Utah. J. of Arid Environ. **29**(2), 239–251 (1995)

C.A. Busso, J.H. Richards, Fenología y crecimiento en dos especies de gramíneas: Efectos del estrés hídrico. Rev. Fac. Agron. **10**, 127–138 (1989)

C.A. Busso, G.L. Bonvissuto, Soil seed bank in and between vegetation patches in arid Patagonia, Argentina. Environ. Exp. Bot. **67**, 188–195 (2009a)

C.A. Busso, G.L. Bonvissuto, Structure of vegetation patches in northwestern Patagonia, Argentina. Biodivers. Conserv. **18**, 3017–3041 (2009b)

C.A. Busso, R.M. Bóo, D.V. Peláez, Fire effects on bud viability and growth of *Stipa tenuis* in semiarid Argentina. Ann. Bot. **71**, 377–381 (1993)

C.A. Busso, H.D. Giorgetti, O.A. Montenegro, G.D. Rodríguez, Perennial grass species richness and diversity on Argentine rangelands recovering from disturbance. PHYTON, Int. J. Exp. Bot. **73**, 9–27 (2004)

C.A. Busso, G.L. Bonvissuto, Y.A. Torres, Germination and seedling establishment of grasses and shrubs in arid Patagonia, Argentina. Land Degrad. Dev. **23**, 116–129 (2012)

C.A. Busso, D. Bentivegna, O.A. FernÃ¡ndez, A review of invasive plants in rangelands of Argentina. Interciencia **38**, 95–103 (2013)

M. Cabido, C. González, A. Acosta, S. Díaz, Vegetation changes along a precipitation gradient in Central Argentina. Vegetatio **109**, 5–14 (1993)

A.L. Cabrera, La vegetación de la Puna Argentina. Rev. Invest. Agric. **11**, 317–412 (1957)

A.L. Cabrera, Ecología vegetal de la Puna. Colloq. Geographicum. **9**, 91–116 (1968)

A.L. Cabrera, La vegetación del Paraguay en el cuadro fitogeográfico de América del Sur. Bol. Soc. Argent. Bot. **11**, 121–131 (1970)

A.L. Cabrera. in *Enciclopedia Argentina de Agricultura y Jardinería*, ed by E.F. Ferreira Sobral. Regiones fitogeográficas Argentinas (ACME, Buenos Aires, 1976), pp. 1–85

A.L. Cabrera, A. Willink, *Biogeografía de América Latina*, Ser. Biol. Monogr. 13, 2nd edn. (Organización de Estados Americanos, Washington, DC, 1980)

M.V. Campanella, M.B. Bertiller, Leaf growth dynamics in four plant species of the Patagonian Monte, Argentina. J. Plant Res. **126**, 497–503 (2013)

E. Cano, Pastizales en la región central de la provincia de La Pampa. INTA-IDIA **331–333**, 1–15 (1975)

E. Cano, *Pastizales naturales de La Pampa. Descripción de las especies más importantes* (Talleres gráficos Mariano Mas, Buenos Aires, 1988)

N.V. Cerda, M. Tadey, A.G. Farji-Brener, M.C. Navarro, Effects of leaf-cutting ant refuse on native plant performance under two levels of grazing intensity in the Monte Desert of Argentina. Appl. Veg. Sci. **15**, 479–487 (2012). doi:10.1111/j.1654-109X.2012.01188.x

N. Ciano, Revegetación de ambientes degradados en la Patagonia Sur: experiencias en la Cuenca del Golfo San Jorge. II Jornadas de Herbario y Biodiversidad. UNPSJB. Comodoro Rivadavia (2010)

N. Ciano, V. Nakamatsu, J. Luque, M.E. Amari, O. Mackeprang, C. Lisoni, Establecimiento de especies vegetales en suelos disturbados por la actividad petrolera. En 3ras Jornadas de Preservación de agua, aire y suelo en la industria del petróleo y del gas, 14 al 17 de septiembre de 1998, Comodoro Rivadavia-Chubut (1998), pp. 435–445

N. Ciano, V. Nakamatsu, J. Luque, M. E. Amari, M. Owen, C. Lisoni, Revegetación de áreas disturbadas por la actividad petrolera en la Patagonia extrandina (Argentina). Convenio YPF S.A.- INTA EEA- INTA Chubut (Trelew). Actas de la 11° Conferencia de la Organización Internacional de la Conservación del Suelo. Buenos Aires. 22 al 27 de Octubre del 2000 (2000)

L. Concostrina-Zubiri, E. Huber-Sannwald, I. Martínez, J.L. Flores, J.A. Reyes-Agüero, A. Escudero, J. Belnap, Biological soil crusts across disturbance – recovery scenarios: effect of grazing regime on community dynamics. Ecol. Appl. **24**, 1863–1877 (2014.) http://dx.doi.org/10.1890/13-1416.1

F.R. Coronato, M.B. Bertiller, Precipitation and landscape related effects on soil moisture in semi-arid rangelands of Patagonia. J. Arid Environ. **34**, 1–9 (1996)

G. Covas, in *IV Reunión nacional para el estudio de las regiones áridas y semiáridas*. Árboles y arbustos forrajeros nativos de la Provincia de La Pampa, Resúmenes Comunicaciones Universidad Nacional La Pampa (Santa Rosa, Argentina, 1971), pp. 24–25

D. Dalmasso, Revegetación de áreas degradadas con especies nativas. Bol. Soc. Argent. Bot. **45**(1–2), 149–171 (2010)

D. Dalmasso, E. Martínez Carretero, O. Console, Revegetación de áreas degradadas. Bol. Extensión científica. IADIZA. Mendoza **46**, 2 (2002)

L.A. De Falco, T.C. Esque, S.J. Scoles-Sciulla, J. Rodgers, Desert wildfire and severe drought diminish survivorship of the long-lived Joshua tree (*Yucca brevifolia;* Agavaceae). Am. J. Bot.. 2010 **97**, 243–250 (2010)

G.E. Defossé, R. Robberecht, M.B. Bertiller, Seedling dynamics of *Festuca* spp. in a grassland of Patagonia, Argentina, as affected by competition, microsites, and grazing. J. Range Manag. **50**, 73–79 (1997)

H.F. Del Valle, N.O. Elissalde, D.A. Gagliardini, J. Milovich, Status of desertification in the Patagonian region: Assessment and mapping from satellite imagery. Arid Land Res. Manag. **12**(2), 95–121 (1998)

R.A. Distel, R.M. Bóo, in *Proceedings of the Fifth International Rangeland Congress*. Vegetation states and transitions in temperate semiarid rangelands of Argentina (Salt Lake City, 1995), pp. 117–118

R.A. Distel, R.M. Bóo, Unidad demostrativa de cría bovina en el sur del Caldenal. Resúmenes del Congreso Argentino de Producción Animal (Buenos Aires, 2002), p. 334–335

R.J. Donohue, M.L. Roderick, T.R. McVicar, G.D. Farquhar, Impact of CO_2 fertilization on maximum foliage cover across the globe's warm, arid environments. Geophys. Res. Lett. **40**, 3031–3035 (2013)

P.H. Etchevehere, *Mapa de Suelos. República Argentina* (Instituto Nacional de Tecnología Agropecuaria, Buenos Aires, 1971)

E. Ezcurra, C. Montana, S. Arizaga, Architecture, light interception, and distribution of *Larrea* species in the monte desert, Argentina. Ecology **72**, 23–34 (1991)

F. Farinaccio, A.E. Rovere, D.R. Pérez, *Rehabilitación con Pappostipa speciosa (Poaceae), en canteras abandonadas por actividad petrolera en zonas áridas de Neuquén, Argentina* (Rehabilitación en la Diagonal Árida de la Argentina, Autonomous City of Buenos Aires, 2013), pp. 343–354

O.A. Fernández, Los pastizales naturales del Caldenal. Anales de la Academia Nacional de Agronomia y Veterinaria **57**, 68–92 (2003)

O.A. Fernández, C. Busso, in *Proceedings of the Rangeland Desertification International Workshop*. Arid and semiarid rangelands: Two Thirds of Argentina. Case studies of rangeland desertification (Agriculture Research Institute, Reykjavik, 1999), pp. 41–60

R.J. Fernández, J.M. Paruelo, Root systems of two Patagonian shrubs: a quantitative description using a geometrical method. J. Range Manag. **41**, 220–223 (1988)

O. A. Fernández, R. M. Brevedan, A. O. Gargano (eds.), *El Pasto Llorón: Su Biología y Manejo* (Universidad Nacional del Sur, Bahía Blanca, 1991), p. 394

R.J. Fernández, R.A. Golluscio, A.J. Bisigato, A. Soriano, Gap colonization in the Patagonian semidesert: seed bank and diaspore morphology. Ecography **25**, 336–344 (2002)

O.A. Fernández, D.V. Peláez, M.D. Mayor, Aún estamos a tiempo. AgroUNS **10**, 14–19 (2008)

O.A. Fernández, M.E. Gil, R.A. Distel, The challenge of rangeland degradation in a temperate semiarid region of Argentina: the Caldenal. Land Degrad. Dev. **20**, 431–440 (2009)

O.A. Fernández, R.E. Brevedan, H. Laborde, M.G. Klich, C.A. Busso, Colección Lecturas de Cátedra, in *Universidad Nacional de Río Negro. Los territorios áridos y semiáridos de la Argentina*, (Universidad Nacional de Río Negro, Río Negro, 2016), pp. 1–21

Fischlin, A., G.F. Midgley, J.T. Price, R. Leemans, B. Gopal, C. Turley, M.D.A. Rounsevell, O.P. Dube, J. Tarazona, A.A. Velichko, in *Climate Change 2007: Impacts, Adaptation and Vulnerability. Contribution of Working Group II to the Fourth Assessment Report of the Intergovernmental Panel on Climate Change.* ed by M.L. Parry, O.F. Canziani, J.P. Palutikof, P.J. van der Linden, C.E. Hanson. Ecosystems, their properties, goods, and services (Cambridge University Press, Cambridge, 2007), pp. 211–272

A.C. Flemmer, C.A. Busso, O.A. Fernández, T. Montani, Effects of defoliation under varying soil water regimes on aboveground biomass of perennial grasses. Arid. Soil. Res. Manag. **17**, 139–152 (2003)

D.E. Fresnillo Fedorenko, O.A. Fernández, C.A. Busso, in *Proceedings IVth International Rangeland Congress*. Forage production of the annual legume *Medicago minima* (1991), pp. 372–374

D.E. Fresnillo Fedorenko, O.A. Fernández, C.A. Busso, Factores en la germinación de dos especies anuales forrajeras de la región semiárida Argentina. Turrialba **44**, 95–99 (1994)

D.E. Fresnillo Fedorenko, O.A. Fernández, C.A. Busso, The effect of water stress on top and root growth in *Medicago minima*. J. Arid Environ. **29**, 47–54 (1995)

H. Giorgetti, O.A. Montenegro, G. Rodríguez, C.A. Busso, T. Montani, M.A. Burgos, A.C. Flemmer, M.B. Toribio, S.S. Horvitz, The comparative influence of past management and rainfall on range herbaceous standing crop in east-central Argentina: 14 years of observations. J. Arid Environ. **36**, 623–637 (1997)

H.D. Giorgetti, C.A. Busso, O.A. Montenegro, G.D. Rodríguez, N.M. Kugler, Cattle raising in central, semiarid rangelands of Argentina. Rangelands **28**, 32–36 (2006)

H. Gitay, S. Brown, W. Easterling B. Jallow, in *Climate Change 2001: Impacts, Adaptation and Vulnerability*, ed by J.J. McCarthy, O.F. Canziani, N.A. Leary, D.J. Dokken, K. S. White. Ecosystems and their goods and services (Cambridge University Press, Cambridge, UK, 2001), pp. 237–242

M.F. González, D.R. Pérez, A.E. Rovere, *Reintroducción de especies nativas de dos grupos funcionales en zonas degradadas del área natural Protegida Auca Mahuida (Neuquén, Argentina)* (Rehabilitación en la Diagonal Árida de la Argentina, Autonomous City of Buenos Aires, 2013), pp. 355–370

M. González-Roglich, J.J. Swenson, D. Villarreal, E.G. Jobbágy, R.B. Jackson, Woody plant-cover dynamics in argentine savannas from the 1880s to 2000s: the interplay of encroachment and agriculture conversion at varying scales. Ecosystems **18**, 481–492 (2015)

D. Granados-Sánchez, M.Á. Hernández-García, A. Vázquez-Alarcón, P. Ruíz-Puga. Los procesos de desertificación y las regiones áridas. *Revista Chapingo Serie Ciencias Forestales y del Ambiente***XIX**(1), 45–66 2013

J.C. Guevara, J.B. Cavagnaro, O.R. Estevez, H.N. Le Houérou, C.R. Stasi, Productivity, management and development problems in the arid rangelands of the central Mendoza plains (Argentina). J. Arid Environ. **35**, 575–600 (1997)

L. Hauman-Merck, Etude Phytogéegraphique de la Region du Rio Negro Inférieur. Anal. Mus. Nac. Hist. Nat. Buenos Aires **24**, 289–444 (1913)

T. Hoffman, C. Vogel, Climate change impacts on African rangelands. Rangelands **30**, 12–17 (2008)

U. Holtz. La Convención de las Naciones Unidas de Lucha contra la Desertificación (CNULD) y su dimensión política (2003) http://www.unccd.int/Lists/SiteDocumentLibrary/Parliament/2003/PDUNCCD(spa).pdf. Access date: 31 Octubre 2015

J.H. Hunziker, B.O. Saidman, C.A. Naranjo, R.A. Palacios, L. Poggio, A.D. Burghardt, Hybridization and genetic variation of argentine species of *Prosopis*. For. Ecol. Manag. **16**, 301–315 (1986)

INTA, Provincia de La Pampa, Universidad Nacional de La Pampa, *Inventario integrado de los recursos naturales de la Provincia de La Pampa* (INTA, Buenos Aires, 1980), p. 493

IPCC, *Technologies, Policies and Measures for Mitigating Climate Change*, ed by R.T. Watson, M.C. Zinyowera, R.H. Moss. Technical Paper I (Geneva, 1996), p. 84

IPCC, Cambio climático 2007: Informe de síntesis. Contribución de los Grupos de trabajo I, II y III al Cuarto Informe de evaluación del Grupo Intergubernamental de Expertos sobre el Cambio Climático [Equipo de redacción principal: Pachauri, R.K. y Reisinger, A. (directores de la publicación)] (IPCC, Ginebra, Suiza, 2007), 104 págs

E.G. Jobbágy, J.M. Paruelo, J.C.R. León, Vegetation heterogeneity and diversity in flat and mountain landscapes of Patagonia (Argentina). J. Veg. Sci. **7**, 599–608 (1996)

H. Laborde, O.A. Fernández, M.N. Fioretti, S.S. Baioni, R.E. Brevedan, *Steppe Ecosystems: Biological Diversity, Management and Restoration*, ed by M. B. Morales and J. T. Díaz. Chapter 15: Sustainable development in the Caldenal (Argentina) (NOVA Science Publishers, USA, 2013)

C. Larregy, A.L. Carrera, M.B. Bertiller, Effects of long-term grazing disturbance on the belowground storage of organic carbon in the Patagonian Monte, Argentina. J. Environ. Manag. **134**, 47–55 (2014)

T.M. Letcher (ed.), *Climate Change: Observed Impacts on Planet Earth* (Elsevier, Oxford, 2016)

E. Manghi, M. Taurian, N. Griffiths, S. García Alvarez, G. Sañudo, J. Bóono. Actas 4to Congreso Forestal Argentino y Latinoamericano (2013)

M.J. Mazzarino, M.B. Bertiller, C.L. Sain, F. Laos, F.R. Coronato, Spatial patterns of nitrogen availability, mineralization, and immobilization in Northern Patagonia, Argentina. Arid. Soil. Res. Rehabil. **10**, 295–309 (1996)

C. Montenegro, M. Strada, J. Bono, I. Gasparri, E. Manghi, E. Parmuchi, M. Brouver, UMSEF Unidad de Manejo del Sistema de Evaluación Forestal, Dirección Bosques, Secretaría de Ambiente y Desarrollo Sustentable (2005)

C. Montenegro, M. Strada, G. Parmuchi, E. Mangui, UMSEF Unidad de Manejo del Sistema de Evaluación Forestal, Dirección Bosques, Secretaría de Ambiente y Desarrollo Sustentable (2007)

C. Morales, *Evaluación económica de la desertificación y degradación de tierras en LAC. Project Note* (Proyecto Conjunto CEPAL/Mecanismo Mundial, Santiago, 2012)

J. Morello, La provincia fitogeográfica del monte. Opera Lilloana **2**, 3–155 (1958)

J. Morello, Las grandes unidades de vegetación y ambiente del Chaco Argentino. Primera parte. Objetivos y metodología. Ser. Fitogeogr. **8**, 1–125 (1968)

J. Morello, Modelo de relaciones entre pastizales y leñosas colonizadoras en el Chaco Argentino. Inf. Invest. Agric. **276**, 31–52 (1970)

J.H. Morello, C. Saravia Toledo, El Bosque chaqueño. I. Paisaje primitivo, paisaje natural y paisaje cultural en el oriente de Salta. Rev. Agron. Noroeste Argent. **3**, 5–81 (1959)

J.H. Morello, N.E. Crudelli, M. Saraceno, Los vinalares de Formosa. Ser. Fitogeogr. **11**, 1–159 (1971)

J. Morello, S. Matteucci, A. Rodriguez, M. Silva, *Ecorregiones y complejos ecosistémicos argentinos* (Orientación Gráfica Editora, Buenos Aires, 2012), p. 752

A. Morris, S. Ubici, Range management and production on the fringe: the Caldenal, Argentina. J. Rural. Stud. **12**, 413–425 (1996)

J.J. Morrison, La ganadería en la región de las mesetas australes del territorio de Santa Cruz. Thesis Facultad de Agronomía y Veterinaria, Buenos Aires, Argentina, 1917

D.S. Ojima, W.J. Parton, D.S. Schimel, J.M.O. Scurlock, T.G.F. Kittel, Modeling the effects of climatic and CO_2 changes on grassland storage of soil C. Water Air Soil Pollut. **70**, 643–657 (1993)

L.R. Parodi, Las plantas indígenas no alimenticias. Rev. Argent. Agron. **1**, 165–212 (1934)

L.R. Parodi, in *Enciclopedia Argentina de Agricultura y Jardinería*, vol. 2, ed by ACME. Las regiones fitogeográficas argentinas (Buenos Aires, 1964), pp. 1–14

J.M. Paruelo, J.P. Guerschman, S.R. Verón, Ciencia Hoy **15**, 14–23 (2005)

D.V. Peláez, R.M. Bóo, O.R. Elía, Emergence and seedling survival of calden in the semiarid region of Argentina. J. Range Manag. **45**, 564–568 (1992)

D.V. Peláez, R.M. Bóo, O.R. Elía, M.D. Mayor, Effect of fire on growth of three perennial grasses from central semi-arid Argentina. J. Arid Environ. **55**, 657–673 (2003)

D.V. Peláez, R.J. Andrioli, O.R. Elía, F.R. Blázquez, Distintas frecuencias de fuego controlados en el sur del Caldenal: efectos sobre la vegetación leñosa. EdiUNS **18**, 13–17 (2012)

D.R. Pérez, A.E. Rovere, F.M. Farinaccio, *Rehabilitación en el desierto: Ensayos con plantas nativas en Aguada Pichana, Neuquén, Patagonia* (Vázquez Mazzini Editores, Buenos Aires, 2010)

R.E. Quiroga, R.J. Fernández, R.A. Golluscio, L.J. Blanco, Differential water-use strategies and drought resistance in *Trichloris crinita* plants from contrasting aridity origins. Plant Ecol. **214**, 1027–1035 (2013)

A.E. Ragonese, J.C. Castiglioni, La vegetación del parque chaqueño. Bol. Soc. Argent. Bot. **11**, 133–160 (1970)

A.E. Ragonese, B.C. Piccinini, Límite entre el Monte y el semi-desierto Patagónico en las provincias de Rio Negro y Neuquén. Bol. Soc. Arg. Bot. **11**, 299–302 (1969)

M.E. Roggers, A.D. Craiga, R.E. Munns, T.D. Colmer, P.G.H. Nichols, C.V. Malcolm, E.G. Barrett-Lennard, A.J. Brown, W.S. Semple, P.M. Evans, K. Cowley, S.J. Hughes, R. Snowball, S.J. Bennett, G.C. Sweeney, B.S. Dear, M.A. Ewing, Aust. J. Exp. Agric. **45**, 301–329 (2005)

F.A. Roig, Flora y vegetación de la reserva forestal de Ñacuñán. Deserta **1**, 25–232 (1970)

C.M. Romero, L. Noe, A. Abril, E.A. Rampoldi, Resilience of humification process to evaluate soil recovery in a semiarid agroecosystem of Central Argentina. J. Soil Sci. **4**(3), 211–224 (2014)

C.M. Rostagno, H.F. del Valle, Mound associated with shrubs in aridic soils of northeastern Patagonia: characteristics and probable genesis. Catena **15**, 347–359 (1988)

A. Ruiz Leal, Flora popular mendocina. Deserta **3**, 9–296 (1972)

B. Ruthsatz, Los arbustos de las estepas andinas del noroeste argentino y su uso actual. Bol. Soc. Argent. Bot. **16**, 27–45 (1974)

O.E. Sala, R.A. Golluscio, W.K. Lauenroth, A. Soriano, Resource partitioning between shrubs and grasses in the Patagonian steppe. Oecologia **81**, 501–505 (1989)

E.D. Schulze, H.A. Mooney, O.E. Sala, E. Jobbágy, N. Buchmann, G. Bauer, J. Canadell, R.B. Jackson, J. Loreti, M. Oesterheld, J.R. Ehleringer, Rooting depth, water availability, and vegetation cover along an aridity gradient in Patagonia. Oecologia **108**, 503–511 (1996)

D.J. Sherman, B. Li, S.M. Quiring, E.J. Farrell, Geography of Climate Change, ed. by R. Aspinall. Benchmarking the war against global warming (Routledge, New York, 2012), p. 301

A. Soriano, El limite entre las provincias botánicas Patagónica y Central en el territorio del Chubut. Lilloa 20, 193–202 (1949)

A. Soriano, La vegetación del Chubut. Rev. Argent. Agron. **17**, 30–66 (1950)

A. Soriano, in *Temperate Deserts and Semi-Deserts*, ed by N.E. West. Desert and semi-deserts of Patagonia (Elsevier Scientific Publishing Company, Amsterdam, 1983), pp. 440–453

A. Soriano, O.E. Sala, Ecological strategies in a Patagonian arid steppe. Vegetatio **56**, 9–15 (1983)

A. Soriano, O.E. Sala, Emergence and survival of *Bromus setifolius* seedlings in different microsites of a Patagonian arid steppe. Isr. J. Bot. **35**, 91–100 (1986)

A. Soriano, R.A. Golluscio, E.H. Satorre, Spatial heterogeneity of the root system of grasses in the Patagonian arid steppe. Bull. Torrey. Bot. Club **114**, 103–108 (1987)

S. Tabeni, I.A. Garibotti, C. Pissolito, J.N. Aranibar, Grazing effects on biological soil crusts and their interaction with shrubs and grasses in an arid rangeland. J. Veg. Sci. **25**, 1417–1425 (2014). doi:10.1111/jvs.12204

Y.A. Torres, C.A. Busso, O.A. Montenegro, L. Ithurrart, H.D. Giorgetti, G.D. Rodríguez, D. Bentivegna, R.E. Brevedan, O.A. Fernández, M.M. Mujica, S.S. Baioni, L. Entío, M. Fioretti, G. Tucat, Plant growth and survival of five introduced and two native/naturalized perennial grass genotypes exposed to two defoliation managements in arid Argentina. Grass Forage Sci. **69**, 580–595 (2014). doi:10.1111/gfs.12071

L. Torres, E.M. Abraham, C. Rubio, C. Barbero-Sierra, M. Ruiz-Pérez, Desertification research in Argentina. Land Degrad. Dev. **26**, 433–440 (2015). doi:10.1002/ldr.2392

M.A. Turic, J.C. Ferrari, *La exploración de petróleo y gas en la Argentina: el aporte de YPF* (S.A., Buenos Aires, 2000), pp. 1–178

W. Ulrich, S. Soliveres, F.T. Maestre, N.J. Gotelli, J.L. Quero, M. Delgado-Baquerizo, M.A. Bowker, D.J. Eldridge, V. Ochoa, B. Gozalo, E. Valencia, M. Berdugo, C. Escolar, M. García-Gómez, A. Escudero, A. Prina, G. Alfonso, T. Arredondo, D. Bran, O. Cabrera, A.P. Cea, M. Chaieb, J. Contreras, M. Derak, C.I. Espinosa, A. Florentino, J. Gaitán, V.G. Muro, W. Ghiloufi, S. Gómez-González, J.R. Gutiérrez, R.M. Hernández, E. Huber-Sannwald, M. Jankju, R.L. Mau, F.M. Hughes, M. Miriti, J. Monerris, J. Muchane, K. Naseri, E. Pucheta, D.A. Ramírez-Collantes, E. Raveh, R.L. Romão, C. Torres-Díaz, J. Val, J.P. Veiga, D. Wang, X. Yuan, E. Zaady, Climate and soil attributes determine plant species turnover in global drylands. J. Biogeogr. **41**, 2307–2319 (2014)

F. Valladares, L. Balaguer, I. Mola, A. Escudero, V. Alfaya, *Restauración ecológica de áreas afectadas por infraestructuras de transporte. Bases científicas para soluciones técnicas* (Fundación Biodiversidad, Madrid, 2011). Fundación Biodiversidad 2011

L.R. Walker, R. Del Moral, *Primary Succession and Ecosystem Rehabilitation* (Cambridge University Press, New York, 2003)

M. Wiesmeier, in *Environmental indicators*, ed by R.H. Armon y Osmo Hänninen. Chapter 14: Environmental indicators of dryland (Springer, Dordrecht 2015), p. 239–250

B.J. Wigley, W.J. Bond, M.T. Hoffman, Thicket expansion in a South African savanna under divergent land use: local vs. global drivers? Glob. Chang. Biol. **16**, 964–976 (2010)

A.D. Wilson, A review of browse in the nutrition of grazing animals. J. Range Man. **22**, 23–28 (1969)

A. Zarrilli, ¿Una agriculturización insostenible? La provincia del Chaco, Argentina (1980–2008). Historia Agraria **51**, 143–176 (2010)

The Impact of Climate Variability on Land Use and Livelihoods in Australia's Rangelands

14

David J. Eldridge and Genevieve Beecham

14.1 Introduction

Variable, unpredictable climates and human-induced climate change have emerged as the greatest threats to human populations over the last few centuries. The extent to which altered climates are predominantly human induced or part of natural temporal variability is still hotly debated. Nevertheless, current predictions for global climate change are that many areas will become hotter, drier, and more variable, while others, such as northern Europe, will experience more temperate climates. For large areas of Earth, changing climate is likely to be most strongly felt in the rangelands: arid or semiarid environments that occupy about 40% of the global land mass and support 38% of the human population (Reynolds et al. 2007). Rangelands are often referred to as 'drylands' and are used predominantly for grazing or dryland cropping. Rangelands are often highly degraded and home to many cultures, generally from low socio-economic groups. These groups are often marginalised and are most likely to be affected by increasing aridity and a more variable temperatures and rainfall.

About 70% of the Australian landmass can be considered rangelands (Fig. 14.1). The bulk of Australia's rangelands have low and variable rainfall, occur on generally infertile soils, and support arid and semiarid pastoral enterprises. Other land uses in the rangelands include mining, indigenous land (traditional ownership), conservation, and military use. Climate change predictions for Australia suggest an increase in warming of up to 5 °C, with a continent-wide general pattern of lower rainfall, which is strongest in the south-west of the continent, but with fewer effects

---

D.J. Eldridge (✉)
Centre for Ecosystem Science, School of Biological, Earth and Environmental Sciences, University of New South Wales, Sydney, NSW 2052, Australia
e-mail: d.eldridge@unsw.edu.au

G. Beecham
ANU College of Law, Australian National University, Acton, ACT 2601, Australia

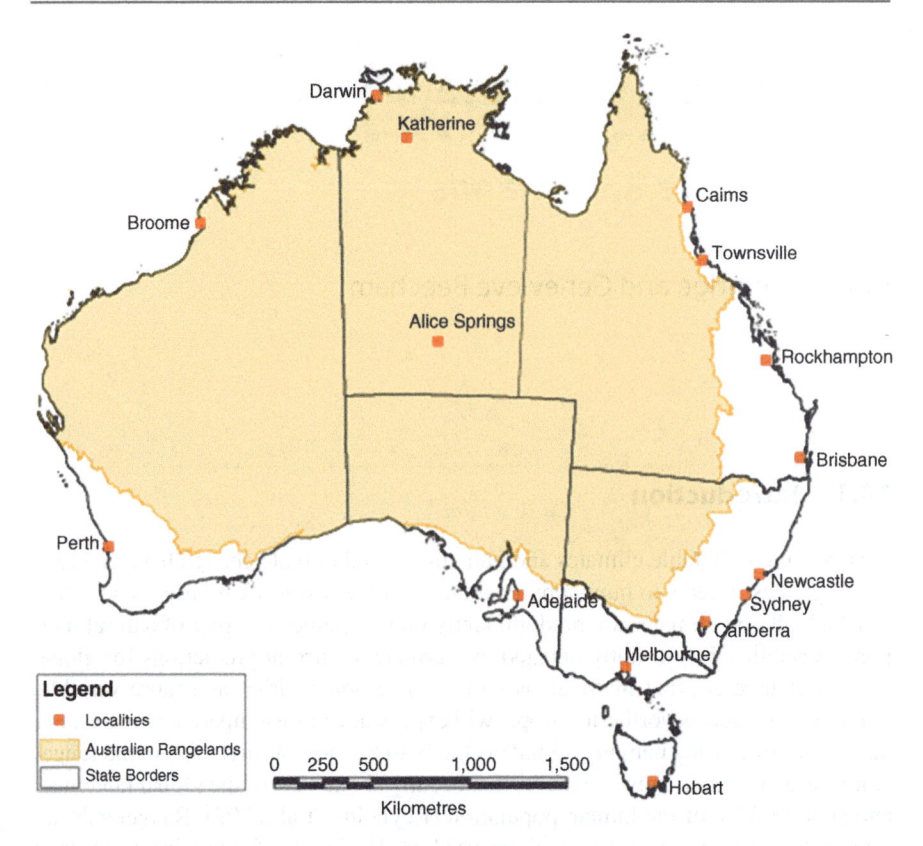

Fig. 14.1 Distribution of Australia's rangelands (*Source*: Commonwealth of Australia, Australian Government Department of the Environment and Heritage 2005)

in northern tropical areas (Stokes et al. 2008a). Projected climate shifts over Australia's rangelands will present substantial challenges for primary producers, land managers, and governments over the next 50 years, with predicted declines in pastoral production and forage quality, increased livestock stress, more frequent droughts and high-intensity rainfall events, and increased land degradation and invasion by exotic pests and diseases. Falling incomes derived from dryland cropping, increasing reliance on government assistance, and lower economic sustainability will create substantial social problems for rural communities.

In this chapter we describe the main climatic drivers that affect continent-wide climatic patterns over Australia and how projected changes in climate and increases in climatic variability are likely to influence the physical and biological environment, its land use, and its peoples across Australia's rangelands. We use a case study of ecosystem decline in the Riverina Region of south-eastern Australia to describe the effects of changing climate on the physical and biological environments. We examine how increases in rainfall variability and heat stress will impact upon Australian pastoral enterprises, and how climate uncertainty is likely to affect the social and economic livelihoods of rural communities. We describe a range of farm-level adaptations to changing climate that could assist producers to manage their enterprise in a climatically variable environment. We conclude with a discussion of the technical, social, and policy barriers to climate change adaptation.

14.2 Climate Variability in Australia: The Current Situation

Overall, Australia's climate is highly variable. Average annual rainfall is less than 500 mm (semiarid) over about three quarters of the land area, and 40% of the land area receives less than 250 mm (arid) (Stafford Smith and Morton 1990). However, before we consider how climatic variability affect ecosystems, their biota, and land uses, we must first consider current climatic variability, particularly variability in rainfall and temperature, which are the main drivers of ecosystem processes in rangelands (Noy-Meir 1973).

Climate variability on the Australian continent is strongly related to a number of key phenomena or drivers (CSIRO and BOM 2015). The El Niño-Southern Oscillations (ENSO) phenomenon is perhaps the most important and is associated with major continental air shifts between the Asian and east Pacific regions. Negative values of the ENSO index (El Niño) result in dry periods and drought, while positive values (La Niña) result in periods of high rainfall. The ENSO phenomenon describes how moist tropical air is collected across the equator. Other major drivers are: (1) The Indian Ocean Dipole, which brings moisture from the Indian Ocean; (2) The Southern Annular Mode, which influences the strength and frequency of cold fronts; (3) The Subtropical High-Pressure Ridge; (4) East coast lows, which are intense low-pressure systems on the eastern seaboard; (5) Monsoonal systems, which are associated with a change in wind direction in northern Australia from the east to the north-west. These systems draw in moist air from the ocean, resulting in periods of high rainfall; (6) Trade winds, which are part of general atmospheric circulation. Moisture-rich trade winds blow across most of the southern hemisphere, collecting moisture, and depositing it in tropical and subtropical areas on the east coast. In inland Australia, however, they are associated with drier conditions; and (7) Tropical depressions, which are associated with high levels of rainfall in tropical northern Australia.

The main driver of variability across Australia's rangelands is the relationship between sea surface temperature and sea level pressure induced by the ENSO phenomenon, the inter-decadal Pacific oscillation (Power et al. 1999), and the Indian Ocean Dipole (Ummenhofer et al. 2009). Temperature predictions for Australian rangelands suggest that the northern subtropical rangelands have a 90% chance of temperatures, 27 °C or lower over the next 50 years, and a 90% chance of temperature lower than 33 °C (CSIRO and BOM 2015). Comparing these with the 10% decile for temperature indicates a 10% chance of temperatures lower than or equal to 24 °C in the Australian rangelands or 30 °C for the northern subtropical region. Rainfall variability data are more clearly defined. Australian rangelands are characterised as moderate to very high rainfall variability. Variability, measured as the

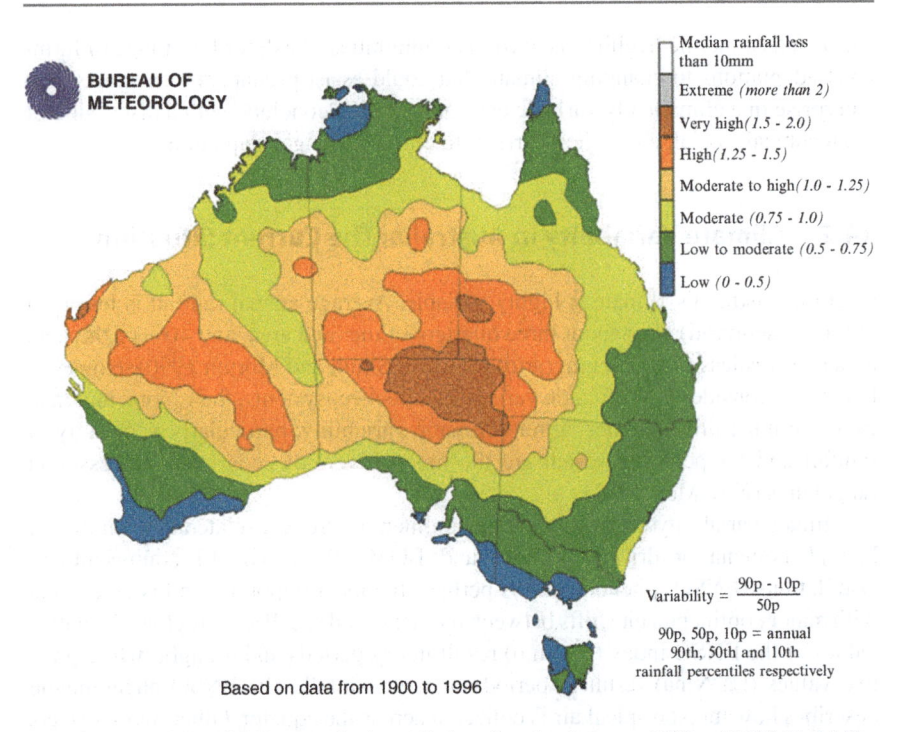

Fig. 14.2 Rainfall variability map for Australia based on data between 1900 and 1996. Variability is calculated as the difference between the 10th and 90th percentile divided by the 50th percentile (*Source*: CSIRO and BOM 2015)

difference between the 90% and 10% rainfall decile divided by 50% decile, ranges from 0.75 to 2.0 for dryland Australia, with the highest variability being in the centre of Australia (approximately 60%; Goudie and Wilkinson 1977; Fig. 14.2). This variability is clear if one looks at the size distribution of rainfall amounts in arid areas. For example, median rainfall in rangelands is generally less than the mean because of high-intensity, local frequency events that are typical of dryland areas (Stafford Smith and Morton 1990; Dunkerley 2010). Long periods of aridity are broken by low-frequency heavy rainfalls (Morton et al. 2011). Similarly, rangelands are characterised by a frequency distribution of rainfall skewed towards very small events. So, for example, examination of a rainfall event at Cobar in western New South Wales over a 2-day period revealed that 70% of rain occurred in falls of <2.8 mm (Jackson 1958). This is a typical pattern of rainfall distribution in rangelands (Whitford 2002; Dunkerley 2010).

14.3 Climate Projections for Australia's Rangelands

In brief, temperatures across Australia's rangelands are likely to increase by 1–5 °C, depending on location and different modelling scenarios (Stokes et al. 2008a). By 2070, this rise will be greatest in Queensland, Northern Territory, and northern part of Western Australia (CSIRO and BOM 2015) and will be associated with increasing frequency of hot days and warm nights, with a doubling of temperatures over 35 °C and more frequent dust storms (CSIRO and BOM 2015). Rainfall is also predicted to change substantially, with a general increase in the number of rain-free days at the continental scale. The greatest reductions are predicted for southwestern Australia, but the northern subtropical and eastern areas of the continent will also be affected, to a lesser degree. Rainfall seasonality will be affected. For example, a 10–20% decline in rainfall is predicted for the winter-dominant rainfall areas of Southern Australian (Stokes et al. 2008a), and the effect of this will increase with increasing latitude (OEH 2011). By 2070 the frequency of drought in Australia is likely to increase, by 40% in the east and 80% in the south-west (Mpelasoka et al. 2008). Overall, therefore, the continent will be hotter, with less but more variable rainfall, and more frequent droughts. This will have substantial effects on soil and ecosystem processes, human and animal populations, land use, and sustainable management of rangelands. These changes will require a substantial paradigm shift in human behaviour, outlook, and institutional arrangements (Colloff et al. 2016).

14.4 Climate Effects on the Biotic and Abiotic Environment

Climate has major effects on dryland ecosystems, with direct effects on soils, and indirect effects, mediated by changes in plant and microbial communities (Delgado-Baquerizo et al. 2013). Direct effects of climate change in Australia are likely to be related to changes in (i) nutrient cycling, (ii) carbon retention, (iii) resilience and resistance to erosion, and (iv) the maintenance of function hydrological processes (Eldridge et al. 2011a).

Changes in rainfall will alter soil processes in rangelands directly due to increased drought severity and soil surface temperatures and exposure of the surface to radiation. For example, the management of rangelands indirectly affects soil health and climate by either increasing or reducing rates of carbon sequestration. Changes in the rates of carbon sequestration will have corresponding effects on CO_2 fluxes into the atmosphere and hence on future climates. The effects of a changing climate on soil processes will depend on a large number of interacting factors including soil moisture, temperature, and the degree of surface disturbance. Positive feedback processes could compromise the ability of plants to sequester CO_2 from the atmosphere, leading to further reductions in soil condition. Similarly, increasing temperature may reduce the rate at which carbon is sequestered into the soil and actually release CO_2 back into the atmosphere. Increasing soil insolation could reduce habitat for subterranean termites, one of the main decomposers of organic matter (Whitford 2002). Coupled with reduced plant biomass, this could lead to reduced levels of soil organic matter and declines in soil productive potential. Given the close links between soil carbon and soil aggregation, this is likely to lead to increasing soil destabilisation and erosion, with feedback effects on productivity and stability (Eldridge et al. 2011). Although reduced rainfall will limit plant production and potentially fire hazard, more intense wildfires are likely to become more frequent with increasing climate change.

Across Australia, land clearance and farming of marginal land in rangelands have led to substantial reductions in soil organic carbon. This loss is estimated to be about 51% of carbon in the surface soil after 40–50 years of farming (Luo et al. 2010). These losses are likely to be greater in rangelands, particularly semiarid areas, because the soils are more degraded and their capacity to store carbon is lower. The extent to which climate variability will lead to reductions in soil carbon will depend on feedback effects of CO_2 on plant community composition. Increased atmospheric carbon dioxide will likely enhance carbon assimilation in plants, resulting in greater production of crop residue, and therefore greater soil carbon storage (Jastrow et al. 2005). However, the overall effect of increased atmospheric CO_2 will depend on the balance between water, nutrient availability, microbial activity, and decomposition (Luo et al. 2010). This is likely to vary amongst different crop types and different agricultural systems.

The management of rangelands will also have a direct effect on climate change. For example, where adverse management such as overgrazing enhances wind erosion, any sediment generated could have a direct positive feedback on local or regional climates. Recent research has highlighted the importance of aerosols such as dust as climate change drivers in the Southern Hemisphere. Dust has substantial impacts on short- and long-wave radiation and can be almost as effective as greenhouse gases in warming the atmosphere (Rotstayn et al. 2007). The IPCC has estimated that the global mean radiative forcing of anthropogenic dust is -0.1 (± 0.2) Wm^{-2} (Forester et al. 2007). Modelling suggests that Australia contributes more than 70% of the atmospheric dust loading over most of the Southern Hemisphere (Tanaka and Chiba 2006). The main periods of high dust activity in Australia correspond to droughts (McTainsh et al. 2005). Therefore, as the tendency for droughts to occur in particular parts of Australia increases with changing climate so does the tendency for dust storm frequency to increase. A large part of the dust generated over the Australian continent comes from either grazed rangelands (Cattle et al. 2009) or areas of opportunistic dryland cropping on the Eyre Peninsula. The alluvial and lacustrine sediments of the Lake Eyre and Murray-Darling Basins are considered to be the main sources of fine-grained dust for contemporary dust storms in south-eastern Australia (Cattle et al. 2009).

Increasing climate variability will intensify problems currently facing rangeland managers worldwide (McKeon et al. 2009). Declining pasture productivity, reduced forage quality, increased livestock heat stress, more frequent weed and pest invasions, more frequent droughts, less frequent, but more intense rainfall, and greater soil erosion are all predicted under current climate change projections for Australia (Stokes et al. 2008b; McKeon et al. 2009). These effects will vary greatly across different regions in Australia due to the differences in climate and vegetation communities; they will also pose a range of major implications for rangeland managers. How the rangeland managers adjust to these different scenarios of climate change will have major feedback effects on the extent of climate change.

> **Case Study 1: Climate-Induced Ecosystem Decline in the Grasslands of the Western Riverina**
>
> The south-western grassland community of eastern Australia, colloquially known as the Western Riverina, is one of the most productive pastoral areas in Australia. A mixture of grasslands and shrublands, dominated by *Atriplex vesicaria* and *Austrodanthonia* spp., supports a productive pastoral industry based on Merino sheep production for wool and meat (Eldridge and Stafford 1999). Climatic projections for the Western Riverina are an increase of up to 3 °C in temperature and an increase of 10–20% in summer rainfall but substantial reductions (20–50%) in winter precipitation. Spring and autumn will also be substantially drier. The net effect of this scenario is that precipitation will decline over substantial areas (OEH 2011). Rainfall and droughts are also predicted to be more extreme (www.climatechangeinaustralia.gov.au). As most plant production occurs during the winter months, there will be substantial negative impacts on the natural environment, pastoral production, and the area's primary producers.

Consistent with other regions in eastern Australia, increasing climatic variability is likely to be associated with overall reduced tree cover, but encroachment by shrubs, increases in the cover and abundance of exotic plants, and declines in native forbs (Prober et al. 2012; Lavorel et al. 2015; Dunlop et al. 2016). These changes are likely to put increasing pressure on agriculturally based industries (Prober et al. 2017).

Declines in rainfall in the Western Riverina over the next 50 years are almost certain to reduce pastoral viability and place increasing stress on plant and animal communities. The area will experience major changes in habitat quality, soil condition, and the viability of pastoral enterprises. Perhaps the greatest changes will result from increased grazing pressure resulting from a similar or slightly lower number of livestock grazing an ever declining area of native grassland. Increased grazing pressure will result in a change in plant composition from palatable winter-growing grasses (e.g. *Austrodanthonia, Enteropodon*) to more grazing-resistant species with lower levels of crude protein (e.g. *Aristida, Austrostipa* spp.). In the Western Riverina areas of coarse-textured soils occur as isolated sandhills, source-bordering dunes, or the levees of prior stream. These soils respond more rapidly to lower levels of rainfall than the clay soils and are therefore likely to support considerable livestock grazing, particularly during dry times. However, their soils are highly erodible by wind due to their coarse textures and overgrazing.

Notwithstanding these increased effects of grazing, increased rainfall variability since the mid-1980s has already resulted in declines in stocking rates in the Western Riverina by about 10% since 1990 (Eldridge and Stafford 1999). This decline has corresponded to general regional-wide declines in the cover and productivity of shrubland vegetation (Eldridge and Stafford 1999). Coupled with some reductions in livestock grazing, however, are increases in the densities of European rabbits (*Oryctolagus cuniculus*) as the vegetation changes from native plants to exotic, often short stature grasses that are favoured by rabbits.

Riparian systems dominated by red gum (*Eucalyptus camaldulensis*) and black box (*Eucalyptus largiflorens*) are likely to be inundated less frequently under a more variable climate because of altered flow regimes and pressure from irrigators (Stokes et al. 2010). These communities will be likely to transform from semiaquatic to terrestrial systems (CSIRO and BOM 2015), and vegetation in the floodplain will likely transform to an open woodland with a ground storey dominated by terrestrial rather than semiaquatic species (Stokes et al. 2010). Under severe water restriction, some areas of irrigation are likely to revert to systems based on opportunistic dryland cropping and grazing (Crimp et al. 2010; Colloff et al. 2016).

Overgrazing is also likely to reduce the cover and composition of biological soil crusts: complex assemblages of mosses, lichens, and liverworts found on dryland soils that play an important role in fixing nitrogen, sequestering carbon, providing habitat for soil microbes, and protecting the soil against wind and water erosion (Eldridge et al. 2011b; Delgado-Baquerizo et al. 2013). A reduction in winter rainfall, combined with an increase in summer rainfall, will likely reduce the cover and composition of lichens because photosynthetic performance declines under high temperatures when the lichen thallus is hydrated (Budel et al. 2013). Declines in lichens will reduce soil surface stability and nitrogen production, and the combined effects of reduced crust cover and increased erosion will be to shift soil microbial community composition (Maestre et al. 2015), reduce niches for soil invertebrates, and therefore reduce habitat and resources for soil fauna. Declines in soil crust communities will also result in reduced soil insulation, leading to higher soil temperatures and, potentially, changes in termite populations. Termites are a major decomposer in these systems; thus any changes are likely to initiate negative feedback processes on decomposition and soil nutrient pools.

The combined effects of overgrazing and resultant landscape degradation is a simplification of the plant community structure (OEH 2011), which ultimately results in declining habitat for fauna. Changes in the plant community will likely manifest themselves as an increase in woody, arid adapted species such as Dillon bush (*Nitraria billardierei*) on clay soils or shrubs such as turpentine (*Eremophila sturtii*), narrow-leaved hop bush (*Dodonaea viscosa*), and punty bush (*Senna* spp.) on sandy soils. Endemic plant species from the family *Asteraceae* will undoubtedly decline and be replaced by exotic European weeds such as Patterson's curse (*Echium plantagineum*), ryegrasses (*Lolium* spp.), and horehound (*Marrubium vulgare*). Faunal communities most at risk will be grassland-dependent birds (Harrington et al. 1988) such as the already endangered Plains-wanderer (Pedionomus torquatus). This species is currently threatened because of habitat alteration through overgrazing, conversion of grasslands to rice production, and increasing frequency and severity of droughts. Plains wanderers require a vegetation cover of about 30 cm (OEH 2011) for protection and predator avoidance. Changes in the composition of grassland plant species will likely alter habitat structure for these grasslands species and lead for further declines in an already highly threatened species (Harrington et al. 1988).

Woody (typically shrub) encroachment is a global phenomenon and occurs over extensive areas of the Australian continent, from tropical and temperate savannah woodlands and grasslands (e.g. Hodgkinson 1991; Eldridge et al. 2011; Lunt et al. 2010) to coastal and alpine forests (Lunt et al. 2010). Encroachment changes the balance between grasses and woody plants, resulting in an increase in woody density and changes in the structure and composition of the vegetation communities (Eldridge and Soliveres 2014). Many of the changes induced by climate change are likely to be exacerbated by grazing, particularly where woody encroachment reduces the total cover of available forage for grazing. Increases in woody vegetation are also associated with changes in land management, particularly historic overgrazing (Eldridge and Soliveres 2014). Historic grazing by domestic livestock and high densities of browsing macropods (kangaroos) are thought to contribute, in a large way, to the increase in density and cover of woody plants (Hradsky et al. 2015). For example, overgrazing is likely to increase the competition between grasses and woody plants for soil moisture, typically giving shrubs a competitive advantage over grasses (Eldridge and Soliveres 2014). This effect is likely to be more pronounced during dry seasons when surface moisture is depleted more rapidly (Knoop and Walker 1985). This is also likely to be exacerbated over extended drought periods when there is strong competition for grass by grazing animals. Interactions between grazing and fire can influence the location, shape, and movement of woodland-grassland ecotones (Fensham et al. 2005). Grazing can also reduce fuel loads, limiting the potential for landholders to use controlled fire to control the grass shrub balance (Hodgkinson 1991).

14.5 Effects of Climatic Variability on Socio-economics of Rangeland Enterprises

As discussed above, temperature and variable rainfall are major drivers of ecosystem functions in rangelands (Stafford Smith and McAllister 2008). Changes in climatic conditions will affect not only affect biophysical attributes, as discussed above, but will have a marked effect on the social, economic, and cultural wellbeing of people in rangelands. Over the next half a century, the predictions are for greater climatic variability in rangelands, increased temperatures, more variable rainfall, and longer periods of drought with more extreme events. These changes will present new challenges (Gunasekera et al. 2007; Kingwell et al. 2013) associated with reduced pasture productivity, increased stress on livestock, increased prevalence of pests, weeds and diseases, and an increased risk of land degradation and erosion (Stokes et al. 2008a). Together these changes will reduce the long-term viability of dryland agricultural practices and make it harder to manage for sustainable production. These changes will also operate with a high degree of uncertainty (Addison 2013).

Rural and rangeland economies are highly dependent upon agriculture, and a changing climate will likely exacerbate the existing challenges faced by land managers in rangelands (Ash et al. 2012), adding an additional level of complexity to the interaction between social, economic, and biophysical systems. Therefore variability in climate will likely have major effects on the socio-economic status of rangeland areas (Reisinger et al. 2014). Managers of rangelands enterprises currently use a range of practices, frameworks, and models to deal with the vagaries of a varying climate (Stafford Smith and McAllister 2008; Kingwell et al. 2013). Managing enterprises under these adverse conditions will require resilience and adaptation and the adoption of adaptive management practices to explore novel crop types, animal species, and land uses (Prober et al. 2017). Agricultural producers will therefore need to adapt to climate change and may act to buffer or mitigate the impact of climate change on the social and economic capital of rangelands.

In the following section, we discuss the socio-economic character of rural Australia and the implications of increased climate variability for dryland communities. We then discuss the importance of adapting to a changing climate and explore potential adaptations at the level of the individual farming enterprise. Lastly, we briefly describe some of the barriers of adapting to climate change.

14.5.1 Socio-economic Status, Rural Health, and Climate Change

Australians living in rural areas are at a socio-economic disadvantage compared with those living in non-rural areas. Levels of income and education are lower; employment opportunities are fewer, and there is less access to goods and services (Alston 2004; AIHW 2007). Contributing to the low socio-economic status of rural communities is the steady decline of rural populations and services. Many rural towns, including those in the rangelands, have been experiencing population declines for decades as people move to regional cities in search of better education and employment and better access to services and amenities (Alston 2004). The loss of people from small rural towns has a feedback effect by further reducing the provision of services. This can lead to a decline in public infrastructure such as health services, schools, and banks (Vines 2011). The loss of people and infrastructure from rural towns often increases the socio-economic disparity between rural and non-rural areas.

Socio-economic status is correlated with physical and mental health. On average, the overall health of those in rural communities is lower than those in non-rural communities (Beard et al. 2009). This can be attributed to reduced access to health services, longer working life (Fragar et al. 2008; Vines 2011) and more physically demanding work (e.g. farm work) than city dwellers (Singh et al. 2015). Those living outside major cities are more likely to be overweight or obese, smokers, and drink alcohol in larger qualities (AIHW 2007). Lifestyle, in conjunction with working conditions and a long work lifespan, places rural inhabitants at a greater risk of lifestyle-related diseases and exposure-related conditions (Beard et al. 2009; Singh et al. 2015) than those in non-rural areas. This corresponds with rural areas having higher rates of morbidly and mortality (Beard et al. 2009) and a life expectancy that declines with increasing remoteness (Vines 2011). Mental illness is more prevalent in rural areas, and the rates of suicide are higher than in non-rural areas (Fragar et al. 2008). The health and mental health statistics are much worse in indigenous communities in rural areas (Green and Minchin 2014). Individuals in rural communities whose employment and livelihood is dependent upon agriculture may experience a further decline in health and well-being if climate change disrupts agriculture. Rural people with a low socio-economic status and overall health are particularly vulnerable to climate change (Hanna and Spicket 2011).

By influencing rural economies, changes in climate influence the socio-economic status, health, and well-being rural communities. Climate change is predicted to cause a decline in the production of wheat, sheep, beef, dairy, and sugar across all Australian states (Cobon et al. 2009; Moore and Ghahramani 2013 Reisinger et al. 2014; Race 2015). The decline in rural economies will likely contribute to a reduction in available services and amenities and a loss of social cohesion (Addison 2013), reducing the socio-economic status and well-being of rural inhabitants (Hanna et al. 2011; Buys et al. 2012). Increased temperatures and drought associated with climate change will have impacts on the social and economic capital of rangelands, with potential threats to food production and water quality (Holland et al. 2015).

Climate change will increase the frequency of extreme temperature days and the occurrence and duration of drought periods in rangelands (Cobon et al. 2009; Ash et al. 2012). There will be more daily deaths due to hot days and an increase the incidence of thermal stress (Fritze et al. 2008). Heat stress will have the greatest impact on rural labourers and those living in houses with poor thermal control (Bastin et al. 2014; Race 2015). Drought has been associated with farm closure, the involuntary separation of communities, and increases in stress and mental health disorders amongst farmers and farming families (Fagar et al. 2008; Greenhill et al. 2009; O'Brien et al. 2014; Reisinger et al. 2014; Fleming et al. 2015). Farmer stress may occur for numerous reasons: environmental exposure (Addison 2013), the stress of providing for dependents (Fagar et al. 2008), economic hardship (Hanna et al. 2011), and reduced access to labour (Fritze et al. 2008; Fleming et al. 2015), to name a few. In conjunction with a decline in health and well-being, there is likely to be a decline in engagement with education, employment, and community participation (Race 2015). As a consequence of climate-induced change, there may be a heightened pull to regional centres. Drought conditions will make farming more challenging, and dryland communities may collapse as a consequence of environmental conditions failing to keep agriculture viable (Alston 2004; Hanna et al. 2011). Climate change in conjunction with reduced local economies and reduced access to services may cement a perpetual cycle of decline in rural towns. Dryland land managers able to maintain enterprises alongside climate change may reduce socio-economic decline in rangeland towns.

14.6 Adaption to Increasing Climate Variability

Different combinations of temperature, rainfall, and carbon dioxide have different implications for farm management, and this generates considerable uncertainty regarding the most appropriate practices to adopt in order to mitigate the effects of changing climates (Race 2015). Certain climate change-induced conditions may advantage rangeland enterprises. For example, graziers may benefit from increases in forage production associated with increased temperature and increased rainfall (Webb et al. 2013), due to greater atmospheric CO_2. Other combinations of change may not be beneficial; for example, the positive effect of increased carbon on the growth of grain crops will likely be restricted by warming (Howden et al. 2008b), and hotter, drier conditions will disadvantage graziers as they will likely experience reduced forage production or a greater number of crop failures (Webb et al. 2012; Table 14.1). Adding to the complexity of predicting the impact of climate change is the heterogeneous nature of landscape health, soil fertility, soil water storage capacities, and plant water use across rangelands. Thus climate change will impact dryland enterprises differently (Stafford Smith and McAlister 2008; Webb et al. 2012; Reisinger et al. 2014). The number of possibilities for future climate combined with local variation in resource availability and quality requires farmers to tailor adaptive measures to individual enterprises (see Table 14.1).

The ability of dryland farmers to adapt to climate change is dependent on their desire and means, including available resources and the ability to mobilise these resources (Fleming et al. 2015; Nelson et al. 2010). The physical, financial, social, natural, and human capital accessible to rangeland managers determine their adaptive capacity (Shaw and Witt 2015). Deficiencies in these forms of capital may preclude any effective adaptation.

Options for adaptation for grazing and cropping in response to various environmental scenarios under climate variability and climate change are presented in Table 14.2. Many of these adaptation measures currently adopted in the Australian rangelands in response to short-term variability in climate are relevant to adapting to long-term climate change. Indeed, it has been suggested that climate change will not present major problems for dryland agriculture as agriculture has persisted for decades under a highly variable climatic regime (Stokes et al. 2008b; Ash et al. 2012; O'Reagain and Scanlan 2013; Brown et al. 2015; Race 2015). Nevertheless, a degree of adaptation will be required in the next half century as Australian rangelands move towards more variable climate with more frequent periods of drought and high temperatures. Individual adaptation options are unlikely to help producers to maintain the status quo, and different types of adaptation strategies will need to be implemented at the same time using a whole of systems level approach (Ferguson 2012; Fitzhardinge 2012; Bastin et al. 2014).

Fundamental to adapting grazing enterprises to climate change is maintaining landscape, soil, and plant health in order to reduce potential declines in forage production in hot dry conditions (Webb et al. 2013). Key to maintaining pasture

Table 14.1 Summary of the short-term and long-term effects of increased climate variability on farming and grazing enterprises

Anticipated environmental change	Farming enterprise affected	Effect of environmental change on farm system	Potential farm level adaptation		References
			Short term (climate variation)	Long term (climate change)	
Temperature increase	Grazing	Increased animal heat stress	Provide shelter for animals	Diversify land use if appropriate	Thornton et al. (2009), Webb et al. (2013), Ash et al. (2012), Moore and Ghahramani (2013), and Stafford Smith and McAlister (2008)
			Provide shelter near water points	Change to a hardier breeds of cattle	
			Alter breeding time to reduce young animal exposure	Modify existing genetics (e.g. genes from tropical breeds such as *Bos indicus*)	
		Less lower-quality forage	Substitute feed paddocks in wet season	May need to substitute feed seasonally	O'Reagain and Scanlan (2013), Bastin et al. (2014), Race (2015), McKeon et al. (2009), and Howden et al. (2008a)
				Spell paddocks in wet season	
				Maintain safe stocking rate	
			Safe stocking rate responsive to season, may have to destock	Evolve pasture species, increase drought and heat-tolerant plants	
		Increase in weeds pests and pathogens (e.g. fly strike, ticks)	Reduce animal susceptibility, e.g. shear animals prior to fly emergence	May be appropriate to change to hardier cattle breed	Wall and Ellse (2011), Thornton et al. (2009), and Stokes et al. (2008b)
		Longer life cycle of pest and pathogens outside of host			
		Reduced livestock immunity			
	Cropping	Heat stress on plants	Maximise water storage, if possible irrigate	If appropriate look to diversification of land use	Reisinger et al. (2014), Stokes et al. (2008b), and Race (2015)
		Inefficient plant water use			
		Increased evapotranspiration		Change variety of crop	
		Increase in pests and weeds	Change sowing time to match seasonal change in rainfall	Change crop every few seasons	

(continued)

Table 14.1 (continued)

Anticipated environmental change	Farming enterprise affected	Effect of environmental change on farm system	Potential farm level adaptation		References
			Short term (climate variation)	Long term (climate change)	
Decrease in rainfall	Grazing	Less water availability	Strategic water storage on property	Maintain the same practices as for environmental variability	Stafford Smith and McAlister (2008), Bastin et al. (2014)
			Cover dams to reduce evaporation		
			Replace or supplement dams with reliable bore water		
		Less forage and lower-quality forage	Use flexible stocking rates	Rest paddocks to promote vegetation growth	Stafford Smith and McAlister (2008), Bastin et al. (2014)
			Rest paddocks to reduce further degradation		
			Use conservative stocking rates		
Increase in extreme temperatures	Grazing cropping	As above	Same as temperature increase	Same as temperature increase	Stafford Smith and McAlister (2008), Bastin et al. (2014)
	Cropping	Less water availability	Strategic water storage on property	Possibly change species of crop	Bastin et al. (2014)
			Cover dams to reduce evaporation	May need to leave land that is too small to be viable, may be able to combine resources with adjoining properties	
			Supplement dams with reliable bore water		
Increased drought frequency and rainfall variability	Grazing	As above for decrease in rainfall	More variable rainfall and increase in drought periods will exacerbate the effect of increased temperature and decreased rainfall on farm systems; coping with these environmental changes will require a number of the adaptive responses identified above		
	Cropping				

Increased high intensity rainfall events	Structures	Erosion, especially around structures	Fence off erosion prone area to allow vegetation to regenerate	Build structures that won't increase likelihood of land eroding	Webb et al. (2013), Stokes et al. (2008b), Bastin et al. (2014), and Howden et al. (2008a)
	Grazing		Strategically locate water points to reduce erosion risk	If necessary relocate structures to reduce erosion risk and damage to structures in high rainfall	
	Cropping	Increase in water logging, flooding, weather damage, and pests	Sow at a different time of year		Howden et al. (2008b)
Increased atmospheric carbon	Grazing	May increase water use efficiency of some vegetation types	Use flexible stocking rate management	Maintain the same practices as for environmental variability	Thornton et al. (2009), Webb et al. (2012), Stokes et al. (2013), Howden et al. (2008b), Howden et al. (2008a), Cullen et al. (2009), and Prober et al. (2017)
			Protect soil resources		
		May increases forage growth dependant on rainfall and temperature	Maintain safe stocking rates		
		May decrease forage quality	Substitute feed		
		Favours growth of woody plants (less forage)	Use fire to remove woody plants	Use fire to manage woody plants	
Changes to fire regime	Cropping	May favour growth if temperature and rainfall not limiting	Sow with rainfall and temperature in mind		Howden et al. (2008b)
	Farm structures grazing	Excess fuel increases the likelihood of fire	Fuel management necessary	If possible establish buildings with reduced flammability	Reisinger et al. (2014), Stokes et al. (2008b), Stafford Smith and McAlister (2008), and Prober et al. (2017)
	Cropping	Time of year for fires may shift	May need to use fire to manage tree-grass balance to maintain forage	Maintain the same practices as for environmental variability	
		Fires frequency may reduce			

Adapted in part from Healy (2015)

Table 14.2 Predicted declines (%) in agricultural production in Australia for selected enterprises by 2050

Product	NSW	VIC	WA	QLD	NT	SA	ACT	TAS
Wheat	11.6	13.4	13.4	–	–	–	–	–
Beef	3.0	6.5	–	33.5	33.2	–	–	–
Sheep meat	13.2	12.9	13.4	–	–	–	–	–
Dairy	11.3	10.0	–	–	–	–	–	–
Sugar	–	–	–	17.0	–	–	–	–

Source: http://www.environment.gov.au/climate-change/climate-science/impacts (Accessed March 20, 2016)
– No data available

health is matching livestock numbers to rainfall levels and using low-risk conservative stocking (McKeon et al. 2009). Retaining non-breeding (excess) livestock in periods of low rainfall or drought increases stresses on available water and forage resources, increases the risk of erosion, and may reduce the long-term market value of livestock (O'Reagain and Scanlan 2013). Conservative stocking and risk adverse management must occur alongside the promotion of plant and soil health in order to maximise any landscape benefits of grazing (McKeon et al. 2009; Race 2015). Similar to grazing, adapting crops to a changing, more variable climate will require the maintenance of landscape health, particularly soil health. Soil degradation, either physical or chemical, will reduce the viability of farming and grazing enterprises. It is important, therefore, to match different crop varieties with different seasonal conditions (Eldridge et al. 2011a). For both grazing and cropping enterprises, there is potential to increase adaptive capacity by introducing new species adapted to climate change and using new technologies to monitor and predict climatic conditions (Bastin et al. 2014; Race 2015). Having methods in place to adapt to climate change reduces the vulnerability of dryland enterprises to failure and may have flow on effects for the socio-economic status of dryland communities.

14.7 Technical, Social, and Policy Barriers to Climate Change Adaptation

Adaptation will depend on a number of technical advances or innovations. For example, new agronomic information will be required to enable new crops and pasture plants to be sown that are able to grow and flourish under a drier, hotter environment (Table 14.2). It will also be necessary to explore potential species that are able to maintain adequate ground cover to protect the soil against wind and water erosion under drier conditions (e.g. Crimp et al. 2010; Moore and Ghahramani 2013). Agronomic technologies such as new methods of seed harvesting, seed coating, and sowing protocols will be needed to ensure effective plant establishment in a period of more variable climate (Prober et al. 2017). As the area of viable productive land declines and is replaced by dryland farming, there will be a greater flexibility in the management of water (Colloff et al. 2016). Grazing management may also need to change. Appropriate stocking levels will be needed to prevent erosion or pasture decline. This will require knowledge of the relative effects of native and domestic (European) herbivores so that control of one group does not lead to compensatory grazing by the other. This will lead to a need for information on low risk or conservative grazing practices (e.g. O'Reagain et al. 2011; Webb et al. 2012).

Methods will need to be developed to cope with the projected increases in weedy, woody species, either exotic shrubs, such as *Acacia nilotica*, or native shrubs such as *Eremophila* and *Senna* spp. that currently compete with ground-storey plants. Increased atmospheric CO_2 is likely to give these shrubs a competitive advantage over ground-storey plants, making it more difficult for pastoralists across large areas of rangelands to maintain profitable grazing enterprises (Eldridge and Soliveres 2014). Two additional technical constraints are related to woody plant management. The first is how to develop appropriate technologies to use fire to control these species. Fire is currently used to control woody regrowth (e.g. Hodgkinson 1991), but it will be increasingly difficult under a drier climate to have enough ground fuel to carry a fire. The second additional technical constraint is that more information is needed on the relative ecosystem services benefits of woody plants. Considerable benefits accrue from the retention of woody regrowth on pastoral properties (Eldridge and Soliveres 2014) ranging from ecosystem benefits (greater hydrology, increased plant facilitation, habitat for predators of crop pests) to production benefits (refugia for lambs during cold periods). There is considerable information needed on how best to manage woody plants, and the needs will be more critical with increased climatic variability.

The restoration of degraded lands will need to be given a greater priority as rainfall declines and temperate increases in amount and variability. Appropriate methods to revegetate degraded landscapes using different plant species and local provenance species are currently a large technical gap. Finally, policy and social constraints to adaptation will also need to be a priority for governments. For example, increases in stress experienced by producers, poor social networks, low levels of education, and entrenched views will be challenges that will need to be overcome if we are to help producers stay on the land (Table 14.3). Small family producers may be less flexible to make the hard decisions required to affect meaningful change.

Table 14.3 Social and economic barriers to former adaptation to climate change in rural communities

Constraint type	Issue	Mechanism	Reference
Social	Physical and emotional stress	Reduces the capacity of farmers to adapt to change	Fleming et al. (2015)
	Poor recognition of producer knowledge	Producer may feel disenfranchised if their knowledge is not used in policy formulation	Brown et al. (2015), Webb et al. (2013), and Reisinger et al. (2014)
	Poor social networks	Reduced community support	Stokes et al. (2012) and Race (2015)
	Low level of education	Reduces the level of environmental awareness thereby restricting adoption of novel practices	Stokes et al. (2012)
	Strong attachment to work and place	Greater likelihood of retaining the *status quo*	Marshall and Stokes (2014)
	Community attitudes and long-held beliefs	Rural residents may perceive that drought and climate variation are a normal part of rural life. This scepticism can perpetuate among farmers, reducing their ability to act	Buys et al. (2012), Mazur et al. (2008) and Prober et al. (2017)
	Rigid policy	Policy must be flexible, risk based, and provide options for different scenarios	Stokes et al. (2008b) and Prober et al. (2017)
Economic	Short-term loss for long-term gain	The loss of short-term profitability may be a sacrifice that farmers are not prepared to make	Webb et al. (2013)
	Flow-on effects	Drought results in feed shortages, in turn increasing the cost of feed and reducing the viability of maintaining grazing enterprises	Howden et al. (2008a) and Hanna et al. (2011)
	Small holdings lack the capacity to change	High stocking rates are associated with higher profits, but maximum income can occur at stocking rates below the biological maximum. Larger enterprises have a greater capacity to manage risk by altered management	O'Reagain et al. (2011), Race (2015), and Stokes et al. (2012)

References

J. Addison, *Impact of Climate Change on Health and Wellbeing in Remote Australian Communities: A review of Literature and Scoping of Adaptation Options*. CRC-REP Working Paper CW014 (Ninti One Limited, Alice Springs, 2013)

M. Alston, You don't want to be a check-out chick all your life: the out-migration of young people from Australia's small rural towns. Aust. J. Soc. Issues **39**(3), 299–313 (2004)

A. Ash, P. Thornton, C. Stokes, C. Togtohyn, Is proactive adaptation to climate change necessary in grazed rangelands? Rangel. Ecol. Manag. **65**(6), 563–568 (2012)

Australian Institute of Health and Welfare, *Rural, Regional and Remote Health: A Study on Mortality*, 2nd edn. (Australian Institute of Health and Welfare, Canberra, 2007). AIHW Cat. no. PHE 95

G. Bastin, C. Stokes, D. Green, K. Forrest, *Australian Rangelands and Climate Change – Pastoral Production and Adaptation* (Ninti One Limited and CSIRO, Alice Springs, 2014)

J.R. Beard, N. Tomaska, A. Earnest, et al., Influence of socioeconomic and cultural factors on rural health. Aust. J. Rural Health **17**(1), 10–15 (2009)

P.R. Brown, Z. Hochman, K.L. Bridle, N.I. Huth, Participatory approaches to address climate change: perceived issues affecting the ability of South East Queensland graziers to adapt to future climates. Agric. Hum. Values **32**(4), 689–703 (2015)

B. Budel, M. Vivas, O.L. Lange, Lichen species dominance and the resulting photosynthetic behavior of Sonoran Desert soil crust types (Baja California, Mexico). Ecol. Process. **2**(1), 6 (2013). doi:10.1186/2192-1709-2-6

L. Buys, E. Miller, K. van Megen, Conceptualising climate change in rural Australia: community perceptions, attitudes and (in)actions. Reg. Environ. Chang. **12**, 237–248 (2012)

S. Cattle, R. Greene, A. McPherson, The role of climate and local regolith-landscape processes in determining the pedological characteristics of aeolian dust deposits across South-Eastern Australia. Quat. Int. **209**, 95–106 (2009)

D.H. Cobon, G.S. Stone, J.O. Carter, et al., The climate change risk management matrix for the grazing industry of Northern Australia. Rangel. J. **31**, 31–49 (2009)

M.J. Colloff, S. Lavorel, R.M. Wise, et al., Adaptation services of floodplains and wetlands under transformational climate change. Ecol. Appl. **26**, 1003–1017 (2016)

S.J. Crimp, C.J. Stokes, S.M. Howden, et al., Managing Murray-Darling Basin livestock systems in a variable and changing climate: Challenges and opportunities. Rangel. J. **32**, 293–304 (2010)

CSIRO and Bureau of Meteorology, *Climate Change in Australia. Information for Australia's Natural Resource Management Regions: Technical Report* (CSIRO and Bureau of Meteorology, Australia, 2015)

B.R. Cullen, I.R. Johnson, R.J. Eckard, et al., Climate change effects on pasture systems in South-Eastern Australia. Crop Pasture Sci. **60**, 933–942 (2009)

M. Delgado-Baquerizo, F.T. Maestre, A. Gallardo, et al., Decoupling of soil nutrient cycles as a function of aridity in global drylands. Nature **504**, 672–667 (2013)

D. Dunkerley, Ecogeomorphology in the Australian drylands and the role of biota in mediating the effects of climate change on landscape processes and evolution. Geol. Soc. London Spec. Publ. **346**(1), 87–120 (2010)

M. Dunlop, R. Gorddard, P. Ryan, et al., *Exploring Adaptation Pathways in the Murray Basin* (CSIRO, Australia, 2016)

D.J. Eldridge, S. Soliveres, Are shrubs really a sign of ecosystem degradation? Disentangling the myths and truths of woody encroachment in Australia. Aust. J. Bot. **62**, 594–608 (2014)

D.J. Eldridge, M.R. Stafford, *Rangeland Health in the Western Riverina: 1990 to 1997* (Department of Land and Water Conservation, Sydney, 1999)

D.J. Eldridge, R.S.B. Greene, C. Dean, Climate change the rangelands: implications for soil health and management in *Chapter 13. Soil Health and Climate Change. Soil Biology Series*, ed. by B. Singh, A. L. Cowie, K. Yin Chan, (Springer, London, 2011a), pp. 237–255

D.J. Eldridge, J. Val, A.I. James, Abiotic effects predominate under prolonged livestock–induced disturbance. Aust. Ecol. **36**, 367–377 (2011b)

R.J. Fensham, R.J. Fairfax, S.R. Archer, Rainfall, land use and woody vegetation cover change in semi-arid Australian savanna. J. Ecol. **93**(3), 596–606 (2005)

J. Ferguson, A sustainable future for the Australian rangelands. Rangel. J. **34**(1), 27–32 (2012)

G. Fitzhardinge, Australia's rangelands: a future vision. Rangel. J. **34**(1), 33–45 (2012)

A. Fleming, A. Dowd, E. Gaillard, et al., Climate change is the least of my worries: stress limitations on adaptive capacity. Rural Soc. J. **24**(1), 24–41 (2015)

P. Forster, V. Ramaswamy, P. Artaxo, et al., Changes in atmospheric constituents and in radiative forcing, in *Climate Change 2007: The Physical Science Basis. Contribution of Working Group*

I to the Fourth Assessment Report of the Intergovernmental Panel on Climate Change, ed. by S. Solomon, D. Qin, M. Manning, et al., (Cambridge University Press, Cambridge, UK, 2007)

L. Fragar, A. Henderson, C. Morton, K. Pollock, *The Mental Health of People on Australian Farms* (Australian Centre for Agricultural Health and Safety, 2008)

J.G. Fritze, G.A. Blashki, S. Burke, J. Wiseman, Hope, despair and transformation: climate change and the promotion of mental health and wellbeing. Int. J. Ment. Heal. Syst. **2**(1), 13 (2008)

A. Goudie, J.C. Wilkinson, *The Warm desert environment* (Cambridge University Press, Cambridge, 1977)

D. Green, L. Minchin, Living on climate-changed country: Indigenous health, well-being and climate change in remote Australian communities. EcoHealth **11**(2), 263–272 (2014)

J. Greenhill, C. MacDougall, D. King, A. Lane, *The Resilience and Mental Health and Well-Being of Farm Families Experiencing Climate Variation in South Australia* (National Institute of Labour Studies, Flinders University, Adelaide, 2009)

D. Gunasekera, Y. Kim, C. Tulloh, M. Ford, Climate change – impacts on Australian agriculture. Aust. Commodities: Forecast. Issues **14**(4), 657–676 (2007)

E.G. Hanna, J.T. Spickett, Climate change and human health: building Australia's adaptation capacity. Asia-Pac. J. Public Health **23**(2), 7S–13S (2011)

E.G. Hanna, E. Bell, D. King, R. Woodruff, Climate change and Australian agriculture: a review of the threats facing rural communities and the health policy landscape. Asia-Pac J. Public Health **23**(2), 105S–118S (2011)

G.N. Harrington, P.N. Maher, D.J. Baker-Gabb, The biology of the plains-wanderer *Pedionomus torquatus* on the Riverine Plain of New South Wales during and after drought. Corella **12**, 7–13 (1988)

M. A. Healy (ed.), *It's Hot and Getting Hotter. Australian Rangelands and Climate Change – Reports of the Rangelands Cluster Project* (Ninti One Limited and CSIRO, Alice Springs, 2015)

K.C. Hodgkinson, Shrub recruitment response to intensity and season of fire in a semiarid woodland. J. Appl. Ecol. **28**, 60–70 (1991)

J.E. Holland, G.W. Luck, C.M. Finlayson, Threats to food production and water quality in the Murray-Darling Basin of Australia. Ecosyst. Serv. **12**, 55–70 (2015)

S.M. Howden, S.J. Crimp, C.J. Stokes, Climate change and Australian livestock systems: impacts, research and policy issues. Aust. J. Exp. Agric. **48**(7), 780–788 (2008a)

S.M. Howden, R.G. Gifford, H. Meinke, 'Grains', in *An Overview of Climate Change Adaptation in Australian Primary Industries – Impacts, Options and Priorities*. Report prepared for the National Climate Changes Research Strategy for Primary Industries (CSIRO, Canberra, 2008b), pp. 44–70

B.A. Hradsky, J.C. Loschiavo, M.A. Hradsky, J. Di Stefano, Shrub expansion alters forest structure but has little impact on native mammal occurrence. Aust. Ecol. **40**, 611–624 (2015)

E.A. Jackson, *A Study of the Soils and Some Aspects of the Hydrology of Yudnapinna Station, South Australia. Soils and Land use Series No. 24* (CSIRO Division of Soils, 1958)

J.D. Jastrow, M. Miller, R. Matamala, et al., Elevated atmospheric carbon dioxide increases soil carbon. Glob. Chang. Biol. **11**, 2057–2064 (2005)

R. Kingwell, L. Anderton, N. Islam, V. Xayavong, A. Wardell- Johnson, D. Feldman, J. Speijers, *Broadacre Farmers Adapting to a Changing Climate* (National Climate Change Adaptation Research Facility, Gold Coast, 2013), p. 171

W.T. Knoop, B.H. Walker, Interactions of woody and herbaceous vegetation in a Southern African savanna. J. Ecol. **73**, 235–253 (1985)

S. Lavorel, M.J. Colloff, S. McIntyre, et al., Ecological mechanisms underpinning climate adaptation services. Glob. Chang. Biol. **21**, 12–31 (2015)

I.D. Lunt, L.M. Winsemius, S.P. McDonald, et al., How widespread is woody plant encroachment in temperate Australia? Changes in woody vegetation cover in lowland woodland and coastal ecosystems in Victoria from 1989 to 2005. J. Biogeogr. **37**, 722–732 (2010)

Z. Luo, E. Wang, O.J. Sun, Soil carbon change and its responses to agricultural practices in Australian agro-ecosystems: a review and synthesis. Geoderma **155**, 211–223 (2010)

F.T. Maestre, M. Delgado-Baquerizo, T.C. Jeffries, et al., Increasing aridity reduces soil microbial diversity and abundance in global drylands. Proc. Natl. Acad. Sci. U. S. A. **112**(51), 15684–15689 (2015)

N. Marshall, C.J. Stokes, Identifying thresholds and barriers to adaptation through measuring climate sensitivity and capacity to change in an Australian primary industry. Clim. Chang. **126**(3), 399–411 (2014)

N. Mazur, A. Curtis, R. Thwaites, D. Race, *Rural Landholder Adaptation to Climate Change: Social Research Perspectives* (Institute for Land, Water and Society, Charles Sturt University, Albury, 2008)

G.M. McKeon, G.S. Stone, J.I. Syktus, et al., Climate change impacts on Northern Australian rangeland livestock carrying capacity: a review of issues. Rangel. J. **31**(1), 1–29 (2009)

G.H. McTainsh, Y.C. Chan, H.A. McGowan, et al., The 23rd October, 2002 dust storm in Eastern Australia: characteristics and meteorological conditions. Atmos. Environ. **39**(7), 1227–1236 (2005)

A.D. Moore, A. Ghahramani, Climate change and broadacre livestock production across Southern Australia. 3. Adaptation options via livestock genetic improvement. Anim. Prod. Sci. **54**, 111–124 (2013)

S.R. Morton, D.M. Stafford Smith, C.R. Dickman, et al., A fresh framework for the ecology of arid Australia. J. Arid Environ. **75**(4), 313–329 (2011)

F. Mpelasoka, J.R. Hennessy, R. Jones, B. Bates, Comparison of suitable drought indices for climate change impacts assessment over Australia towards resource management. Int. J. Climatol. **28**(10), 1283–1292 (2008)

R. Nelson, P. Kokic, S. Crimp, The vulnerability of Australian rural communities to climate variability and change: part II—integrating impacts with adaptive capacity. Environ. Sci. Pol. **13**(1), 18–27 (2010)

Noy-Meir, Desert ecosystems: environment and producers. Annu. Rev. Ecol. Syst. **4**, 25–51 (1973)

L.V. O'Brien, H.L. Berry, C. Coleman, I.C. Hanigan, Drought as a mental health exposure. Environ. Res. **131**, 181–187 (2014)

P.J. O'Reagain, J.C. Scanlan, Sustainable management for rangelands in a variable climate: evidence and insights from Northern Australia. Animal **7**(1), 1–11 (2013)

P.J. O'Reagain, J. Bushell, B. Holmes, Managing rainfall variability: long-term profitability of different grazing strategies in Norther Australia tropical savanna. Anim. Prod. Sci. **51**, 210–224 (2011)

OEH, *NSW Climate Impact Profile Technical Report: Potential Impacts of Climate Change on Biodiversity* (Department of Environment and Climate Change, Sydney, 2011)

S. Power, T. Casey, C. Folland, et al., Inter-decadal modulation of the impact of ENSO on Australia. Clim. Dyn. **15**, 319–324 (1999)

S.M. Prober, D.W. Hilbert, S. Ferrier, et al., Combining community-level spatial modelling and expert knowledge to inform climate adaptation in temperate grassy eucalypt woodlands and related grasslands. Biodivers. Conserv. **21**, 1627–1650 (2012)

S.M. Prober, M.J. Colloff, N. Abel, et al., Informing climate adaptation pathways in multi-use woodland landscapes using the values-rules knowledge framework. Agric. Ecosyst. Environ. **241**, 39–53 (2017)

D. Race, *The Impacts of, and Strategies to Ameliorate, the Intensity of Climate Change on Enterprises in Remote Australia. CRC-REP Working Paper CW020* (Ninti One Limited, Alice Springs, 2015)

A. Reisinger, R.L. Kitching, F. Chiew, et al., Australasia, in *Climate Change 2014: Impacts, Adaptation, and Vulnerability. Part B: Regional Aspects. Contribution of Working Group II to the Fifth Assessment Report of the Intergovernmental Panel on Climate Change*, ed. by V. R. Barros, C. B. Field, D. J. Dokken, et al., (Cambridge University Press, Cambridge, UK, 2014), pp. 1371–1438

J.F. Reynolds, D.M. Stafford Smith, E.F. Lambin, et al., Global desertification: building a science for dryland development. Science **316**, 847–851 (2007)

M. Robertson, R. Murray-Prior, Five reasons why it is difficult to talk to Australian farmers about the impacts of, and their adaptation to, climate change. Reg. Environ. Chang. **16**(1), 189–198 (2014)

L.D. Rotstayn, W.J. Cai, M.R. Dix, et al., Have Australian rainfall and cloudiness increased due to the remote effects of Asian anthropogenic aerosols? J. Geophys. Res. 112(D9) (2007). doi:10.1029/2006JD007712

E. Shaw, G.B. Witt, Climate change and adaptive capacity in the Western Australian rangelands: a review of current institutional responses. Rangel. J. **37**(4), 331–344 (2015)

S. Singh, E.G. Hanna, T. Kjellstrom, Working in Australia's heat: health promotion concerns for health and productivity. Health Promot. Int. **30**(2), 239–250 (2015)

M. Stafford Smith, R.R.J. McAllister, Managing arid zone natural resources in Australia for spatial and temporal variability – an approach from first principles. Rangel. J. **30**(1), 15 (2008)

D.M. Stafford Smith, S.R. Morton, A framework for the ecology of arid Australia. J. Arid Environ. **18**, 255–278 (1990)

C.J. Stokes, A. Ash, S.M. Howden, Climate change impacts on Australian rangelands. Rangelands **30**(3), 40–45 (2008a)

C.J. Stokes, S. Crimp, R. Gifford, et al., Broad acre grazing, in *An Overview of Climate Change Adaptation in Australian Primary Industries – Impacts, Options and Priorities. Report prepared for the National Climate Changes Research Strategy for Primary Industries* (CSIRO, 2008b), pp. 229–258

K.E. Stokes, K.A. Ward, M.J. Colloff, Alterations in flood frequency increase exotic and native species richness of understorey vegetation in a temperate floodplain eucalypt forest. Plant Ecol. **211**, 219–233 (2010)

C. Stokes, N. Marshall, N. Macleod, *Final Report: Developing Improved Industry Strategies and Policies to Assist Beef Enterprises Across Northern Australia Adapt to a Changing and More Variable Climate* (Meat & Livestock Australia Limited, 2012)

T.Y. Tanaka, M. Chiba, A numerical study of the contributions of dust source regions to the global dust budget. Glob. Planet. Chang. **52**(1), 88–104 (2006)

P.K. Thornton, J. van de Steeg, A. Notenbaert, M. Herrero, The impacts of climate change on livestock and livestock systems in developing countries: a review of what we know and what we need to know. Agric. Syst. **101**(3), 113–127 (2009)

C.C. Ummenhofer, A.S. Gupta, A.S. Taschetto, M.S. England, Modulation of Australian precipitation by meridional gradients in East Indian Ocean sea surface temperature. J. Clim. **22**, 5597–5610 (2009)

R. Vines, Equity in health and wellbeing: why does regional, rural and remote Australia matter? Psych.: Bull. Aust. Psychol. Soc. Ltd. **33**(5), 8–11 (2011)

R. Wall, L.S. Ellse, Climate change and livestock parasites: integrated management of sheep blowfly strike in a warmer environment. Glob. Chang. Biol. **17**(5), 1770–1777 (2011)

N.P. Webb, C.J. Stokes, J.C. Scanlan, Interacting effects of vegetation, soils and management on the sensitivity of Australian savanna rangelands to climate change. Clim. Chang. **112**(3), 925–943 (2012)

N.P. Webb, C.J. Stokes, N.A. Marshall, Integrating biophysical and socio-economic evaluations to improve the efficacy of adaptation assessments for agriculture. Glob. Environ. Chang. **23**(5), 1164–1177 (2013)

W.G. Whitford, *Ecology of Desert Ecosystems* (Academic, London, 2002)

Additional Reading

B.K. Henry, T. Danaher, G.M. McKeon, W.H. Burrows, A review of the potential role of greenhouse gas abatement in native vegetation management in Queensland's rangelands. Rangel. J. **24**, 112–132 (2002)

B.K. Henry, G.M. McKeon, J. Syktus, J.O. Carter, K. Day, D. Rayner, Climate variability, climate change and land degradation, in *Climate and Land Degradation*, ed. by M. V. K. Sivakumar, N. Ndiang'ui, (Springer, New York, 2007), pp. 205–221

G.M. McKeon, S.M. Howden, D.M. Stafford Smith, The management of extensive agriculture: greenhouse gas emissions and climate change, in *Assessing Technologies and Management Systems for Agriculture and Forestry in Relation to Global Climate Change*, (Intergovernmental Panel on Climate Change, Response Strategies Working Group, Australian Government Publishing Service, Canberra, 1992), pp. 42–47

D.M. Stafford Smith, G.M. McKeon, I.W. Watson, et al., Learning from episodes of degradation in variable Australian rangelands. Proc. Natl. Acad. Sci. U. S. A. **104**, 20690–20695 (2007)

M. Stafford Smith, L. Horrocks, A. Harvey, C. Hamilton, Rethinking adaptation for a 4°C world. Philos. Trans. Royal Soc. A **369**, 196–216 (2011)

R.J. Williams, R.A. Bradstock, G.J. Cary, et al., *The Impact of Climate Change on Fire Regimes and Biodiversity in Australia—A Preliminary Assessment. A CSIRO unpublished report to the Australian Government* (Department of Climate Change, Canberra, 2009)

Part V
Summary, Synthesis and Concluding Remarks

The two chapters here attempt to synthesize the key issues while focussing on those issues and salient points raised in the preceding chapters of the book.

Mahesh Gaur and *Victor Squires* try to sum up and catalogue the problems and the opportunities (prospects) faced by aridland/dryland dwellers throughout the world. Climate change impact and other upheavals attributed to global change are discussed. The water, soil, food, energy nexus and its implications are highlighted. The long-term future of traditional land uses, especially arid zone pastoralism, are discussed in the light of environmental, economic, social and political changes.

The final chapter by *Victor Squires* and *Mahesh Gaur* is a synthesis of the whole book and a summing up based on the inputs of 21 contributors from eleven countries. Of course we draw heavily on our colleagues in the *invisible college* whose writings have educated us and inspired us. We are grateful to them. The likely impact of global change (including climate change) on the lands and its peoples will be profound, but as a *slow-onset* phenomenon, there may be time to adapt, but the pace of change and the dire circumstances in which many now live may make this hope fade. Initially, we will see coping mechanisms brought into play but ultimately adaptation must occur. We conclude that the pace of change and the already degraded state of much of the world's aridlands leave little room for optimism about the land and its peoples over the next 20–50 years. We identify research priorities and conclude with a fervent hope that the various initiatives such as the land degradation neutral world can achieve its targets.

Drylands Under a Climate Change Regime: Implications for the Land and the Pastoral People They Support

Mahesh K. Gaur and Victor R. Squires

15.1 Introduction

Drylands,[1] a major focus of this book, occupy about one half of the world's land surface (see also Sidahmed, this volume). There is a misconception among the modern societies that deserts (below 200 mm rain) which dominate the dryland areas in many countries are wastelands of no use. However, for the inhabitants of the desert lands, this is not true. Historically, human communities have lived in the drylands and used productively one of its most valued resources – the rangelands.[2] Because the pastoralists were marginalized and exposed to influences of infrastructural and policy changes, unwise exploitation and use of the rangelands resulted in progressive – and in some cases what was thought as irreversible – deterioration.

By the end of the twentieth century, the rangelands of most African and Middle Eastern countries had been subjected to extraordinary pressures. This is reflected in an obvious degradation of rangeland resources and increasing desertification hazards. Much current academic and political interest has focused on these problems, and many people are interpreting these problems as the result of overuse and mismanagement of vegetation and livestock by traditional pastoralists.

During the last five decades, awareness of the need to develop and restore the rangelands of these countries has increased and has been expressed in many development activities. Various programs and projects with general and specific

---

[1] Drylands is an internationally accepted term to cover subhumid, semiarid, and arid regions of the world.

[2] Rangeland ecosystems include deserts, shrublands, grasslands, and open forests and only exclude commercial forests, cultivated lands, ice-covered regions, and areas covered by solid bare rocks.

M.K. Gaur (✉)
ICAR-Central Arid Zone Research Institute, Jodhpur, India
e-mail: maheshjeegaur@yahoo.com

V.R. Squires
Institute of Desertification Studies, Beijing, China

objectives were carried out in some rangeland areas located in various ecological zones. Some programs were simple with only a few components, while others were very ambitious and involved some packages with the following components: establishing regulations and by-laws on rangeland utilization and protection (grazing rotations), artificial revegetation, digging boreholes for permanent water supply, sedentarizing nomads, veterinary care, constructing water-harvesting and water-spreading facilities, establishment of feed reserves, creating pastoral and fattening cooperatives, and subsidizing feeds and livestock production. Reviewing the outputs of many of these development programs showed largely disappointing results when judged by the criteria of improving forage production, the economic rate of return, the welfare of land users, and the conservation of rangeland resources.

Evaluation of development policies carried out in some African, Central and South Asian, and Middle Eastern countries during the last four decades demonstrated that the policies themselves are the root cause of many difficulties and problems encountered in rangelands and that these policies themselves have contributed to the deterioration of rangeland resources and the acceleration of desertification hazards (Behnke and Mortimer 2015).

Rangelands amount to nearly 40% of the earth's land surface and are considered the primary land type of the world. If all the land resources presently grazed by domestic animals and all the uncultivated land with the potential to support livestock are taken into account, rangelands comprise about 60% of the earth's land area, of which a major part lies within the world's dry regions. On a worldwide basis, rangelands contribute to 70% and over 95% of the feed needs of domestic and wildlife ruminants, respectively. Forage for livestock is a major contribution, providing between 60 and 85% of the total feed needs of their domestic ruminants. But rangeland values go far beyond grazing, to include water, food, fuel, recreation, and home to ethnic minorities or sites for antiquities as well as are repositories of ancient system of gene pool conservation (Gaur 2014).

Drought is a common feature of the rangelands of the arid zone (Squires, 2017), and pastoralists (i.e., nomads, transhumants, and settlers) are highly adapted to variable water deficits and shortage of forage supplies. However, for thousands of years, these rangelands had traditionally supported some extensive forms of animal production, somewhat in harmony with the indigenous wildlife population and the more intensive cropping systems of adjacent higher potential lands. The functioning and survival of rangeland systems of these countries are dependent on complex interrelationships between people, domestic animals, vegetation, other wild organisms, and the physical environment. Similarly, the ability of pastoralists to survive in these marginal lands is attributed to their opportunistic mobility, their large numbers of animals, and diversified livestock husbandry, as well as to a wide spectrum of adaptive strategies. Some of these strategies are ecologically based, while others depend upon socioeconomic relations (Squires and Feng, this volume).

The regions in question are characterized by the predominance of communal rangelands which were equally available to the herds and flocks of all members of the community and are utilized and managed by villagers, nomads and transhumants according to agreed roles and traditions. In drought years, accessibility to communal grazing could be granted to neighboring tribes or groups, and the donors of such a grant receive, in return, an equal treatment when the need arises.

For centuries, traditional livestock movement from dry to wet season or from low- to high-altitude ranges formed a simple but very efficient rotation, in which pastures are utilized for a specific period and then livestock are moved to another season's pasture. The availability of water from temporary dugouts, natural ponds, and springs, the occurrence of biting insects, and the prevalence of low temperature govern the period for which livestock utilize forage in a particular site; hence, rangeland plants in different sites have a rest period which allows them to recover, store carbohydrates needed for regrowth, and set seed.

By the turn of the twentieth century, rangelands of most African, Middle Eastern, and Southwest Asian countries had been subjected to extraordinary pressures due to the obvious increase of human and livestock population (which caused a serious degradation of range resources). During the last five decades, many development projects and various activities (based on technology and rangeland management principles developed in the more advanced countries) had been carried out in the regions to solve degradation problems and restore range resources. Unfortunately, most of these projects failed to overcome the problems and failure was attributed to many different reasons (Squires et al. 2010).

Until relatively recent times, the trend of international development effort was technology-led (i.e., formulated in terms of highly advanced technology, investment, and management). Socioeconomic factors were not considered, with the implicit assumption that the social relations that constitute the societies in question would rearrange or reform themselves in adaptation to the new exogenous techno-environmental condition.

During the last four decades, large-scale interventions represented in many development projects had been carried out in African and Asian rangelands to overcome the degradation problems of their resources. Most of these projects were based on technology and range management principles developed in the USA, Europe, Australia, the former Soviet Union, and South Africa and responded to the problems by attempting to institute radical management changes. These projects included the following strategies: sedentarizing nomads, digging boreholes for permanent water supply, veterinary care, subsidizing feeds and livestock production, artificial revegetation of depleted ranges, constructing water-harvesting and water-spreading facilities, establishing feed reserves, creating pastoral and fattening cooperatives, and establishing regulations and by-laws to optimize utilization of rangeland resources. Unfortunately, most of these development projects failed to overcome the degradation problems, and the failure was often attributed to one or more of the following reasons: the traditionalism of pastoralists; inadequacies of administrations; problems of land tenure and rights of rangeland utilization; conflicts between pasture production and other agriculture production systems, as well as between individual and common interests of pastoralists; rural and urban communities; underestimation of environmental constraints of the pastoral areas; and lack of adequate information for the selection of appropriate technology in the approach to development.

In fact, failure of such projects was due to the development policies themselves which disturbed the traditional utilization and management of these rangelands and caused more degradation of resources. Uninformed manipulation of these pastoral societies had produced unpredicted and ramifying changes both in the environment and in the pastoralists' capacity to exploit it. A major factor in the social tragedy that followed the Sahelian drought was the recent loss of flexibility in resource use, which is a very important aspect of any land use system particularly in arid and semiarid regions (Sidahmed, this volume). In retrospect, massive interventions based on technology and range management principles developed in advanced countries may not provide the optimal solutions to the problems of African and Asian pastoralists because of the significant differences in their ecological and cultural histories. However, it is now generally accepted that technological change does not occur (and cannot be induced) in isolation from economic change and general social change. It has also become clear that, in order to induce change successfully in any social system, it is necessary first to investigate the dynamics of the existing system. It is unscientific to expect to change particular social practices without first ascertaining what generates the social formation underlying those practices.

15.2 Importance of Rangelands

Rangelands play an important role in the livelihood of local communities and wildlife. In this chapter, however, the three main roles of rangelands in low-rainfall areas will be discussed in the following section: (i) as a feed source for livestock production; (ii) a base of survival for local communities, their institutions, and management practices they developed to overcome environmental variability; and (iii) as a means of avoiding conflicts between herders and farmers.

Much of the development thrust toward resource improvement and conservation has been technically oriented. This approach has been one of the shortcomings of rangeland research and policy-making; economic efficiency of rangeland resources continues to be the guiding principle of government intervention in rangelands. Rangeland users, nomads, and transhumants often continue to be perceived as the causes of rangeland problems.

The human dimension of rangelands (i.e., indigenous communities, local institutions, and resource management systems) often unfortunately continues to be disregarded in policy formulation (Squires 2009). The human dimension is a key factor for sustainable development policies on rangelands. It is timely that policy-makers and researchers start to talk about the interaction between people and resources, not only in terms of efficiency but also in terms of equity and sustainability. The role and rights of local communities on rangeland resources need to be well-defined as these are necessary to their livelihood. As primary beneficiaries of these rangeland resources, local communities ought to be made responsible for their management and long-term conservation. Such a policy will not be a complete innovation, only a restitution of many traditional rights and management roles that pastoral communities used to exercise on rangelands. The strength of local communities is indicated by community members' continued adherence to customary rules and institutions. The capacity of local institutions depends on (i) the existence of their former territory, (ii) their social legitimacy to enforce customary rules, and (iii) their recognition by the state as viable management institutions. An important role of rangelands that is generally missed is in conflict avoidance between farmers and herders. The use of rangelands during the cropping season enhanced the relations between herders and farmers by securing the welfare of each party.

15.3 Pressures Affecting Rangelands

Pastoralists, policy-makers, and researchers have been talking alarmingly about the reduction and degradation of native pastures. The recent debate around rangeland problems has shifted rangeland development from economic efficiency, which for many years concentrated on settling and transforming the pastoral population into farmers (regardless of the long-term environmental impacts), to the sustainable use of rangeland resources. Several factors are identified as being causes of resource misuse and impediments to sustainable management of rangelands. Some of these causes will be highlighted to provide background and explore the solutions that are being promoted.

The debate around rangeland development should focus now, not only on the areas that are considered as native pastures but also on the settled areas. It is only through such an integrated land use approach that rangelands could be tackled adequately for sustainable resource use. These areas are a continuum used by the same rural communities who developed different coping strategies in response to nature and to government development policies. Thus, the new interest in rangelands should be oriented toward understanding present communities and their resource management systems. Any approach that seeks sustainable resource use but does not integrate local communities and management systems in the design process is doomed to failure. There is an ongoing debate around community participation but, in general, commitment to developing local production systems is still missing. This does not mean that local production systems should be used blindly, but they should be studied with respect to their changes and the remaining resource management practices and rules that could be beneficial to future rangeland development (Squires 2009). Efforts for rangeland management are usually confined either to improvement of existing areas or introduction of suitable exotics. As such, sound management plans for the holistic development of rangelands and protection of native patches of grasslands which are unique and harbor rich fauna, are more or less missing (Gaur 2014).

15.3.1 Land Tenure Confusion

The tenure situation in rangelands is the most confusing. Two systems continue to claim legitimate ownership of rangelands: state and local communities. In many arid zone countries, legislation that makes the state the owner of rangelands was enacted. The rights of local communities were reduced to "use rights." This appropriation of rangelands by the state has many implications regarding rangeland management, because it reduces the capabilities of local communities to control and manage the use of resources.

The main issue regarding the question of appropriation is to determine the trade-off between efficiency and sustainability. The state, claiming that local communities are not efficient resource users, takes the responsibility to set up new rules of access and create resource control mechanisms. With rare exceptions, governments have generally not been efficient controllers of rangeland management. *Any policy based solely on state absolute property regardless of prior claims risks failing partially or even totally.* The failure of state control mechanisms is due to the high cost of patrolling a very rough and large area and the lack of community participation. The main question that emerges under state control is how to get local participation for sustainable resource use.

Local communities often continue to view rangelands as their territory and continue to control access on an informal basis. Customary management rules are often no longer being enforced; this is one of the major impediments caused by state appropriation. Neighboring groups (local institutions) will otherwise continue to use their social networks to demand reciprocal access for grazing from one another. They grant each other access as a means of confirming their claims and strengthening their traditional social relations with other communities. Importantly, these arrangements enhanced their risk management strategies during drought years (Behnke et al. 1993).

Pastoral communities have maintained some of their customary claims by adapting their strategies to state development policies. For example, local communities were the major beneficiaries of land allocation in settlement schemes. The only difference has been the change from common to individual resource control. As a result, community members claim two types of rights: (i) individual rights of ownership they derive from their community membership and that are confirmed by the state and (ii) common ownership rights not recognized by the state, which they continue to claim on unsettled rangelands.

This dichotomy is found in all the North African countries (and others) with the exception of Morocco and Tunisia, where tribal rights were recognized on collective lands. In these two countries, however, land allocation to community members has favored land fragmentation and reduced common community pastures. In the remaining countries, we find different degrees of this duality (Mohamed, this volume). It is important, however, to note that regardless of government land policies, local institutions continue to view their rights over rangelands as superior to state claims. Such claims are even asserted on improved state rangeland reserves.

These opposing claims between state and local communities have resulted in poorly defined tenure rights on rangeland resources. The confusion between who manages and enforces rules of use and who grants access to rangelands has fostered a situation of "no control" which is called "open access." Some argue that instability of life and lack of property rights are the real causes of overgrazing and misuse. In addition, such tenure confusion raises many equity issues because wealthy community members, who have the political means to defend those holdings despite their questionable legal status, enclose large grazing areas at the expense of poor community members. Such a situation is all too common in post-Soviet Central Asia (Halimova 2012; Kreutzmann 2012).

15.3.2 Collapse of Traditional Institutions and Management Systems

The establishment of national borders, the appropriation of rangelands by the state, and the confinement of herding communities into smaller grazing areas narrowed their traditional grazing access-options. Some argue that the legal assault on property rights seems to share one common objective – overthrowing the customary rights and breaking the traditional organization of the pastoral society. The collapse of traditional migration patterns has put great pressure on community pastures and increased the use of purchased feeds and crop residues. All too often, traditional practices and management systems in rangelands, which were developed by local communities in response to their different constraints, have broken down.

Tribal control of rangelands, virtually "states-within-states," was revoked in many countries. The unintended result of this was to take rangelands out of traditional common property management and move them to open access and subsequent uncontrolled use and heavy degradation. In addition, the power loss of tribal institutions fosters the individualization of many common resources.

15.3.3 Competition Between Pastoralism and Farming

Expansion of agricultural production has shifted the boundaries of rangelands. In a desperate pursuit of food self-sufficiency, the governments of the some countries have encouraged the production of staple food crops and small ruminants even in high-risk areas, regardless of environmental damage. The development of the transport system, and cheap fuel, allows greater mobility of herds, feeds, and water. This new access to, and availability of, water has permitted livestock herders to stay much longer on the range. Modern transport of animals and water has disturbed the traditional flock movements and caused overgrazing.

15.3.4 Privatization of Rangelands

The persistence of rangeland degradation in many arid zone countries prompted several approaches for better resource use. Generally, poor or destructive resource use is perceived as a consequence of a lack of well-defined rights. In order to promote sustainable resource management, it therefore becomes necessary to grant secure access rights to resource users. Provision of tenure security to resource users promotes better resource use and encourages investment in resources. This is because holders of such rights can reasonably hope to enjoy the benefits of their investments in good stewardship. Ellis and Swift (1988) noted that "the assumption is that some form of privatization will alleviate the imbalances supposedly induced by communal grazing." As such, privatization of common resources is thought to be one of the most practical solutions to environmental degradation. In this case, it is very important to depart from the narrow view of private property, confined solely to the individual.

The main feature of private property is that it is legally respected by the state and is easily marketable. As such, three types of private property rights are distinguished by (i) community private rights under collective management (i.e., cooperative), (ii) community private rights under individual management (i.e., tribal system), and (iii) private rights under individual management. The first two types are forms of corporate ownership of private property.

As a strategy against rangeland degradation, private property could be part of the answer in any of the three forms wherever they are viable. There are still some communities, for example, with strong local leaders who continue to use efficient traditional resource management systems. For such communities, the best strategy may not be individual private property but, through recognizing and strengthening traditional rights, to allow them access to credit with borrowing on reasonable terms.

15.4 Constraints to Pastoralism in the Arid Zone

Under the conditions of highly variable rainfall, traditional pastoral economies are in continuous disequilibrium. Rainfall, rather than grazing pressure, determines the following year's primary production (Behnke et al. 1993). This disequilibrium conserves soil and vegetation, because grazing pressure is adjusted to the quantity of feed available. As long as mobility and flexibility of grazing are conserved, the annual vegetation especially can maintain its resilience. This resilience is the result of the way arid rangelands have been traditionally used.

The key constraints which have plagued pastoralism in the past include both the technical (animal health and nutrition) and the sociopolitical (land tenure, policy issues, religion) and economic (marketing). These constraints deserve some elaboration, although the reader should refer to the excellent papers in Part IV of this volume for more detailed examples.

New socioeconomic issues have arisen which impact on the way in which traditional societies view the future. Demands for education, better health, higher expectations for their children, and a desire for a more technologically-based lifestyle (radio, TV, satellite communications, motorized transport) have shifted priorities.

Changes in land tenure, security of access, reductions in internecine disputation, the emergence of stronger central governments, and rising nationalism as globalization of trade and commerce takes hold have all played their part. These shifts call for a different set of institutions, markets, and policies. They also call for the development and adaptation of new technologies to make livestock production environmentally more benign – the scope is enormous and so is the task.

Emerging problems include the conservation of the resource base (including biodiversity issues), the globalization of the world economy, the breakdown of tradition, and the potential impacts of climate change. The challenge is to find ways of managing drylands that are more environmentally sustainable, economically viable, and socially equitable than at present.

15.5 Driving Forces in Arid Grazing Systems

Over the longer term, the fundamental driving force on natural resources is population pressure, especially from outside the arid rangelands and their pastoral users. While population growth of the pastoral peoples has been rather low, the growth of non-pastoral groups in the arid and semiarid regions has been among the highest in the world. This growth of other groups causes an increasing encroachment by arable farmers on the pastoral "key resource" sites and constrains the critical mobility necessary to adjust to the disequilibrium conditions.

The increased population pressure also leads to water development and settlements in arid rangelands. Also within the system, the population pressure mounts and causes land degradation. Thus, in spite of the resilience of the system, many pastoralists face a downward spiral of increased crop encroachment, increased fuelwood requirements, and decreased grazing availability. These forces contribute to impoverishment of the pastoral population and to land degradation. This trend is being exacerbated by drought, and vulnerability to drought is one of the main indicators of long-term environmental and social sustainability of these arid grazing systems.

15.5.1 World Demand for Food, Especially Meat Products

The massive appetite of the growing urban populations for meat and milk (and other livestock products) often translates into environmental damage and disruption of traditional patterns of livestock raising. At the same time, livestock producers are forced into resource degradation where population pressure and poverty coincide, such as in marginal pastoral areas. On a global basis, production based on pure grazing systems is relatively unimportant and grows also at the lowest rate. Pastoral system production grows at 1% per annum, mixed farming at 3%, and industrial production at more than 7%. Despite their cost-effectiveness, traditional pastoralists will not supply much of the increased demand for animal products but the more favored margins of the rangelands will be encroached upon by croplands. This can often lead to land degradation.

15.6 Land Degradation in Arid Regions

Livestock production, mainly as a result of pressures in this process, has become an important factor in environmental degradation. Large land areas have become degraded through overcropping, overgrazing, and the concomitant loss of vegetation. The most important degradation of land and vegetation is around settlements and water points. They are mostly in a radius of about 1–5 km of the actual water point. However, assuming an average distance of about 10–30 km between water points, the degraded area would amount, at most, to 5–10% of the total area.

Until recently, arid rangelands, more than any other system, have been associated with land degradation (Heshmati and Squires 2013). The concept of "desertification" originated from a perception of degrading fringes of arid rangelands and advancing deserts. In addition, the arid grazing systems and pastoral production modes of the developing world have been described as inefficient and backward production systems.

These views have changed radically. Firstly, there is now evidence that in the arid zones, the extent of land degradation is greatly exaggerated and that we are dealing with a highly resilient ecosystem. Secondly, there is convincing evidence that traditional mobile production systems on arid rangelands are highly efficient (Ellis and Swift 1988). Production of protein per ha of traditional nomadic pastoralists in Mali and Botswana is two- or threefold higher (and at a much lower cost in nonrenewable fuel resources) than production from sedentary production systems or ranching, respectively, under similar climatic conditions in Australia and the USA. In addition, arid grazing systems have often multiple uses, with wildlife and other plant products (including medicinal plants) being important products.

Land degradation of semiarid lands in the West Asia and North Africa (WANA) region, and in Africa and India, is caused by a complex set of factors involving pastoralists and their stock, crop encroachment in marginal areas, and fuelwood collection. Changes in land tenure arrangements, settlement, and incentive policies (e.g., subsidized barley in WANA region) have undermined traditional land use practices and contributed to degradation.

However, some would question whether irreversible degradation is occurring at all (Harris 2010; Behke and Mortimer 2015; Reynolds and Stafford Smith 2002), and just what the role of livestock role is in that process? Livestock do not move, produce, or reproduce without human intervention. Livestock do not degrade the environment – humans do. As a result of these misconceptions about livestock development, institutions and governments continue to miss opportunities which would permit the livestock sector to make its full contribution to human welfare and economic growth. It should be noted that livestock are an important source of gaseous emission, contributing to global warming, which is projected to increase by 2 °C (or more) worldwide over the next few decades. All these pressures on the environment are the result of a metamorphosis, where the role of livestock has altered due to rising and changing demands for livestock commodities and to a different perception of the environment. In essence, the conflict between livestock and environment is a conflict between different human needs and expectations. In many places, livestock production is growing out of balance with the environment or is under so much pressure that it leads to environmental degradation.

Finding the balance between increased food production and the preservation of the world's natural resources remains a major challenge (IFPRI 1995). It is clear that food will have to be produced at less cost to the natural resource base than prevails at present.

Biodiversity may also be affected by extensive livestock production, although there are a large number of cases, showing increases in plant biodiversity in well-balanced grazing systems, especially those using multispecies. The interaction between wildlife and livestock in these ecosystems is complex. Firstly, there is increasing evidence of complementarity and only limited competition of wildlife and livestock in grazing. The grazing "overlap" between most wildlife species and livestock is rather limited. The combination of livestock raising and wildlife management often results in an equal or better species wealth than any of these activities done individually. Furthermore, in national parks in Kenya, where livestock are not permitted, biodiversity is decreasing, with an increase in unpalatable species and bush encroachment. On the other hand, there are many degraded areas in Kenya, due to combined wildlife-livestock pressure.

The driving forces leading to losses in animal biodiversity are habitat destruction, species introduction, and hunting. In addition, hunting and culling of wildlife were encouraged in the past, because wildlife in general was considered to be a reservoir of diseases, such as rinderpest and malignant catarrhal fever, and vectors of disease, such as East Coast fever and trypanosomiasis. The control of the above-mentioned diseases has improved considerably, and there is a much better understanding of which particular species harbor specific diseases, opening better opportunities for wildlife-livestock integration.

Therefore, the understanding of the interaction between the technical and socioeconomic constraints was mostly inadequate and led to implementation of inappropriate, uniform projects which not only overlooked the goals and strategies of the pastoral communities but also failed to involve them in planning. The environmental factors (i.e., temperature, water availability, diseases and insects, local flooding, and/or feed shortage) make livestock keeping in one area more or less risky than in another area (based on a regular and predictable seasonal timetable). Government officials (e.g., veterinarians-doctors, administrators) dislike the mobility of nomads because it spreads diseases and re-infects areas once cleared, imposes difficulties in providing them with social services, causes many clashes between them, and results in many complications at the international borders which may pose a threat to national security. Therefore, their settlement is desirable because centralized control can be increased, taxes can be levied, a herd's size can be limited in line with some ideal of predetermined carrying capacity, and herders can be encouraged to commercialize their production.

This line of reasoning is justified by the argument that the pastoral areas are experiencing a crisis in terms of overpopulation and overgrazing, producers are "backward" because of their apparent failure to respond to price incentives, and it is time for people to take individual responsibility for limiting their herd size. Sedenterization is thought to make it easier for governments to provide services such as health care and education. It is, therefore, viewed as a necessary precondition for development.

In cases where settlement has been imposed on nomads compulsorily or by government decree or by other means, the results have been disastrous. However, settlement impoverishes pastoralists; it has caused almost massive losses of livestock through starvation and diseases. Furthermore, it has changed the nature of society's structure and organization with some consequent loss of feeling of identify and continuity. Finally, it does not improve social services due to the low density of population per area.

15.7 Technology and Policy Options

The prevailing combination of poverty and high population growth, which characterizes many countries of the semiarid regions of the world, cannot be easily broken. The overriding need is to stop the building up of further human pressure in arid zones, through adequate population control and alternative employment generation policies.

As the second priority, external interventions in the system need to acknowledge the disequilibrium status of the pastoral systems in the arid zones and respond to their need for flexibility and mobility. This means that attempts to regulate stocking rates need be stopped. Even apart from the technical flaws in the estimation of the carrying capacity, experience has shown that it is almost always impossible to enforce those stocking rates.

The third priority should be to empower traditional pastoral institutions and develop effective comanagement regimes, forging partnerships between the state and a wide variety of users, with the state carrying the overall responsibility for arbitrating conflicting interests at the national level and facilitating negotiation between the multiple stakeholders. Access to land is to be based on customary resource user rights, however, avoiding rigid territorial boundaries.

The fourth priority is the identification of effective drought management policies.

Finally, the fifth priority should be to establish appropriate incentive policies by:

(i) Increasing the costs of grazing on the range which can reduce animal pressure by promoting an earlier offtake.
(ii) Full-cost recovery, especially for water supply and animal health services. Water has been in many cases a free good supplied by the public sector (and frequently financed by the international donor community).
(iii) Removing of price distortions for other agricultural inputs, in order to reduce the conversion of pastoral key resources into marginal crop land.

15.8 Research Needs

The key areas where research is urgently needed emerge from the above recommendations. They include the identification of:

(i) Appropriate indicators, which provide reliable information on the resource trends in the arid areas
(ii) Appropriate methodologies for economic appraisals on the investment in converting such "key resources"
(iii) The factors which lead to strong pastoral institutions
(iv) Sustainable drought preparedness plans, with particular emphasis on decentralized management and the design of appropriate banking and insurance schemes, and appropriate conflict resolution schemes.

15.8.1 Whither Nomadic Pastoralism?

At a time when traditional lifestyles are under threat throughout the world, it seems appropriate to ask "what future is there for traditional (noncommercial) range/livestock production systems? The strongholds of nomadic and seminomadic pastoral systems are undergoing rapid evolution as market-based systems overtake them (Box 15.1).

At the same time, concerns about the sustainability of the resource base have emerged with the ratification, by many nations, of international conventions on *biodiversity* and on *desertification and drought*. Many people are left to wonder if the environmental damage done to rangelands, especially on desert margins, will bring about the demise of traditional pastoralism. In general, ecosystems are grazed because they are not sufficiently productive or reliable to be cropped. This means that management must cope with low or unreliable production, complex seminatural systems, large management units, and greater economic risk.

In the light of the above observations, we must conclude that traditional nomadic pastoralism will become less and less important. There will always be those who wish to utilize (exploit) the otherwise unusable forage and water resources of the drylands, but social, economic, and political pressures will see the demise of this way of life. The Beduoin, who traditionally occupy the arid zones of West Asia and North Africa, have adapted to changing conditions but are under increasing pressure. Their situation is described in Box 15.1.

15.8.2 Changing Status of Pastoral Nomads

A change in the balance between settled agriculture and nomadic pastoralism appears to have been ongoing in the region since the early years of the last century. The process of change has not been due to any perception of ecological damage caused by the pastoral nomads to rangelands through excessive use leading to land degradation but due to the political reality of governments preferring settled farmers who could be taxed and conscripted to the nomads who are outside the political community and might presumably pose a danger to order.

There has also been a second factor at work. In the interface between agriculture and pastoralism, an increasing sophistication and diversity in the consumption patterns and preferences of the population led inevitably to a decreasing demand for the main products of the rangelands or rather to a shrinking profits from them, as compared to those from agricultural crops. For example, the market for camels began to shrink with the advent of modern transportation (the coming of railways, network of roads, automated vehicles), which struck a substantive blow at the traditional use of camels for transport. The demand for sheep and goats continued, and may have increased as the population grew, but capital was profitably invested in the growing of crops. As a result, the numbers of livestock in proportion to human population have decreased dramatically over many regions, even though 99% of the pastoral households today own sheep for cash value and to meet their family needs.

The nomads have lived a largely self-contained existence with strict observance of traditional rites and obligations. Nevertheless, during the 1920s and 1930s,

> **Box 15.1 The Future of Nomadism: An Example from the Bedouin in Saudi Arabia**
> The Al-Taysiyah Bedouin, like other nomad groups, face a difficult future. On the one hand, public policy and the impacts of modernization have protected their livelihood from extinction, and they are able to survive at reasonable levels of economic well-being. They have responded to current economic incentives and have been able to overcome many of the constraints associated with the traditional livestock livelihood system. In many ways, modernization has transformed the Bedouin into a tent-dwelling rancher, with a capital-intensive enterprise that hires outside labor and depends critically on market conditions as much as on range conditions. This transformation, however, has also changed the nature of nomad decision-making. Herd management strategies and grazing practices reflect the demands of short-term economic maximization rather than long-term conservation. Herd sizes are large, water is free, range access is open, and cash is needed to pay the expenses. This combination favors the short-term option on natural resource management that is observed among the Al-Taysiyah Bedouin.
>
> At the same time, there is increasing competition for the Bedouin in a modern world. Urban residents with available capital are moving into livestock herding for profit. These capitalists seek the economic benefits without the rigors of the lifestyle. Other interests are competing for the desert, such as city recreationists, agriculturalists, and now the conservation interests such as the wildlife conservation agency, which is in the process of claiming rangeland for the recuperation of wildlife species. Such pressures, combined with the exodus of young Bedouin to urban areas, will continue to place the Bedouin existence in jeopardy.
>
> For policy-makers, the choices are equally challenging. In a society that values the Bedouin lifestyle for its cultural contribution to national identity, new solutions are needed to ensure the sustainability of this livelihood. It is inevitable that such sustainability will require a redefinition of the Bedouin place in their desert and in their society.
>
> *Sources*: Al Gain (1985)

according to a noted historian of the region, Albert Hourani, "nomadic pastoralism virtually disappeared as an important factor in Arab society." This does not mean that nomads are no longer to be found or that they have been fully integrated into the mainstream of economic life. The development and management of livestock production continues to be followed through reliance on "opportunistic" stock movements. The techniques utilized for the purpose have evolved with time. A generation ago, the more talented young men of the extended family were sent out to scout for rainfall areas and vegetative cover; today, the young men are working in oilfields or construction sites, sometimes across the border in other countries, and there is a reliance on trucks for transporting livestock and on telephones for tracking climatic and other conditions. In this process, the pastoral nomads are actually settling down to a more profitable existence with a family house in their ancestral village and access to schooling for their children.

Thus, while techniques have changed, the long-standing strategies for livestock management have not changed. These strategies are based on the need to respond as rapidly as possible to changing climatic and vegetative conditions, through enhanced mobility and such means of information gathering as may be available.

15.8.3 Grazing Systems

Grazing systems cannot be replaced easily by prescriptions to reduce land degradation through the control of excessive livestock numbers. Such prescriptions are usually ignored. Carrying capacity and critical loads are useful concepts in broad scientific analysis. These are equilibrium concepts, but equilibrium is not the normal state of the ranges in drylands. The key to the success and survival of the pastoral nomad lies in the keenness of his observation of variations in vegetation and precipitation in time and over different parts of the rangelands and on his successful (or unsuccessful) exploitation of the observations (Finan and Al-Harani 1988). Information exchange and transportation have always been and still remain the major instruments for rangeland management. For this reason also, there is little doubt that in due course the pastoral communities will become more involved and participate directly in the adaptation and use of the new technologies for rangeland monitoring (Feng and Squires 2017).

Meanwhile, policy and program interventions are needed to help pastoral nomads overcome a number of emerging concerns that have made effective and efficient (in the economic sense) livestock management more difficult for them. These concerns include the rapid increase in human populations in pastoral communities, a more sedentary life on the pasturelands, the increasing need for technology to deal with emerging problems, and changing political, economic, and social conditions. These concerns are not static. They are changing in nature and impact as their incidence becomes more burdensome and they interact with one another. Time is not on the side of the nomads and early action is indicated, if we are to avoid an accelerated deterioration in economic and environmental conditions. Bad management decisions stemming largely from the lack of local participation and poor foreign advisory services (e.g., through unqualified experts) often contribute to a worsening of these concerns.

15.9 Summary and Conclusions

Climate change impact and other upheavals attributed to global change are inevitable. Models suggest that drylands may be among those regions most affected. Emerging problems include the conservation of the resource base (including biodiversity issues), the globalization of the world economy, the breakdown of culture and tradition, and the constraints imposed by systems of governance. Arid zone dwellers are under increasing pressure. The pastoralist lifestyle that has stood the test of time over millennia is under threat. New solutions are needed to ensure the sustainability of this livelihood. It is inevitable that such sustainability will require a redefinition of the pastoralists' place in the arid lands and in modern society (Humphries and Sneath 1999). This is especially so for those who depend on migration in search of forage and water for their flocks/herds. Those who herd in the vast rangeland regions of the earth face a precarious situation as they struggle to respond to the momentous political and economic changes of recent years.

References

A. Al Gain. Integrated resource survey in support of nomads in Saudi Arabia – a proposal. Proc. International Research and Development Conf. Arid Lands Today and Tomorrow (Tucson, 1985), pp. 1213–1221

R. Behnke, M. Mortimer. *The End of Desertification? Disputing Environ-mental Change in the Drylands* (Springer, 2015)

R. H. Behnke Jr., I. Scoones, C. Kerven (eds.), *Range Ecology at disequilibrium. New Models of Natural Resource Variability and Pastoral Adaptation in African Savannas* (Overseas Development Institute, London, 1993)

J.E. Ellis, D.M. Swift, Stability of the African pastoral ecosystems: alternate paradigms and implications for development. J. Range Manag. **41**, 450–459 (1988)

T.J. Finan, E.R. Al-Harani, in *Drylands: Sustainable Use of Rangelands into the Twenty-First Century*, ed. by V.R. Squires, A.E. Sidahmed. Modern Bedouins: the transformation of nomad society in the Al-Taysiyah region of Saudi Arabia (IFAD, Rome, 1988)

N. Halimova, in *Rangeland Stewardship in Central Asia: Balancing Improved Livelihoods, Biodiversity Conservation and Land Protection*, ed. by V. Squires. Land tenure reform in Tajikistan: Implications for land stewardship and social sustainability: a case study (Springer, Dordrecht, 2012), pp. 305–332

G.A. Heshmati, V.R. Squires, *Combating Desertification in Asia, Africa and the Middle East: Proven Practices* (Springer, Dordrecht, 2013)

C. Humphries, D.S. Sneath, *The End of Nomadism?: Society, State, and the Environment in Inner Asia* (Duke University Press Central Asia Book Series 368 p, 1999)

International Food Policy Research Institute (IFPRI), *A 2020 Vision for Food, Agriculture and the Environment* (International Food Policy Research Institute, Washington, DC, 1995)

H. Kreutzmann, *Pastoral Practices in High Asia: Agency of 'development' Effected by Modernisation, Resettlement and Transformation* (Springer, Dordrecht, 2012)

J. Reynolds, D.M. Stafford Smith, *Global Desertification: Do Humans Cause Deserts?* (Dahlem University, Berlin, 2002).

V.R. Squires, in Range and Animal Sciences and Resources Management. Encyclopedia of Life Support Systems (EOLSS) developed under the auspices of UNESCO, ed. by V.R. Squires. People in rangelands: their role and influence on rangeland utilization and sustainable management (EOLSS Publishers, Oxford, 2009) pp. 34–42

V.R. Squires, X. Lu, Q. Lu, T. Wang, Y. Yang, *Rangeland Degradation and Recovery in China's Pastoral Lands* (CABI, Wallingford, 2010.) p. 264

Further Readings

H.Y. Feng, V.R. Squires, *Climate Variability and Impact on Livelihoods in the Cold Arid Tibet Plateau* (Springer, Cham, 2017), pp. xx, this volume

R.B. Harris, Rangeland degradation on the Qinghai-Tibetan plateau: a review of the evidence of its magnitude and causes. J. Arid Environ. **74**, 1–12 (2010)

A.E. Sidahmed, Recent trends in drylands and future scope for advancement (Springer, Cham, 2017) pp. xx, this volume

Unifying Concepts, Synthesis, and Conclusions

16

Victor R. Squires and Mahesh K. Gaur

16.1 Introduction

Approximately one third of the global population lives in and depends on drylands (Fig. 16.1) for their livelihoods (Gaur and Squires 2017). Rangelands are by far the most widespread land-use type in the dry areas (Table 16.1) and are home to most of its poorest inhabitants. Having low productivity and potential, the people they support remain largely marginalized. It is said that they are "living on the edge" – geographically, socially, and politically. A large area of the world region is arid or semiarid with shallow- and low-fertility soils and poor plant cover and pastoral/agropastoral use are dominant production systems. Many dryland regions around the world are affected by rapid change in vegetation cover, plant community composition, hydrologic conditions, or soil properties, which results in an overall loss of ecosystem services and poses serious threats to sustainable livelihoods. The process underlying these changes is often termed "desertification." There is a significant interaction with drought. Both *desertification* and *drought* are ill-defined concepts, as explained in this book (see also Squires 2017a). Desertification is commonly associated with changes that persist for several decades and are presumably permanent and irreversible, at least within the time scales of a few human generations (Heshmati and Squires 2013). While land degradation can occur as a result of natural processes, there is a widespread opinion that it mostly happens as a result of the impact of users' activity on the land and is often a "social problem," which can be prevented if the underlying causes are addressed properly (Vlek et al. 2010). The increasing demand for food, feed, fuels (including biofuels), and fodder linked to an

V.R. Squires (✉)
Institute of Desertification Studies, Beijing, China
e-mail: dryland1812@internode.on.net

M.K. Gaur
ICAR-Central Arid Zone Research Institute, Jodhpur, India
e-mail: maheshjeegaur@yahoo.com

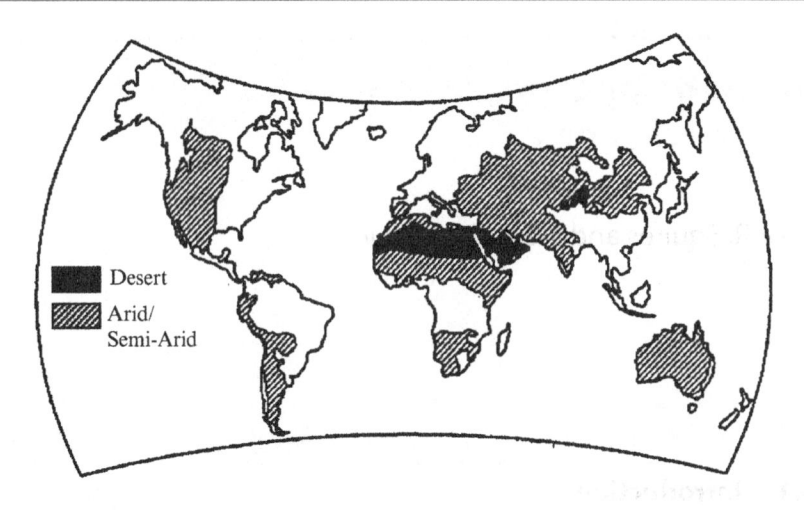

Fig. 16.1 World drylands

Table 16.1 World drylands (excluding hyper arid lands) in millions of hectares (Dregne and Tucker 1988)

	Arid	Semiarid	Dry subhumid	Total
Africa	504	514	269	1,287
Asia	626	693	353	1,672
Australia	303	309	51	663
Europe	11	105	184	300
North America	82	419	232	733
South America	45	265	207	517

increase in human population and a conversion of land through deforestation, environmental services, irrigation, and pollution, among other issues, contributes significantly to land degradation. In addition, land degradation is often defined as the "reduction in the capacity of the land to provide ecosystem goods and services over a period of time" which emphasizes the time aspect of this process. This definition is especially useful when the mechanisms at play in land degradation is discussed.

16.2 Combating Land Degradation and Mitigating Drought and Other Hazards

Land-use patterns influence the long-term productivity of agroecosystems and result from socioeconomic as well as biophysical and climatic drivers (Walker et al. 2004). The DPSIR model is widely used now to characterize the interplay between key factors. The determination of the likely causes (drivers and pressures: DP) of the actual physical states (S) of land (R) to reverse the degradation at a local scale is

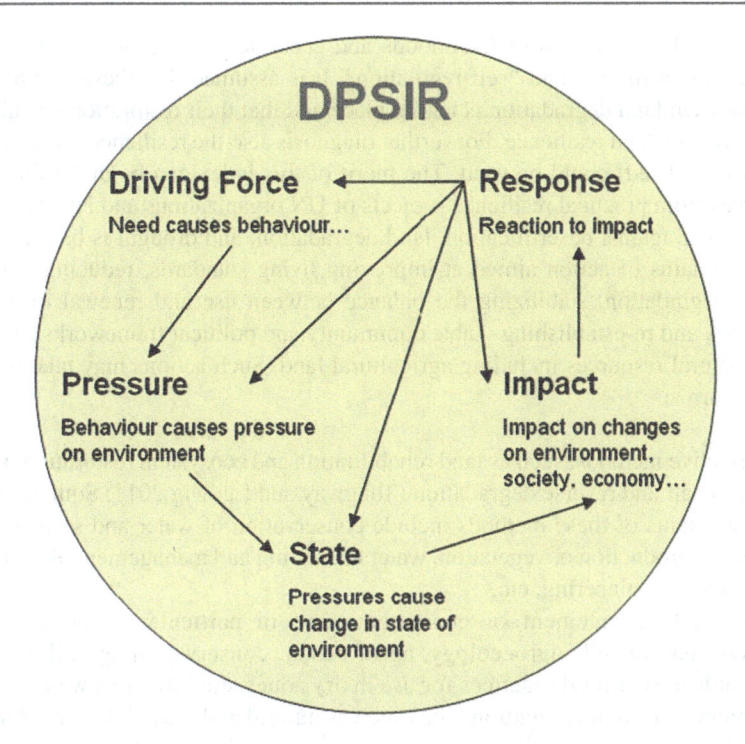

Fig. 16.2 A schematic showing key elements of the DPSIR model

fundamental for identification of impacts (I) and the design of appropriate responses process. Linking indicators of drivers and pressures to state indicators on their presumed cause-effect relationship are an essential component of the DPSIR approach to the design of appropriate responses and remedial actions (Fig. 16.2).

Besides the drivers, which are commonly mentioned in the land degradation literature – population growth, poverty, climate change – other parameters should be considered. These are resilience and inequality. They also lack explicit focus on issues of justice and equity (Pelling and Manuel-Navarrete 2011). Closely related to the resilience element of self-regulation of socio-ecological systems are inequality factors. While examination of the processes of land degradation highlights the relations between poverty and desertification and land degradation, it could be argued that poverty is related to poor land condition, but that land degradation and desertification are more related to inequality. This argument is based on the theories of environmental justice and underpinned by regression analyses of certain parameters of injustice with land degradation, but there are coefficients of land and income distribution, inequalities of purchasing power, and power of decision-making that are analogous to Gini coefficients.

The components of *resilience* as described by Holling (1973) are growth, conservation, release, and reorganization. To translate them into practically applicable parameters which constitute adaptive capacities, we designate them as "adaptive learning," "diversification of livelihoods and ecosystems," "sustainable technologies and innovations," and "self-regulation." It is assumed that these components are lost when land degradation is taking place and that their restoration can also be used to restore land resilience. For further diagnosis also the resilience index framed by FAO (undated) could be used. The merit of this index can be underpinned by examples from practical resilience projects of UN organizations and NGOs.

The fight against desertification, land degradation, and drought is based on different domains of action aimed at improving living standards, reducing environmental degradation, stabilizing the balance between use and renewal of natural resources, and re-establishing viable community and political frameworks for managing natural resources including agricultural land. Such actions may take the following forms:

- Corrective methods, such as land rehabilitation and ecosystem restoration, which aim to halt and reverse degradation (Tongway and Ludwig 2011; Squires 2012). Components of these methods include conservation of water and soils, protection or introduction of vegetation, water harvesting and management of droughts, ecological engineering, etc.
- Improved management of ecosystems and in particular agroecosystems. These may include agroecology, agroforestry, conservation agriculture, sustainable agricultural practices for use in dry zones, etc., But also water-related matters in aridland irrigation areas such as natural and artificial oases (Squires 2017)
- Development of models for integrated management systems for shared natural resources. These would address such issues as access to information, negotiation of access and user rights between local and national organizations, exchange of data, conflict management, etc.
- Implementation of favorable institutional and policy settings that promote sustainable practices that achieve livelihood goals. These can include access to market for products from dry zones, diversification of the economy, payment for environmental services, land ownership rights, access to credits, training for farmers, insurance systems, and public-private partnerships. Still other questions relating to sources of information that need answers are:
  What are the major contributions from traditional and local practices and scientific research?
  How are they related to specific settings?
  How can they be generalized so that they can be adapted and applied to broader settings?
  What are the obstacles to their more widespread use?
  How can adaptive capacities be developed or maximized in the short, medium, and long term?

For all these actions, a "standard" or academic model often cannot be applied effectively in the field or be used to outscale practices (technological transfer) from one region to another (Squires 2013). Combining scientific research with traditional and local knowledge should be encouraged through participatory research to facilitate and enhance adoption of innovations so that they are well adapted to different social, economic, political, and ecological contexts (Hua and Zhang 2012). Toward this end, focus should be placed on family/householder farming/pastoral systems and achievement of a better understanding of socio-ecological dynamics at micro- and landscape levels.

- A significant problem militating against clearer understanding of desertification as a tangible process relates to its confused relationship with the terms *drought*, *climatic variation*, *climate change*, and *climatic fluctuation* which are all used interchangeably in the literature.
- Both cyclical (climatic) and anthropogenic changes are evident in most drylands worldwide. Climate is inherently variable at all scales. The cyclicity of rainfall impacts on a plant community and exacerbates the negative conditions imposed on vegetation as a result of sustained anthropogenic activity. However, the difficulty of differentiating between the effects of normal cyclical changes and anthropogenic changes is still a matter of debate. Anthropogenic and environmental pressures on the earth's ecosystems have led to substantial land degradation in all parts of the world (see this volume for specific regional studies) throughout the centuries. For example, the so-called "fertile crescent" in the Tigris-Euphrates basin was the cradle of agriculture 10,000 years ago. Agriculture and sedentarization were based on domestication of food grains and livestock such as sheep, cattle, etc. Nowadays it is one of the most severely degraded regions in the world and has been dubbed the "desertification crescent." The scale of this land degradation has been known for some time[1] but the complex socio-ecological drivers behind land degradation have only been acknowledged in recent years (Reynolds and Stafford Smith 2002; Walker et al. 2004; Brunckhorst and Trammell 2016). It might be said that it is through the ubiquity of communication technologies that the complexities that are involved with all of the global challenges and changes from the warming of the planet through to geopolitical instabilities are coming to be known and appreciated for their complexity.
- Confusion also arises in the literature relating to the corresponding adaptive vegetation changes that has led to unreasonable attempts to exclude vegetative indicators from studies of desertification. Depending on the driver and the geographic setting, desertification can result in an increase in bare soil (up to complete denudation of the soil surface), loss of soil resources (e.g., loss of nutrients, fine soil grains, and water-holding capacity), increase in soil salinity and toxicity, or shifts in vegetation composition (e.g., from perennial to annual species, from palatable to unpalatable grasses, or from grassland to shrubland (Squires 2017; Busso and Fernandez 2017)). Badripour and Solaymani (2017) discuss these changes in the context of watershed/catchment management in, while Squires (2017) summarizes the problems of invasive plants in the drier western half of the United States of America, and Ajai and Dhinwa 2017 and Dhir (2017) discuss the situation in India, while Ma (2017) describes the aridlands of China and their management.

---

[1] http://news.nationalgeographic.com/news/2001/05/0518_crescent.html

16.3 The Water, Soil, Food, and Biodiversity (WSF&B) Nexus

Land degradation, desertification, and drought represent significant challenges to the water, energy, and food (WEF) security of dryland peoples (DDC 2013). Supporting WEF security, and therefore sustainable development in drylands, requires holistic approaches (Reynolds et al. 2007; Bawden 2017). Maintaining WEF security is a high priority. The linkages among water, soil, food, and biodiversity (WSF&B) reduce human security from a dual focus as freedom from want and freedom from disaster impacts. This WSF&B nexus has increased both the social and the environmental vulnerability of the poor and often marginalized rural and urban populations and contributes to their human insecurity. Longer and more intensive droughts related to climate variability increase this dual vulnerability (Squires 2017b). Nexus approaches have been developed to highlight relationships and interdependencies between WEF and the need for integrated management to understand the trade-offs and promote synergies (Bazilian et al. 2011). However, they have not yet been widely applied in dryland settings (Van Wyk et al. 2016).

According to Bazilian et al. (2011)

> The areas of energy, water and food policy have numerous interwoven concerns ranging from ensuring access to services, to environmental impacts to price volatility. These issues manifest in very different ways in each of the three "spheres", but often the impacts are closely related. Identifying these interrelationships *a priori* is of great importance to help target synergies and avoid potential tensions. Systems thinking is required to address such a wide swath of possible topics. While environmental issues are normally the 'cohesive principle' from which the three areas are considered jointly, the enormous inequalities arising from a lack of access suggest that economic and security-related issues may be stronger motivators of change. Finally, consideration of the complex interactions will require new institutional capacity both in industrialized and developing countries.

16.3.1 The Water-Food-Energy "Nexus" as a Solution

The concept of the water-food-energy (W-F-E) nexus is defined as "an approach that integrates management and governance across sectors and scales" and which inter alia aims at resource use efficiency and greater policy coherence (Hoff 2009). It is argued that a reduction of negative economic, social environmental externalities in economic planning can lead to a greater overall resource use efficiency and provide additional benefits in the form of securing human rights to water and food. Successfully achieving the implementation of the nexus approach requires breaking down traditional silos in which policymaking takes place. The nexus approach has a number of corollary benefits. It fits well with the approaches for achieving human security – through water and food as human rights – and political security –through economic growth and stability. Such integrated approaches can also catalyze greater regional integration around shared resources; for example, in trans-boundary basins cooperation around water resources becomes more enhanced when it is linked to also sharing energy through hydropower.

The nexus approach also fits well with both mitigation and adaptation strategies under intense discussion at the climate change dialogue. Coupling water, food, and energy security allows climate change negotiators to see beyond narrow benefits and allows for greater integration into national economic policies. Finally, the W-F-E nexus is also relevant in disaster relief situations as well as post-conflict reconstruction situations. Both natural disasters and armed conflict often lead to refugees and internally displaced persons, who in turn are in dire need to receive adequate food, water, and shelter. Provisioning of these services requires availability of adequate energy. It thus becomes advisable that recovery and relief programs in these situations follow a nexus approach. Achieving the W-F-E security nexus in drylands is easier said than done. In part, because it is a new concept and typically government agencies and department are not designed to work across their specific domains, its implementation would offer new challenges. In order to successfully implement the W-F-E nexus, the following three elements are essential.

First, the W-F-E nexus must be connected and integrated into the national economic development narrative. Such integration must start at the top political level and trickle down to line departments and agencies; an entry point could be the ongoing planning around Sustainable Development Goals (SDGs) as part of the post-2015 development agenda. It must also be coupled with mobilizing financial resources and allocating investments within national budgets. It may also be required to formulate an enabling environment in which investments from the private sector are solicited in order to achieve the nexus.

Second, the discussions around the W-F-E nexus must be couched in terms of its human and social dimensions. The obvious links are to food and water as a human right and the role these would play in achieving human health and well-being. Not so obvious links are to job creation and livelihood security, which can be achieved and are bound to be politically attractive. This inter alia means that countries must build human and technological capacity to achieve security around the W-F-E nexus. This capacity development can become an agent of change in the context of understanding and overcoming cross-sectoral divergences.

Third, the W-F-E nexus must be integrated into regional dialogues, most notably in situation where water is a shared resource across national and subnational boundaries. Integrated regional dialogues can boost trans-boundary trade and cooperation and reduce transaction costs for achieving cooperation. It can also help improve institutions dealing with trans-boundary governance and overall policy coherence.

In conclusion, water security in drylands is closely linked to food and energy security. Arguably, these very closely linked domains cannot be planned in isolation. For drylands to overcome their intrinsic resource scarcity, cross-linked policies utilizing the W-F-E approaches offer a way out. We are a long way from achieving water, energy, and food security for all the world's people. In hotspot regions such as India and some other South Asian countries where lack of land is also becoming an issue, and in sub-Saharan Africa, large fractions of the population remain marginalized and deprived of their human rights and development opportunities.

16.4 Climate Change

Climate change is one indication of a lack of sustainability in dominant modes of development (Hulme 2009); other climate indicators of crisis include global biodiversity loss. It has been said that climate change is but one expression of the internal contradictions of capitalism that include also economic inequality and political alienation. Seen in this way analysis of human responses to climate change must engage with social relations of power. Climate change projections provide little information on how this risk might affect rangeland plant communities in specific regions (Eldridge and Beecham 2017; Kellner et al. 2017). However, the farmers' decisions about managing their farming activities in the framework of "sustainability" and "climate change" are plagued with uncertainty. The scientific community is giving increasing importance to the integration of local and scientific knowledge in sustainable environmental decision-making (Kellner et al. 2017).

16.4.1 Adaptation to Global (Including Climate) Change

In the context of climate change, a key challenge is the response capacity – resilience – to reduce the vulnerability of communities (Van Wyk et al. 2016) There is a need to rethink the processes of transformation and socio-environmental transition to more resilient systems that reduce the vulnerability of communities and facilitate the exchange of knowledge and purposeful dialogue between key actors in local and regional transformation: local and regional authorities, universities, research centers and international agencies, social movements emerging from civil society, grassroots organizations, NGOs, media, and corporations/businesses (Feng and Squires 2017).

Globalization has been an important driver of development. It connects and integrates markets, brings investment, and provides access to technology that supports innovation and increased resource use efficiency. International trade has grown rapidly (food trade somewhat more slowly) and the traded percentage of food produced has grown globally from about 10% in 1970 to 18% in 2010. Trade can also mitigate local scarcities. This is evident in the Middle East and North African (MENA) countries (Mohamed and Squires 2017), which increasingly have to rely on food imports (and associated imports of virtual water).

16.4.2 The Land Degradation Paradigm Is It Still Valid?

Several recent publications have questioned the relevance and validity of the conventional paradigm, notably the book edited by Behnke and Mortimer (2015) entitled "The End of Desertification? Disputing Environmental Change in the Drylands" as well a number of papers published in the literature. The dominant paradigm represents "the practical expression of sets of particular beliefs and assumptions that constitute particular approaches to problematic situations in the aridlands by research scientists, policy makers, regulators, educators, NGOs, and so-called

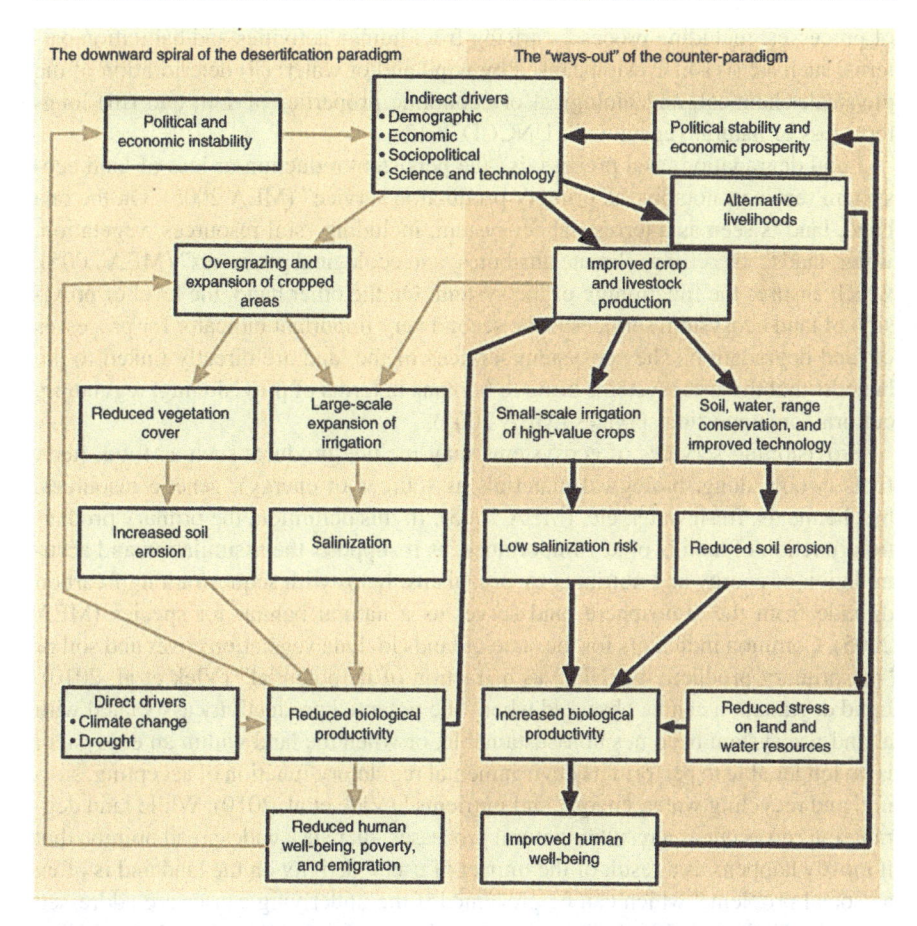

Fig. 16.3 Two contrasting paradigms. On the *left* is the more traditional "desertification," while on the *right* the counter paradigm where indirect drivers such as demography and sociopolitical prevail

"development specialists." Bawden (2017) makes the observation "These worldviews are so global in their adoption and so influential with respect to the paradigm that they have effectively nullified any attempts to replace them even under circumstances, as now, where their relevance is proving to be singularly inadequate and paradigmatically severely limiting (Norgaard 1994). There is a counter paradigm (Fig. 16.3) that places more emphasis on indirect drivers.

Chabay et al. (2015) remind us that land and soil are vital resources that support humans and that we are obliged to minimize/prevent degradation and are under an imperative to restore land wherever we can.

"Land degradation" is "the reduction or loss of the biological or economic productivity and complexity of rainfed cropland, irrigated cropland, or rangeland, pasture, forest and woodlands resulting from land uses or from a process or combination of processes, including processes arising from human activities and habitation patterns, such as: (i) soil erosion caused by wind and/or water; (ii) deterioration of the physical, chemical, and biological or economic properties of soil; and (iii) long-term loss of natural vegetation" (UNCCD 2013).

Land degradation, as a process, is "the persistent reduction or loss of land ecosystem services, notably the primary production service" (MEA 2005). On the one hand, land is seen as a terrestrial ecosystem, including "soil resources, vegetation, water, landscape setting, climate attributes, and ecological processes" (MEA 2005), which ensures the functioning of the system. On the other hand, the level of provision of land ecosystem services is the second very important indicator for processes of land degradation. The ecosystem services of the land are directly linked to the benefits that these ecosystems bring to humans in terms of provisioning, regulating, cultural, or supporting services (MEA 2005).

Provisioning services of ecosystems may include products such as food, fiber, fuels (wood, dung, biological materials as sources of energy), genetic resources, biochemicals, fresh water, etc. (MEA 2005). In this definition, the primary production function of land is of key importance, as it supports the assimilation and accumulation of energy and nutrients by organisms, helps with sequestration of carbon dioxide from the atmosphere, and serves as a natural habitat for species (MEA 2005). Common indicators for the state of lands include vegetation cover and soil or "net primary productivity (NPP) as a fraction of its potential" (Vlek et al. 2010). Land degradation can be observed when "the potential productivity associated with a land-use system becomes non-sustainable, or when the land within an ecosystem is no longer able to perform its environmental regulatory function of accepting, storing, and recycling water, energy, and nutrients" (Vlek et al. 2010). While land degradation can occur as a result of natural processes, there is a widespread opinion that it mostly happens as a result of the impact of users' activity on the land and is often a "social problem," which can be prevented if the underlying causes are addressed properly (Vlek et al. 2010). The increasing demand for food, feed, fuels (including biofuels), and fodder linked to an increase in human population and a conversion of land through deforestation, environmental services, irrigation, and pollution, among other issues, contribute significantly to land degradation. In addition, land degradation is often defined as the "reduction in the capacity of the land to provide ecosystem goods and services over a period of time, which emphasizes the time aspect of this process."

16.5 Rethinking the Transition to Resilient Aridland Systems

The goal of most land restoration efforts and initiatives such as land degradation neutrality is to (i) improve and strengthen dialogue between local governments, academia, and other stakeholders that boost community resilience development, (ii) encourage innovation in socio-ecological systems that are aimed at improving human welfare, and at the same time (iii) increase the production of services from ecosystems. Such programs have a list of expected outcomes. We list a few.

16.5.1 Expected Outcomes of Efforts to Restore Degraded Land

(a) Contribute to clarifying the concept of resilience at different scales and from different disciplinary perspectives.
(b) Conduct a comprehensive review of all methodologies, approaches, and tools for the analysis and the measurement of resilience.
(c) Develop planning methodologies for the transition and showcase good practices of resilience at regional and local level.
(d) Promote knowledge conservation, fundamental in preserving intangible heritage, promoting the use and transfer of appropriate technologies.
(e) Improve environmental and economic resources management, by changing consumption and production systems.
(f) Integrate community development and participatory planning through a coevolutionary approach.
(g) To strengthen communication among stakeholders and between actors.

16.6 Research Priorities

The following research question is a vital one: How could the human security of the people be enhanced that are impacted by the nexus of the water, soil, food, and biodiversity (WSF&B) with climate change impacts.

Other researchable topics are:

Impact on land degradation: soil erosion, sediment redistribution, and slope stability
Impact on livelihoods: food and feed production; agricultural sustainability; soil and water quality; land and water availability for cropping, irrigation, livestock, and forestry systems; energy production and hydropower
Impact on water, runoff, river discharge dynamics
Impact on carbon and nutrient cycling: soil organic carbon dynamics, greenhouse gas emission, and feedback mechanisms
Impact on the cryosphere (snow cover, glacier dynamics, and permafrost) – land-water ecosystem quality interactions

16.7 Summing Up

The opening words to the chapter by Richard Bawden (2017) remind us of the great challenges the aridlands and their peoples are facing now and the even greater stress they will need to endure in the coming decades.

> "It is difficult to imagine a more relevant or useful focus for exploring the consequences of global change in all of their socio-ecological complexity, than that provided by the world's arid lands. As a living example of nature/culture dynamics in the face of global changes in both their biophysical and sociocultural environments, they present far more appropriate images than the…..[images] that dominate the media."

We are reminded of the importance of land by the words of Monique Barbut,[2] Executive Secretary of UNCCD who said:

> As the global population expands dramatically toward a total of more than 9.5 billion people by 2050, demand for the goods and services that the land provides will only get stronger. In the context of a changing climate, as demand starts to outstrip supply, competition for this increasingly scarce and valuable productive resource will only heat up. As the loss of productive land and soil continues at an alarming rate, we expect to see declining food production and more hunger. The natural end point of these trends is mass forced migration, radicalization and deadly conflict in climate change and land degradation hot-spots. The so-called developing world, where natural infrastructure and resource governance mechanisms are often weakest and most vulnerable, is at the greatest risk. It will be hit hardest. However, in an interconnected world, no region or community is immune. No country can rest on its laurels. That is the reality. .... Land and soil are set to be a fundamental part of the Sustainable Development Goals for post-2015 implementation. Achieving land degradation neutrality will be a crucial first step. From the community to the national level, we must stop the loss of healthy and productive land by avoiding degradation wherever we can and by rehabilitating already degraded land. More well-managed land means more food and water available, more of the ecosystems we enjoy, less forced environmental migration, and a greater chance of security and peace in an unstable world. Work on securing our land resources also needs to start now because sustainable land and soil management can buy valuable time in the fight against climate change – time that is desperately needed. Land and soil is the second-largest carbon sink, after the oceans. Getting carbon back in the soil could buy the 30 years that we may need to move to a low-carbon economy. It could also get vulnerable populations, who are already experiencing climate change impacts and resource scarcity, time to adapt.

The likely impact of global change (including climate change) on the lands and its peoples will be profound, but as a "slow onset" phenomenon there may be time to adapt but the pace of change and the dire circumstances in which many now live may make this hope fade. Initially, we will see coping mechanisms brought into play but ultimately adaptation must occur. Adaptation may involve mass migration of ecological refugees. In dry lands, sustaining ecosystem processes in the face of climatic variability requires a sound foundation of monitoring and research, as well as a good working relationship between people and organizations with diverse goals and interests. As a result of the communication revolution, it might be said that "it is through the ubiquity of communication technologies that the complexities that are involved with all of the global challenges and changes from the warming of the planet through to geopolitical instabilities, are coming to be known and appreciated for their complexity" (Bawden 2017).

Sidahmed (2017) presents a wide-ranging overview and highlights the opportunities for future advancement. Many challenges lie ahead in the pluralistic societies that are becoming the norm.

This book brings together experiences from drylands across the globe. If the information and analysis presented here helps to further our understanding of the issues and challenges faced and the action being taken then we will be truly rewarded.

---

[2] Taken from the foreword of the book "Land Restoration: Landscapes Reclaiming for a Sustainable Future. Edited by Ilan Chabay, Martin Frick and Jennifer Helgeson, Elsevier, 2016

References

Ajai, P.S. Dhinwa, in *Climate Variability, Land-Use and Impact on Livelihoods in the Dry Lands*, ed. by M.K. Gaur, V.R. Squires. Desertification and land degradation on the Indian Sub Continent: Issues, present status and future challenges (Springer, New York, 2017), pp. xx.

H. Badripour, R.H. Solaymani, in *Climate Variability, Land-Use and Impact on Livelihoods in the Dry Lands*, ed. by M.J. Gaur, V.R. Squires.(Springer, New York, 2017), pp. xx.

R. Bawden, in Climate Variability, Land-Use and Impact on Livelihoods in the Dry Lands, ed. by M.J. Gaur, V.R. Squires. (Springer, New York, 2017), pp. xx

M. Bazilian, H. Rogner, M. Howells, D. Arent, D. Gielen, P. Steduto, A. Mueller, P. Komor, R.S.J. Tol, K.K. Yumkella, Considering the energy, water and food nexus: Towards an integrated modelling approach. Energ Policy **39**(12), 7896–7906 (2011)

R. Behnke, M. Mortimer, *The End of Desertification? Disputing Environmental Change in the Drylands* (Springer, Dordrecht, 2015)

D.J. Brunckhorst, J.E. Trammell, Restorative futures for social ecological systems in changing times, in *Ecological Restoration: Global Challenges*, ed. by V.R. Squires (Social Aspects and Environmental Benefits. Nova Publishers, New York, 2016), pp. 259–280

C.A. Busso, O.A. Fernandez, in *Climate Variability, Land-Use and Impact on Livelihoods in the Dry Lands*, ed. by M.J. Gaur, V.R. Squires. Arid and semiarid rangelands of Argentina (Springer, New York, 2017), pp. xx

I. Chabay, M. Frick, J. Helgeson (eds.), *Land Restoration: Landscapes Reclaiming for a Sustainable Future* (Elsevier, 2015)

DDC, Global climate change and its impacts on food and energy security in the drylands. Proc 11th Int. Conf on Development of Drylands, Beijing (International Dryland Development Commission, 2013)

H.E. Dregne, C.J. Tucker, Desert encroachment. Desertificat. Cont. Bullet. **16**, 16–19 (1988)

D. Eldridge, G. Beecham, in *Climate Variability, Land-Use and Impact on Livelihoods in the Dry Lands*, ed. by M.J. Gaur, V.R. Squires. The impact of climate variability on land use and livelihoods in Australia's rangelands (Springer, New York, 2017), pp. xx

FAO. (undated) *Resilience Index Measurement and Analysis Model*. http://www.fao.org/3/a--i4102e.pdf

H.Y. Feng and V.R. Squires in *Climate Variability, Land-Use and Impact on Livelihoods in the Dry Lands*, ed. by M.J. Gaur, V.R. Squires Climate variability and impact on livelihoods in the cold arid Tibet Plateau. (Springer,New York, 2017) pp. xx

M.K. Gaur and V.R. Squires in *Climate Variability, Land-Use and Impact on Livelihoods in the Dry Lands*, ed. by M.J. Gaur, V.R. Squires. Drylands under a climate change regime: implications for the land and the pastoral people they support. (Springer, New York 2017) pp.xx

G.A. Heshmati, V.R. Squires, *Combating Desertification in Asia, Africa and the Middle East: Proven Practices* (Springer, Dordrecht, 2013.) 476 p

H. Hoff, Global water resources and their management. Curr. Opin. Environ. Sustain. **1**, 141–147 (2009)

C.S. Holling, Resilience and stability in ecological systems. Annu. Rev. Ecol. Syst. **4**, 1–23 (1973)

L.M. Hua, D. Zhang, Engaging with land users: The first steps on a long road, in *Rangeland Stewardship in Central Asia: Balancing Livelihoods, Biodiversity Conservation and Land Protection*, ed. by V.R. Squires (Springer, Dordrecht, 2012), pp. 333–356

M. Hulme, *Why We Disagree About Climate Change* (Cambridge University Press, Cambridge, UK, 2009)

K. Kellner, G. von Maltitz, M. Seely, J. Atlhopheng, L. Lindeque, A. van Rooyen, in *Climate Variability, Land-Use and Impact on Livelihoods in the Dry Lands*, ed. by M.K. Gaur, V.R. Squires Springer, New York, 2017), pp. xx.

S. Ma, in *Climate Variability, Land-Use and Impact on Livelihoods in the Dry Lands*, ed. by M.J. Gaur, V.R. Squires.(Springer, New York, 2017), pp. xx

Millennium Ecosystem Assessment (MA), Dryland systems, in *Ecosystems and Human Well-Being: Current State and Trends*, ed. by R. Hassan, R.J. Scholes, N. Ash (Earthscan, London, 2005), pp. 623–662

A.H. Mohamed and VR Squires in *Climate Variability, Land-Use and Impact on Livelihoods in the Dry Lands*, ed. by M.J. Gaur, V.R. Squires Aridlands of North Africa and the Mediterranean Basin – Current Status and Future Prospects (Springer ,New York, 2017) pp.xx

R.B. Norgaard, *Development Betrayed: The End of Progress and a Co-evolutionary Revisioning of the Future* (Routledge, London, 1994)

M. Pelling, *Adaptation to Climate Change: From Resilience to Transformation* (Routledge, London, 2011)

Pelling, M., and D. Manuel-Navarrete. 2011. From resilience to transformation: The adaptive cycle in two Mexican urban centers. Ecol. Soc. 16(2): 11. [online] URL: http://www.ecologyandsociety.org/vol16/iss2/art11/

J.F. Reynolds, D.M. Stafford Smith. *Global Desertification: Do Humans Cause Deserts?* (Dahlem Univ. Press, 2002), Dahlem

J.F. Reynolds, D.M. Stafford Smith, E.F. Lambin, B.L. Turner II, M. Mortimore, Global Desertification: Building a Science for Dryland Development. Science **316**(5826), 847–851 (2007). doi:10.1126/science.1131634

A. E. Sidahmed, in *Climate Variability, Land-Use and Impact on Livelihoods in the Dry Lands*, ed. by M.J. Gaur, V.R. Squires. (Springer, New York, 2017), pp. xx

V.R. Squires, *Rangeland Stewardship in Central Asia: Balancing Livelihoods, Biodiversity Conservation and Land Protection* (Springer, Dordrecht, 2012.) 458 p

V.R. Squires 2013. Replication and scaling up: Where to from here?, in *Combating Desertification in Asia, Africa and the Middle East: Proven Practices*, ed. by G.A. Heshmati, V.R. Squires (Springer, Dordrecht), pp. 445–459

V.R. Squires, in *Handbook of Drought and Water Scarcity. Vol. 1 Principles of Drought and Water Scarcity*, ed. by S.S. Eslamian. Desertification and drought (CRC Press, Boca Raton, 2017b), pp. 13–21

V.R. Squires in *Climate Variability, Land-Use and Impact on Livelihoods in the Dry Lands*, ed. by M.K. Gaur, V.R. Squires. Dry Lands of North America – current status and future prospects. (Springer, New York, 2017a) pp. xx

V.R. Squires, H. Feng. Humans as change agents in drylands (with special reference to Qinghai-Tibet Plateau) in *Climate Variability, Land-Use and Impact on Livelihoods in the Dry Lands*, ed. by M.J. Gaur, V.R. Squires. (Springer, New York, 2017), pp. xx

D.J. Tongway, J.A. Ludwig, *Restoring Disturbed Landscapes: Putting Principles into Practice* (Island Press and Society for Ecological Restoration International, Washington, DC, 2011)

UNCCD, Land degradation neutrality. LDN Flyer http://www.unccd.int/Lists/SiteDocumentLibrary/Rio+20/Land%20degradation%20neutrality%202015/LDNFlyer.pdf (2013)

E. Van Wyk, B. Nkhata, C. Breen, W. Friemund, Sustaining ecological restoration through social transformation, in *Ecological Restoration: Global Challenges, Social Aspects and Environmental Benefits*, ed. by V.R. Squires (Nova Publishers, New York, 2016), pp. 17–33

P.L. Vlek, Q.B. Le, L. Tamine, Assessment of land degradation, its possible causes and threat to food security in Sub-Saharan Africa, in *Food Security and Soil Quality*, ed. by R. Lal, B.A. Stewart (CRC Press, Boca Raton, 2010), pp. 57–86

B. Walker, C.S. Holling, S.R. Carpenter, A. Kinzig, Resilience, adaptability and transformability in social–ecological systems. Ecol. Soc. 9(2), 5 (2004). [online] URL: http://www.ecologyandsociety.org/vol9/iss2/art5/

MIX

Papier aus ver-
antwortungsvollen
Quellen
Paper from
responsible sources
FSC® C141904

Druck:
Customized Business Services GmbH
im Auftrag der KNV-Gruppe
Ferdinand-Jühlke-Str. 7
99095 Erfurt